# FOOD PROTEINS

*Proceedings of the Kellogg Foundation International Symposium on Food Proteins held at University College, Cork, Republic of Ireland between 21–24 September, 1981.*

# FOOD PROTEINS

*Edited by*

## P. F. Fox and J. J. Condon

*University College, Cork, Republic of Ireland*

APPLIED SCIENCE PUBLISHERS
LONDON and NEW YORK

APPLIED SCIENCE PUBLISHERS LTD
Ripple Road, Barking, Essex, England

*Sole Distributor in the USA and Canada*
ELSEVIER SCIENCE PUBLISHING CO., INC.
52 Vanderbilt Avenue, New York, NY 10017, USA


**British Library Cataloguing in Publication Data**

Food proteins.
  1. Food—Protein content—Congresses
  I. Fox, P. F.     II. Condon, J. J.
  641.1′2        TX553.P7

ISBN 0-85334-143-5


WITH 88 TABLES AND 52 ILLUSTRATIONS





Printed in Great Britain by Galliard (Printers) Ltd, Great Yarmouth

# Preface

This symposium was organised, with the financial assistance of the Kellogg Foundation, to mark the opening of a new Food Science Building at University College, Cork. Dairy science has been a field of study at the College since 1924 and since about 1970 the curriculum has been extended to include a broad range of nutritional and food sciences. Agriculture and food processing are extremely important to the Irish economy, as discussed in Chapter 1.

The subject chosen for the symposium is a favourite and controversial topic with food scientists. This interest is due to (1) the nutritional significance of proteins which are commonly regarded as the limiting nutrient world-wide, although this thesis may not be fully tenable, (2) the commercial significance of food protein production arising from the limited supply of proteins of high biological value and the cost of producing them, and (3) the importance of proteins in determining the functional properties of foods, e.g. the extensibility and gas retention properties of wheat flour, the heat stability and rennet coagulability of milk and the toughness/tenderness of meat.

In spite of the several available texts on food proteins, we believe that the scope of this symposium is unique. The topics covered include an assessment of the world protein supply; the basic aspects of protein chemistry of importance in food sciences; the chemistry, technology and modification of the three principal food proteins, i.e. dairy products, meat and cereals; technology for the production of selected novel proteins; consideration of the nutritional significance of proteins and possible applications of genetic engineering in food protein production.

We express our sincere appreciation to the many people who contributed to the success of the symposium. In particular we wish to thank the Irish Department of Agriculture, who finance food science education at Cork; the Kellogg Foundation, Battle Creek, Michigan, USA, who, in addition to financing this symposium, have contributed greatly to the development of food science at this college by way of fellowships for faculty training and some items of specialised equipment; all the contributors to the symposium and Applied Science Publishers Ltd, for their cooperation in making possible a permanent record of the symposium proceedings.

J. J. CONDON and P. F. FOX

# Contents

# List of Contributors

ALLEN J. BAILEY

Agricultural Research Council, Meat Research Institute, Langford, Bristol BS18 7DY, UK.

ARNOLD E. BENDER

Department of Nutrition, Queen Elizabeth College, University of London, Campden Hill Road, London W8 7AH, UK.

K. J. BURGESS

Research and Development Division, Milk Marketing Board, Crudington, Telford, Shropshire, UK.

G. COTON

Dairy Crest Research and Technical Services, Milk Marketing Board, Thames Ditton, Surrey, UK.

D. G. DALGLEISH

The Hannah Research Institute, Ayr KA6 5HL, UK.

THAYNE R. DUTSON

Department of Animal Science, Texas A & M University, College Station, Texas 77843, USA.

P. F. FOX

*Department of Dairy and Food Chemistry, University College, Cork, Republic of Ireland.*

P. V. J. HEGARTY

*Department of Food Science and Nutrition, University of Minnesota, St. Paul, Minnesota 55108, USA.*

J. E. KINSELLA

*Institute of Food Science, Cornell University, Ithaca, New York 14853, USA*

B. A. LAW

*National Institute for Research in Dairying, Shinfield, Reading RG2 9AT, UK.*

R. A. LAWRIE

*School of Agriculture, University of Nottingham, Sutton Bonnington, Loughborough LE12 5RD, UK.*

ROLAND J. LEVINSKY

*Institute of Child Health, 30 Guilford Street, London WC1N 1EH, UK.*

SANFORD A. MILLER

*Director, Bureau of Foods, Food and Drug Administration, Washington, DC 20204, USA.*

GERALDINE V. MITCHELL

*Research Chemist, Division of Nutrition, Bureau of Foods, Food and Drug Administration, Washington, DC 20204, USA.*

L. L. MULLER

*Division of Food Research, Dairy Research Laboratory, CSIRO, Highett, Victoria, Australia*

PER O. NETTLI

*Alwatech, A. S., Harbitzalleen 3, Oslo 2, Norway*

FERGAL O'GARA

*Department of Dairy and Food Microbiology, University College, Cork, Republic of Ireland*

D. E. PALMER

*Bio-Isolates Limited, Powell Dubbryn House, Adelaide Street, Swansea SA1 1SE, UK.*

N. W. PIRIE

*Rothamsted Experimental Station, Harpenden, Herts AL5 2JQ, UK.*

A. SIMANTOV

*Food, Agriculture and Fisheries Division, Organisation for Economic Co-operation and Development, 2 rue André Pascal, 75016 Paris, France.*

P. V. TARRANT

*Meat Research Department, An Foras Taluntais, Dunsinea Research Centre, Castleknock, Co. Dublin, Republic of Ireland*

# 1

# The Irish Food Protein Industry

P. F. Fox

*Department of Dairy and Food Chemistry,
University College, Cork, Republic of Ireland*

## INTRODUCTION

The island of Ireland has an area of $\sim 85\,000\,\text{km}^2$ ($8\cdot 5 \times 10^6$ hectares) of which $\sim 83\%$ is in the Republic of Ireland. It lies between 51 and $55°$ north latitude and has a temperate, moist climate which has a significant influence on the type of agriculture practised. The population of the Republic is $\sim 3\cdot 4 \times 10^6$; $60\%$ of the population is classified as urban and $\sim 30\%$ live in the greater Dublin area. Considering that $\sim 65\%$ of the land is classified as arable and much of the remainder is suitable for rough grazing, Ireland is a relatively sparsely populated country.

Ireland is still generally regarded as an agricultural country; however, since about 1960 it has been industrialising rapidly. The proportion of the population directly engaged in agriculture has been declining steadily from $\sim 31\%$ in 1966 to $\sim 20\%$ in 1979 but this is still very high compared with the EEC average of $8\%$. It is estimated[1] that 20–25 % of the work-force engaged in manufacturing industries is involved in food processing or in industries directly connected with agriculture. Obviously, many others are indirectly connected with agriculture (banking, education, transport, research and advisory services, wholesaling and retailing). Thus, while the significance of agriculture in the Irish economy is declining, it is still the largest single sector.

Further evidence of the importance of agriculture in the Irish economy is provided by export statistics. Figure 1 shows the growth in Irish exports, total and agricultural, over the past 15 years (unadjusted for inflation). Total exports have grown rapidly with industrial exports growing at the faster rate; however, agricultural products still represent $\sim 35\%$ of total exports. The data in Fig. 1 represent gross values. Since most Irish industrial products depend on heavy importation of raw materials, the net value of industrial exports is considerably less; agricultural exports, on the other hand, are largely indigenous.

In comparison with that of other European countries, Irish agriculture is very heavily pastoral. Of the 7 million hectares in the Republic of Ireland,

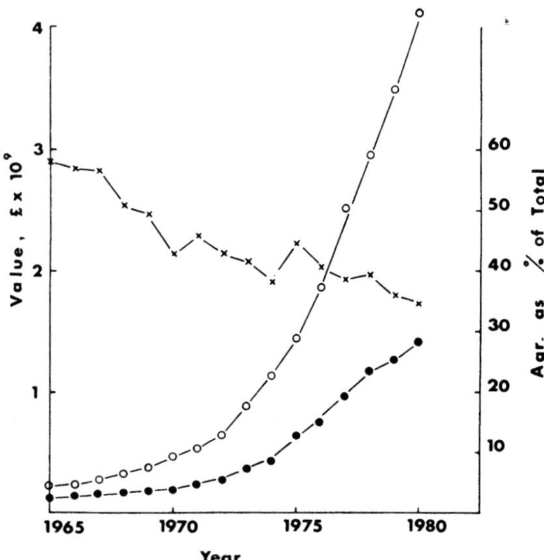


FIG. 1.    Value of Irish exports, 1965–1980. Total, ○. Agricultural, ●. Agriculture as percent of total, ×.

47% is under pasture, 14% is used for hay and silage and only 7% is devoted to tillage crops; the remainder is occupied by woodland, mountain, bog, roads, housing, etc. In Western Europe as a whole, ~50% of the arable land is cultivated. The preponderance of pasture, hay and silage obviously reflects the overwhelming importance of cattle for the production of milk and meat in Irish agriculture (Fig. 2) and their relative importance has increased in recent years. Cattle and milk represent a higher percentage of agricultural output in Ireland than in any other EEC country, except Luxembourg.[2]

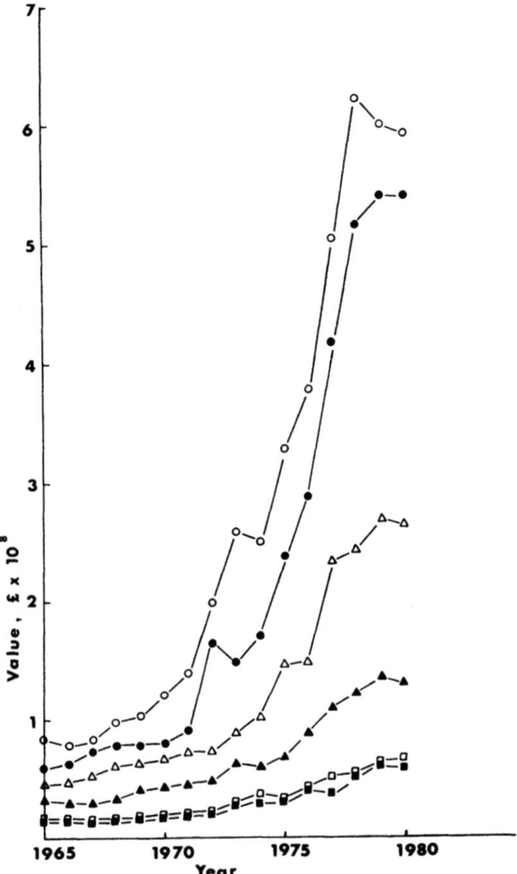

FIG. 2.　Value of Irish farm output. Cattle, ○. Milk, ●. Pigs, ▲. Sheep, ■. Poultry and eggs, □. All crops, △.

## THE IRISH DIET

Information on the Irish diet is, regrettably, rather limited. The only dietary survey undertaken was in 1948; the only other attempt at quantitation of the Irish diet, using consumption data, was made by Cremin and Morrissey.[3] Their data, summarised in Table 1, along with dietary intake for the US and the UK, show that cereals, dairy products and meats are the principal sources of dietary protein. Not surprisingly, these three commodities make similar contributions to dietary protein in

## TABLE 1

PERCENTAGE CONTRIBUTION OF FOOD GROUPS TO TOTAL DIETARY ENERGY, PROTEIN AND FAT[a]

| Group | % Dietary energy | | | % Dietary protein | | | % Dietary fat | | |
|---|---|---|---|---|---|---|---|---|---|
| | *Ireland* | *UK* | *US* | *Ireland* | *UK* | *US* | *Ireland* | *UK* | *US* |
| Milk, cream, cheese | 16 | 14·8 | 11·1 | 19·9 | 23·9 | 22 | 22·9 | 19·7 | 12·5 |
| Meats and fish | 14·6 | 17·2 | 20·0 | 38 | 34·6 | 42·6 | 22·4 | 28·2 | 34·1 |
| Eggs | 2·5 | 1·9 | 1·8 | 4·5 | 4·9 | 4·8 | 4·6 | 2·9 | 2·7 |
| Fats | 11·5 | 15·1 | 18·1 | ~0 | ~0 | ~0 | 30·4 | 36 | 43·3 |
| Sugars and preserves | 9·8 | 9·6 | 17·3 | ~0 | ~0 | ~0 | ~0 | ~0 | ~0 |
| Vegetables and fruits | 12·6 | 9·9 | 11·7 | 8·1 | 9·6 | 12·4 | 0·4 | 1·8 | 4·7 |
| Cereals | 34·8 | 29·7 | 19·2 | 24·6 | 25·1 | 17·6 | 6·9 | 9·9 | 1·3 |

[a] Taken in part from Cremin and Morrissey.[3]

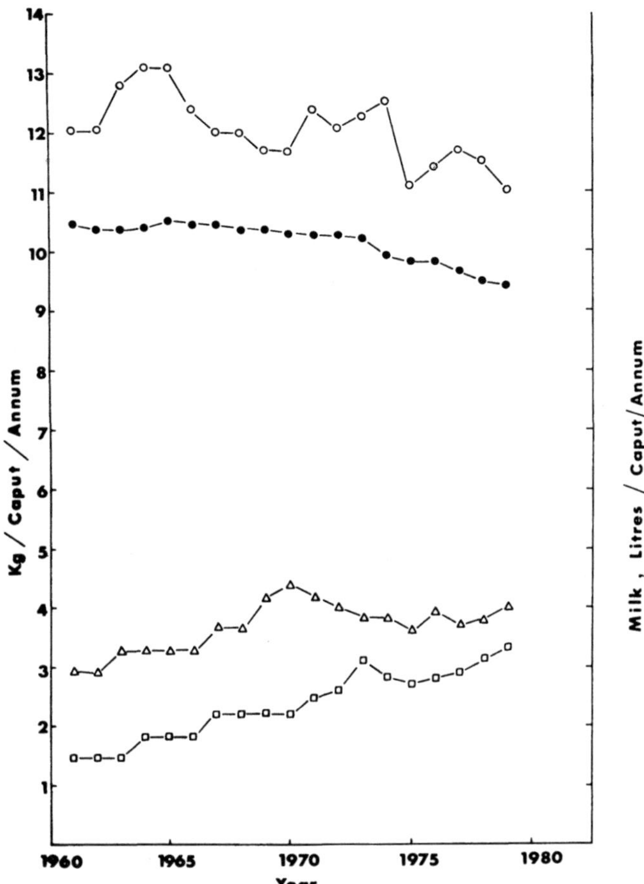



FIG. 3. Consumption of dairy products and margarine in Ireland. Butter, ◯. Cheese, ☐. Liquid milk, ●. Margarine, △.

Ireland and the UK but in the US meat contributes a higher proportion of dietary protein and cereals a lesser amount. In Ireland, we consume $\sim 110$ g of protein per head per day which is $\sim 250\%$ of the recommended dietary intake (45 g); $\sim 70\%$ of Irish dietary protein is of animal origin and therefore of high biological value.

The composition of the Irish diet has changed considerably in relatively recent years. The contribution from animal products has increased while the significance of cereals, and especially of the potato, has declined (Figs 3, 4, 5). This trend no doubt reflects increasing affluence and parallels

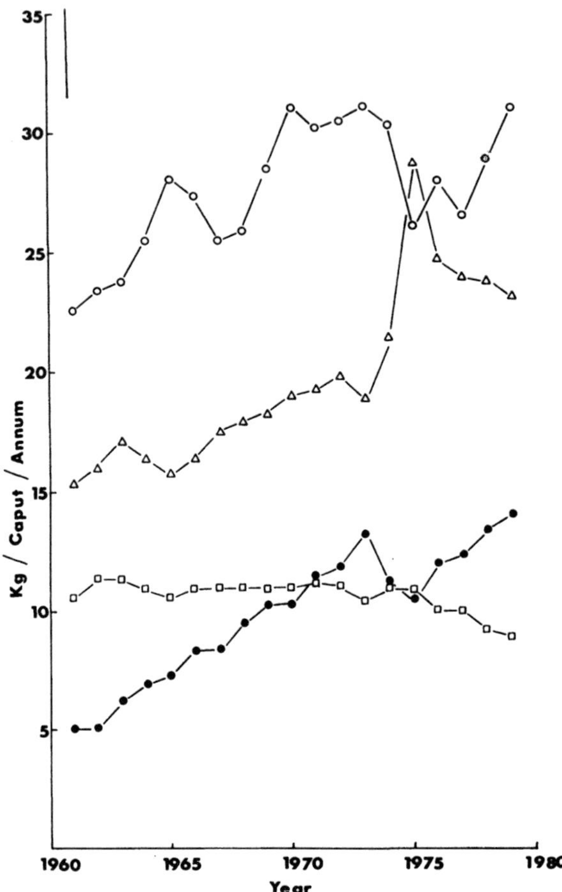



FIG. 4.  Consumption of meats and fish in Ireland. Pig-meat, ◯. Beef, △. Sheepmeat, ☐. Poultry, ●.

changes in most western countries. With the exception of a few countries, notably Argentina and Uruguay, there is a strong correlation between meat consumption and personal income. Apart from economic considerations, the increasing concern of nutritionists about the over-consumption of saturated lipids, mostly from animal sources, and the inadequate levels of dietary fibre in western diets may initiate a reversal of recent trends, especially in the consumption of fatty meats. However, in the Irish situation there is little evidence of such a change to date.

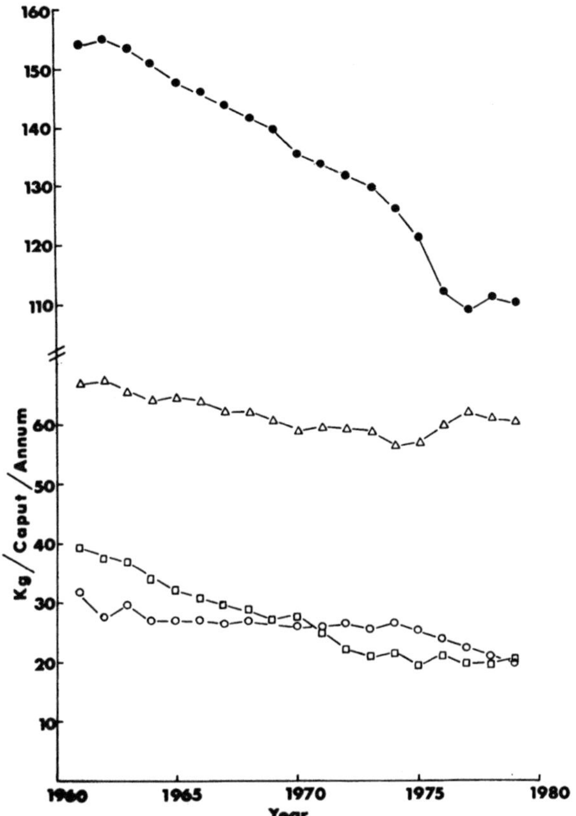



FIG. 5.    Consumption of bread (△), flour (□), sugar (○) and potatoes (●) in Ireland.

## PRODUCTION OF PROTEIN-RICH FOODS IN IRELAND

As shown in Table 1, the principal sources of dietary protein in Ireland and many other western countries are dairy products, flesh foods and cereals. Eggs are relatively minor contributors, while legumes are important sources in many developing countries and, in the form of soy products, are becoming increasingly important in developed countries. The status of each of these five groups in the Irish food industry will be reviewed in the following sections.

### Cereals

Only three cereals, wheat, oats and barley, are grown commercially in

Ireland; recent trends in acreage grown are shown in Fig. 6. Barley, the principal cereal, is used for malting and animal feed compounding and, as such, its protein is not utilised directly for human consumption. Only four species of cereal, wheat, oats, rice and maize, are consumed by humans in Ireland in significant amounts. All the rice consumed in Ireland is imported in processed form; $\sim 3000$ metric tons were imported in 1980 and, at 6 % protein, would contribute $180 \times 10^3$ kg of dietary protein, i.e. 53 g protein per caput per annum or 0·13 % of total dietary protein.

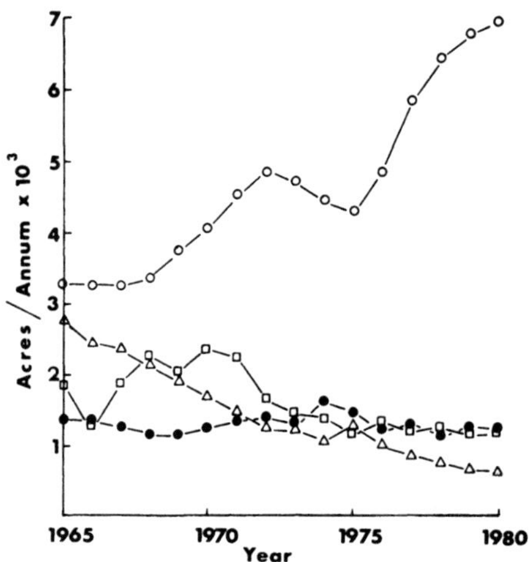

FIG. 6.   Acreage of cereals grown in Ireland. Feeding barley, ◯. Malting barley, ●. Wheat, ◻. Oats, △.

Oats, a popular traditional cereal in Ireland, was grown predominantly as a feed for horses and the rapid decline in the horse population has been paralleled by a decline in oat production (Fig. 6). Today, $\sim 120\,000$ tons of oats are produced in Ireland annually; a further 10 000 tons were imported in 1980. Approximately 12 000 tons of whole oats ($\sim 6000$ tons of flaked oats) are processed annually in Ireland for human consumption; some prepared oat products are also imported. With a protein content of $\sim 12$ %, 6000 tons of flaked oats contribute $\sim 7 \times 10^5$ kg dietary protein per annum, i.e. $\sim 200$ g protein per caput per annum or 0·5 % of total dietary protein.

Oat is unique among cereals in that 80 % of its proteins are albumins and globulins and are easily extractable by aqueous solvents. However, the solubility of the extracted protein in the pH range 2–10 is poor and consequently its functionality is limited. In addition, oat contains $\sim 5\%$ oil which may be extracted, along with much of the protein, by treatment of rolled oats with 0·02 N NaOH and may be recovered by ether extraction. Oat lipid is rich in polyunsaturated fatty acids and could be a significant source of indigenous vegetable oil. An economic assessment of oat as a raw material for the production of functional protein and vegetable oil may be worthwhile.

Although maize is grown in Ireland to a very limited extent for silage production, this cereal does not mature fully under Irish climatic conditions. Some maize is imported as 'corn on the cob' or as a component of canned or frozen mixed vegetables, but the vast majority of maize used in Ireland for human consumption is in the form of cornflakes. Up to 1979 a range of Kellogg products was manufactured under licence in Ireland; that licence was withdrawn in 1979 and all breakfast cereal products are now imported from the UK. Data on the imports and consumption of breakfast cereals (Table 2) show that consumption of these products is increasing rapidly. Data for the consumption of individual products are not available; if it is assumed that the average product contains 10% protein, then breakfast cereals contribute 430 g protein per caput per annum, i.e. $\sim 1\%$ of dietary protein.

As far as I can ascertain, maize is not processed in Ireland for the manufacture of starch and zein. However, maize starch is imported for the

## TABLE 2
### CONSUMPTION OF BREAKFAST CEREALS

| Year | Imports[a] (metric tons × $10^3$) | Total consumption[b] (metric tons × $10^3$) | Per caput consumption per annum (kg) |
|---|---|---|---|
| 1970 | | 6·7 | 2·3 |
| 1971 | | 6·8 | 2·3 |
| 1972 | | 7·8 | 2·6 |
| 1973 | | 8·7 | 2·9 |
| 1974 | | 8·7 | 2·9 |
| 1975 | | 9·9 | 3·2 |
| 1976 | 4·6 | 10·8 | 3·5 |
| 1977 | 4·9 | 11·7 | 3·7 |
| 1978 | 6·8 | 11·8 | 3·7 |
| 1979 | 7·9 | 12·2 | 3·8 |
| 1980 | 15·4 | 14·5 | 4·3 |

[a] From data published by the Central Statistics Office.
[b] From data supplied by the Kellog Company of Great Britain Ltd.

manufacture of glucose and glucose syrups. The scale of this industry at present is not sufficient to justify the manufacture of its own starch requirements from imported maize but perhaps consideration should be given to expanding the operation to include a wider range of starch products (e.g. modified starches, dextrins, glucose and glucose–fructose syrups). The protein by-product might form the basis of a valuable industry. Increased international interest in the production of ethanol as a fuel may offer an opportunity to develop a large-volume starch extraction industry leaving large quantities of zein for a protein products industry.

While Ireland has large areas of suitable soil types for wheat farming, the Irish climate is less than ideal for wheat. Nevertheless, wheat is a fairly important crop (Fig. 6) and, in good years, Ireland is $\sim 70\%$ self-sufficient in wheat. The protein content of Irish wheat is low (Table 3) compared with North American hard wheat ($14\cdot5\%$ protein)[5] and even with British wheats ($\sim 12\%$ protein).[6] Winter wheat has become increasingly popular in

TABLE 3

MEAN PROTEIN CONTENT ($\%$) OF IRISH SPRING
AND WINTER WHEATS[4]

| Year | Spring | Winter |
|------|--------|--------|
| 1974 | 8·6 | 8·5 |
| 1975 | 10·7 | 9·4 |
| 1976 | 10·9 | 9·3 |
| 1977 | 10·1 | 9·0 |
| 1978 | 11·0 | 8·7 |
| 1979 | 9·7 | 9·2 |
| 1980 | 11·3 | 9·5 |

Ireland in recent years because of higher yields, the higher probability of having good weather for harvesting and probably the desire to spread the harvesting season. Undoubtedly, these are attractive features from the agriculturalist's viewpoint, but from the food technologist's aspect, Irish winter wheat has, on average, $\sim 1\%$ less protein than spring varieties (Table 3) and yields flour even less suitable for baking than the already mediocre spring varieties.

The acreage of wheat sown in Ireland, following an increase around 1970, has generally decreased in recent years (Fig. 6). Data on the production and importation of wheat (Table 4) show that Ireland is now $\sim 50\%$ self-sufficient in wheat. Some importation of high-protein North

TABLE 4

PRODUCTION AND IMPORTATION OF WHEAT[a]

| Year | Irish grown wheat ('000 tons) | | | Imported wheat ('000 tons) |
| --- | --- | --- | --- | --- |
| | *Total* | *Millable* | *% Millable* | |
| 1975 | 172 | 165 | 96 | 247 |
| 1976 | 180 | 168 | 93 | 163 |
| 1977 | 227 | 200 | 88 | 175 |
| 1978 | 233 | 183 | 79 | 199 |
| 1979 | 229 | 119 | 52 | 199 |
| 1980 | | | | 298 |

[a] From Central Statistics Office trade statistics.

American wheats is necessary for mixing with Irish wheats to raise the protein content of the flour to a satisfactory level but the production of indigenous wheat could probably be increased. France, Canada and the UK are the principal sources of wheat imports.

Wheat unsuitable for milling, usually because of pre-harvest germination, is normally used in animal feed concentrates. In addition to flour milling and animal feed production, wheat, as far as can be ascertained, is not used in Ireland in the preparation of other products although some imported wheat-based breakfast cereals are used. There are plans to open a plant in Cork to produce gluten and glucose products, largely from imported wheat and unmillable Irish wheat.

### Other Vegetable Proteins

During the past decade there has been considerable interest in the use of vegetable and single cell proteins for human consumption. Most interest has centred on soy bean and, to a lesser extent, on oil seed residues. These crops cannot be produced in Ireland, with the exception of rape seed, small ($\sim$1400 acres in 1980) but increasing acreages of which have been sown recently with very satisfactory results. At present the rape seed is shipped to Britain for processing but if oil-seed rape proves agriculturally attractive it is likely that it will be processed domestically, thus providing the raw material for an indigenous vegetable protein industry although it is more likely to be used for animal feed, at least initially.

There has been some interest in the production of protein from grass but to date, at least, this has been directed toward the production of animal feed. No work appears to have been done on the production of other leaf proteins. Many regions in Ireland are suitable for the growing of peas and beans; approximately 1000 hectares are devoted to these crops. Domestic production and trade in peas and beans are summarised in Table 5. Ireland appears to be 60–70 % self-sufficient with respect to peas but most of the bean requirements are imported, mainly for canning, since kidney beans do not grow here, although it has been suggested that an effort might be made to identify or develop varieties suitable for Irish conditions. There is

TABLE 5

PRODUCTION AND TRADE IN PEAS AND BEANS ('000 metric tons)

| Year | Production | | Imports | | | | Exports | | | |
|------|------------|-------|--------|-------|--------|-------|--------|-------|--------|-------|
| | Peas | Beans | Peas | | Beans | | Peas | | Beans | |
| | | | Frozen | Dried | Frozen | Dried | Frozen | Dried | Frozen | Dried |
| 1975 | 11 | 2·3 | 1·2 | 3·9 | 0·4 | 3·4 | 1·8 | 0·2 | 0·1 | <0·1 |
| 1976 | 11 | 1·2 | 1·0 | 4·6 | 0·4 | 2·2 | 0·4 | <0·1 | <0·1 | <0·1 |
| 1977 | 14·1 | 1·3 | 2·3 | 5·3 | 0·5 | 3·3 | 1·3 | 0·1 | <0·1 | <0·1 |
| 1978 | 10·1 | 1·6 | 4·3 | 5·5 | 0·8 | 3·4 | 0·8 | <0·1 | 0·2 | <0·1 |
| 1979 | 11 | 1·2 | 2·6 | 5·2 | 1·1 | 3·7 | 1·4 | 0·6 | 0·1 | <0·1 |
| 1980 | 5·6 | 0·8 | 2·5 | 3·4 | 1·1 | 2·6 | 0·5 | 1·1 | <0·1 | <0·1 |

considerable interest in the UK in the lupin seed as a source of vegetable proteins, and although only relatively small acreages are grown at present, the crop appears to be economic; economic assessment of this crop in Ireland appears warranted.

Considerable quantities of soy bean are imported and processed in Ireland for animal feed. Although some soy bean products are used here in human food preparations, especially in meat products, soy bean for such applications is not processed in Ireland.

**Eggs**
The annual per caput consumption of eggs in Ireland has decreased slightly from ~250 in 1960 to ~200 in 1978 (~27 g per caput per day). Eggs supply ~3·5 % of total dietary protein.

Data for the production, importation and export of eggs (Table 6) show that Ireland was essentially self-sufficient in eggs and that the extent of foreign trade was small in 1978. However, importation of fresh eggs has now become significant and, in 1980, represented ~30 % of total consumption. This development is apparently due to the establishment of

## TABLE 6
PRODUCTION, IMPORTS AND EXPORTS OF EGGS AND EGG PRODUCTS (TONNES)[a]

|                      | 1978   | 1979   | 1980   |
|----------------------|--------|--------|--------|
| Production           | 35 344 | 32 011 | 31 000 |
| Exports              | 13     | 51     | 47     |
| Imports              | 907    | 3 949  | 9 408  |
| Self-sufficiency (%) | 97·7   | 88·9   | 69·7   |
| Egg products         | 365·4  | 409·1  |        |

[a] Data supplied by Department of Agriculture.

trading relationships between some Irish poultry farms and international companies operating in Northern Ireland.

The processing of eggs in Ireland is confined to the utilisation of broken or cracked eggs for liquid whole egg, liquid white and liquid yolk. Liquid egg products are sold fresh to bakeries but some are frozen. Considerable quantities of fresh and frozen egg and egg fractions are imported and there is a low level of exports (Table 7).

No dehydrated eggs are manufactured in Ireland while 31 000 kg are imported annually. The feasibility of developing a dehydrated egg industry in Ireland has been assessed and it would appear not to be economical because of the low volume of raw material available from widely scattered locations. Considerable quantities of egg white substitute are also imported

## TABLE 7
IMPORT/EXPORT OF EGG PRODUCTS, 1980[a]

| Commodity | Imports | | Exports | |
|-----------|---------|---------|---------|---------|
|           | *Quantity* (metric tons) | *Value* (Ir £) | *Quantity* (metric tons) | *Value* (Ir £) |
| Dried eggs       | 30·9  | 65 631  | —   | —   |
| Liquid eggs      | 363·6 | 258 542 | 3·5 | 886 |
| Liquid egg yolk  | 105·0 | 81 237  | 2·8 | 970 |
| Frozen egg yolk  | 23·4  | 16 781  | 0·4 | 129 |
| Dried egg yolk   | 2·4   | 4 201   | —   | —   |
| Ovalbumin and lactalbumin | 16·0 | 65 173 |   |   |

[a] Data supplied by Department of Agriculture.

                        *P. F. Fox*

but, as far as can be ascertained, no such products are manufactured in Ireland.

## Meat and Meat Products
*Cattle*

Cattle farming has been a major agricultural activity in Ireland for many centuries. In contrast to beef farming systems in most other countries, the practice in Ireland is such that $\sim 70\%$ of beef cattle are the progeny of the dairy herd with only 30% being produced from beef cows. Friesian type cows now strongly predominate in the dairy herd and, consequently, Friesian and Friesian/beef breed crosses predominate (75–80%) among beef cattle. Irish beef is produced mainly on grass although there is a weak trend toward yard fattening on hay/silage and concentrates. There has been a generally upward trend in cattle output over the past 15 years (Fig. 7); at present about 1·5 million beef cattle are reared annually. The disposal and processing of these cattle has been, and still is, a rather contentious issue.

During the 18th and 19th centuries Ireland was a major processor of salted beef. This industry was based mainly in Cork which was a major provisioning port for the British and other navies and for trans-Atlantic ships. The industry arose because the British Cattle Acts, passed in 1665, prohibited the export of live cattle and fresh meat to Britain.

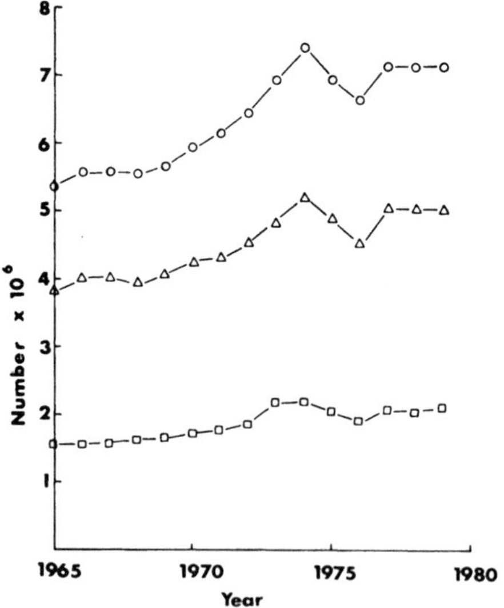


FIG. 7. Cattle numbers in Ireland. All cattle, ○. All non-dairy cattle, △. All cows, □.

The Cattle Acts were gradually and periodically repealed during the period 1757–1776 and completely and permanently repealed in 1776, following which the export of live cattle recommenced and the salt beef industry gradually declined. From the middle of the 19th to the middle of the 20th centuries, essentially all cattle were exported live to Britain either as 'stores' or fat cattle (Fig. 8). The economic folly of this practice was obvious and a few ill-fated attempts were made to develop a meat processing industry. However, it was not until 1926 that the first successful company, Clover Meats, was established at Waterford. This was followed, about 1936, by the establishment of Roscrea Meats, by the Government in collaboration with private interests, to process cull cows, necessitated by the so-called 'Economic War' with Britain during the 1930s. These developments did not significantly alter the pattern of cattle disposal and it

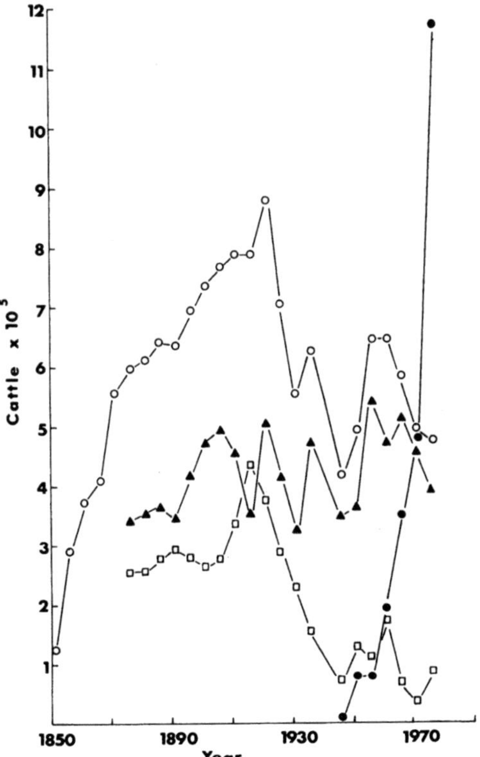

FIG. 8.    Exports of cattle and beef, 1850–1975. Total live, ○. Store cattle, ▲. Fat cattle, □. Fresh beef, cattle equivalent ●.

was not until $\sim 1950$ that a meat processing industry became significant. However, to the displeasure of many, the export of live cattle persists and still represents $\sim 30\%$ of total output (Fig. 9). Britain is still the principal outlet for live cattle but lucrative markets have recently been established in North Africa for fat cattle and in Italy for calves. Farmers argue that the retention of live exports is essential for the buoyancy of the cattle trade but, from the national viewpoint, it is generally considered to be undesirable.

Although the percentage of total cattle exported as meat has greatly increased over the past 15 years, it is agreed that such meat should be more

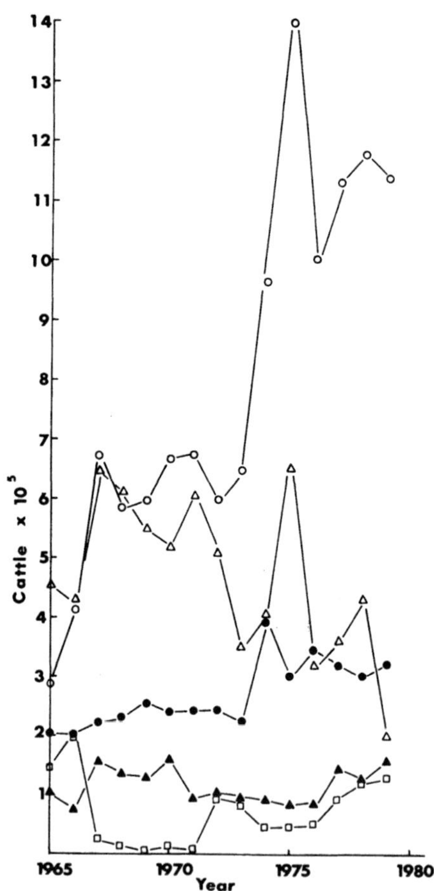

FIG. 9.    Trade in cattle and beef, 1965–1979. Exports of store cattle, $\triangle$. Exports of fat cattle, $\square$. Imports of cattle, $\blacktriangle$. Exports of meat, $\bigcirc$. Home consumption of beef, $\bullet$.

extensively processed domestically. The vast majority of beef is exported as sides or quarters with only relatively small amounts being sold as 'vac-packs' cuts or other high value-added forms (Fig. 10). In recent years, pricing policies of the EEC are at least partially responsible for the slow development of meat processing in Ireland; these policies frequently make it more profitable to dispose of sides into intervention than to break down

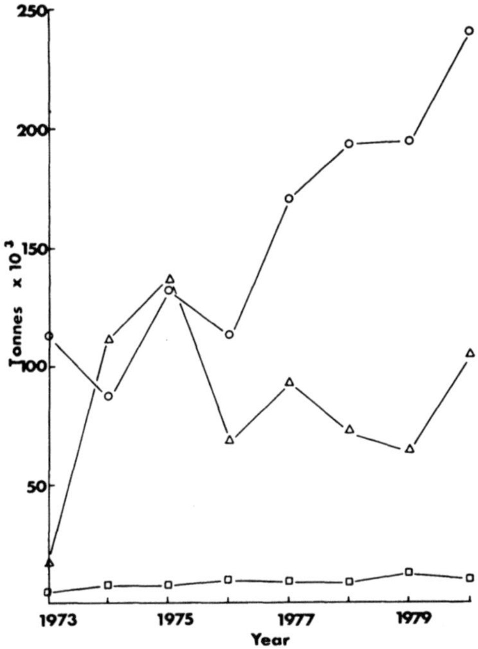


FIG. 10. Exports of meat products. Fresh or chilled (including vacuum packed at 10 000 tonnes), ○. Frozen (with bone or boneless, approximately 1:1), △. Processed (cooked and uncooked), ☐.

the carcass to wholesale or retail cuts. Structural problems within the beef industry are also contributors.

Processing of food items based on beef is largely undeveloped in Ireland. Corned or canned beef is the principal such product but hamburger type products are assuming significant proportions (Table 8). Other traditional products include steak and kidney pie, while pre-cooked, packed meats play a minor role. Most of the major items of edible offal are exported unprocessed in frozen form and presently amount to ~20 000 tons per

## TABLE 8
TRADE IN BEEF PRODUCTS ('000 TONNES)[a]

| Year | Vac-packed and other boneless | Processed | | | |
|---|---|---|---|---|---|
| | | Cooked[b] | | Uncooked[c] | |
| | | Exports | Imports | Exports | Imports |
| 1973 | 6·9 | — | — | — | — |
| 1974 | 10·3 | — | — | — | — |
| 1975 | 9·9 | — | — | — | — |
| 1976 | 8·2 | — | — | — | — |
| 1977 | 8·1 | 3·1 | 1·1 | 2·7 | 0·2 |
| 1978 | 8·1 | 4·7 | 2·0 | 4·2 | 0·1 |
| 1979 | 8·0 | 6·8 | 2·6 | 5·2 | 0·3 |
| 1980 | 10·2 | 4·1 | 2·2 | 4·8 | 0·3 |

[a] Data from Coras Beostoic & Feola.
[b] Corned beef, stewed steak, frozen processed beef.
[c] Hamburgers, etc.

annum. Minor amounts of meat scrap and bone are used in the preparation of soups but most of the bone from boning-out operations is used for animal feed as meat and bone meal, $\sim$40 000 tons of which are produced annually. One company processes and fractionates bovine blood for human consumption; it claims to process $\sim$70% of available blood although processing in the Republic of Ireland appears to be irregular, e.g. the company is not processing in 1980 due to a very large kill in 1979. The same company manufactures decalcified bone for gelatine manufacture but does not produce the finished product.

About 80% of the beef produced in Ireland is exported in one form or another, making Ireland probably the largest exporter of beef per caput in the world. Beef and cattle exports are, in fact, the largest single export item of Irish trade, although the value of exports of dairy products now is of about the same magnitude when EEC subsidies are included in the latter.

### Sheep

Sheep farming is a traditional, although relatively minor, Irish farming enterprise. Sheep numbers have generally declined in recent years (Fig. 11) and, in spite of apparently lucrative outlets for lamb on the French market recently, there is no sign of a recovery. Per caput consumption of sheep-meat in Ireland, at $\sim$10 kg per caput per annum, is high by international

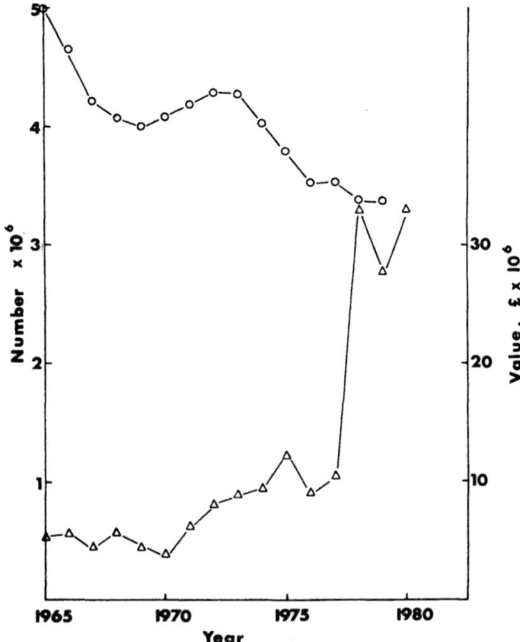

FIG. 11.   Sheep numbers (○) and value of exports of sheep and sheep-meat (△).

standards; domestic consumption represents $\sim 50\%$ of total output. The remainder is exported mainly in carcass form although there are some exports of live lambs to North Africa. Values of exports of sheep and sheep-meat are summarised in Fig. 11; the marked increase in 1978–1980 undoubtedly reflects the opening of the French market to Irish lamb.

## Pig-Meat

Exports of pig-meat from Ireland are recorded as far back as 1650 but values were very small until 1750. Prior to 1750 all exports of pig-meat were of salted pork, rather than bacon and ham, which did not develop as distinct products until $\sim 1760$. There was a considerable increase in exports of both salt pork and bacon during the period 1760–1840 (Fig. 12) and many of today's bacon companies were established toward the latter part of this period. In 1840, Ireland was the principal exporter of bacon to the British market where Irish bacon was highly regarded. Exports were disrupted by the Great Famine (1844–1848) but were resumed afterwards although stiff competition was now encountered from Danish bacon

*P. F. Fox*

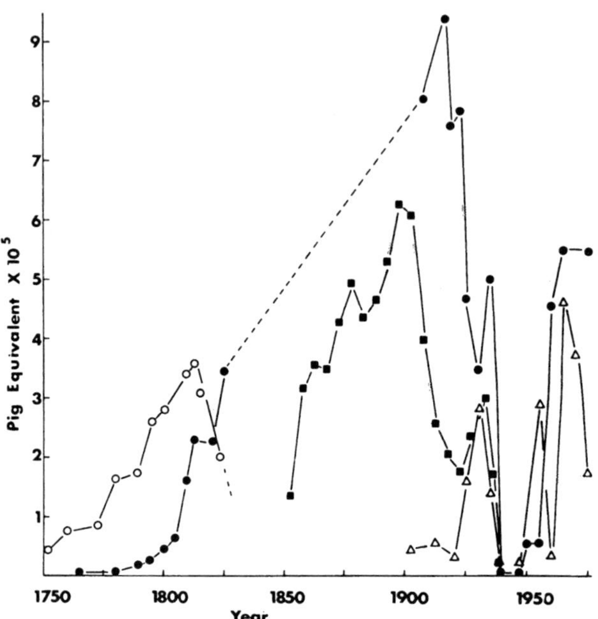

FIG. 12.　Exports of pigs and pig-meat products. Salt pork, ○. Bacon, ●. Fresh pork, △. Live pigs, ■.

(manufacture of Wiltshire bacon did not commence in Denmark until ~1847, with the help of an Irish bacon curer) although, until 1897, Danish bacon commanded a lower price than Irish bacon on the British market. During the 20th century Danish bacon has been much more successful than its Irish counterpart and now commands ~45% of the British market compared with 4–5% for Irish bacon.

The upturn in pig production after 1750 was due to the expansion of potato growing and of dairying, pigs being fattened on skim milk and surplus potatoes. Traditionally, pig production and dairying have been closely associated but, although pig production is still concentrated in dairying areas, the two industries have become more or less separated due to specialisation of both. Traditionally, pigs were reared in small numbers on many farms but today, the pig industry has become highly centralised and specialised on a small number of farms.

The export of salt pork reached a maximum of ~350 000 pigs per annum in 1812 and declined thereafter, being gradually replaced by bacon and ham. An active export trade in live fat pigs to Britain developed from

~1850 onward, reached a peak of ~500 000 pigs per annum in ~1900 but ceased about 1940.

Because of its association with dairying, pig farming was concentrated in roughly equal proportions in Ulster and Munster. However, the development of bacon industries in the two regions followed different paths. In Munster, centralised factories for the slaughter and curing of pig-meat developed early, whilst in Ulster it was traditional, up to the 1930s, for farmers to kill their own pigs and offer the carcass for sale at local markets. While this practice avoided stress-related problems in bacon, e.g. high or low pH, variations in the standard of hygiene could be expected and the bacon industry in Ulster has been factory based since ~1930, permitting better quality control.

The sharp decline in the export of pigs and bacon after ~1920 (Fig. 12) is likely to be due, at least in part, to the partitioning of Ireland at that time since ~50% of the Irish bacon industry was Ulster-based. The sudden increase in exports of pork about 1920 is probably due to a continuation of the traditional movement of fresh pig carcasses from Cavan/Monaghan for curing in factories in what became Northern Ireland. Exports of bacon appear to have been discontinued during World War II but were resumed in 1949 and increased gradually up to 1960. Since 1960, in spite of some fluctuations, there has been a small general increase in pig numbers, and in home consumption of bacon and pork (Fig. 13). Pig-meat is the most popular form of meat in Ireland although beef consumption has increased rapidly in recent years (Fig. 4). However, exports of bacon, and especially of pork, have been rather irregular (Fig. 13); most of the exports of pig-meat products are to the UK, although small amounts go to a variety of markets.

The traditional form of bacon curing in Ireland is the Wiltshire process but some factories have developed sizeable outputs of fabricated, pressed joints and canned ham/bacon products. Sausages are the major line of pork by-products; blood (black) sausage, made from porcine blood, is also very popular.

*Poultry*

Until about 15 years ago, small numbers of free-range poultry, especially chickens, but frequently ducks, geese and turkeys, were reared on most farms for egg and meat production to meet family requirements and as a significant source of cash. Although the Government embarked on a major programme about 1950 to expand the poultry industry, it has never become more than a minor sector of the agricultural industry. Like the pig industry, and even more dramatically the poultry industry, both egg and meat

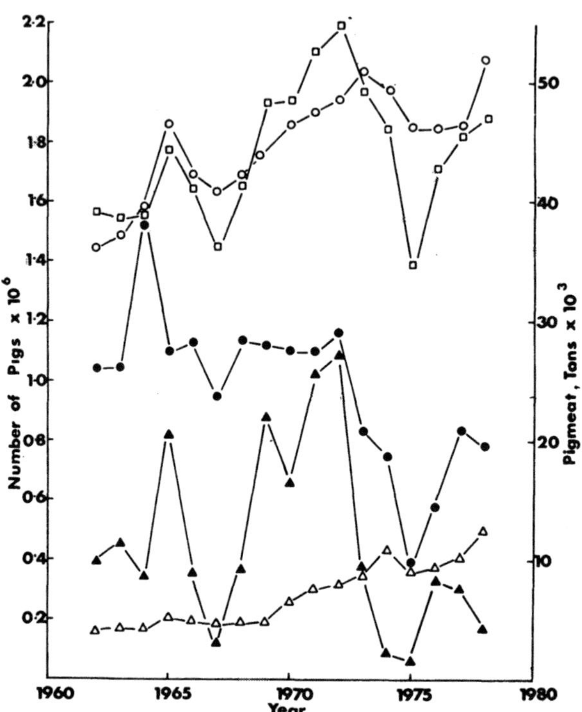



FIG. 13.    Same features of the modern Irish pig industry. Total number of pigs processed, □.
Domestic consumption of bacon, ○. Exports of bacon, ●. Domestic consumption of pork,
△. Exports of pork, ▲.

sectors have switched from a farm-yard enterprise to a highly specialised, well-organised industry involving relatively few companies, many of whom process and market their own production; some even manufacture a range of secondary products.

Ireland is presently about self-sufficient in total poultry meat (Table 9). It is a net importer of broilers but $\sim 90\%$ of the 1·6 million ducks reared are exported. The industry is growing slowly but, at an annual value of £64 million ($\sim 4\%$ of gross agricultural product), is a relatively minor sector, although it is surpassed in value only by cattle, dairying and barley, in terms of farm output.

## Fishing

Although Ireland has a very long coastline proportional to its area and is located in very rich fishing grounds, the Irish national fish catch is smaller

TABLE 9
ASPECTS OF POULTRY MEAT INDUSTRY[a]

|  | 1978 | 1979 | 1980 |
|---|---|---|---|
| Broilers (numbers) | $23 \times 10^6$ | $24 \cdot 3 \times 10^6$ | $24 \cdot 4 \times 10^6$ |
| Ducks (numbers) | — | — | $1 \cdot 6 \times 10^6$ |
| Production (tonnes) | 43 000 | 47 000 | 47 000 |
| Export (tonnes) | 3 820 | 3 700 | 7 200 |
| Imports (tonnes) | 4 197 | 4 626 | 4 961 |
| Self sufficiency | 99·1 | 98·0 | 105·8 |

[a] Data supplied by the Department of Agriculture.

than those of other EEC countries except Luxembourg and Belgium.[7] However, in contrast to most other EEC countries, the Irish catch is increasing slowly and there has been considerable expenditure both on fishing vessels and on shore-based fish processing facilities (Fig. 14).

The fishing industry is a significant employer along the western coast where other sources of employment are few. It is estimated that there are 3485 full-time and 5333 part-time fishermen; 1600 are employed in on-shore processing and 770 in other on-shore jobs.[8] The majority of fishermen operate close to shore but there has been a recent significant increase in the number of larger vessels capable of deep-sea fishing; for example, in 1980, there were nine Irish-owned vessels over 90 ft in length compared with two in 1975.

Actual landings of some of the principal species or groups of species for 1977–1979 are shown in Table 10; clearly, herring and mackerel are the dominant species. The values of landings, exports and imports of fish are summarised in Fig. 14. The value of exports is about twice that of imports but examination of exports and imports according to the degree of processing (Table 11) shows that fresh, chilled or frozen products represent 54·9 % of total exports but only 20·5 % of imports while processed products, excluding fish meal and oils, represent 13 % of exports and 57·8 % of imports. Presumably these data indicate that most of our fish are exported in unprocessed form while we import principally processed products. It is hoped that the increased investment in on-shore processing facilities will redress this imbalance; indeed, the first fish cannery in Ireland was opened in Killybegs in 1977 and the first factory producing fish fingers went into production in 1981 (imports of this product are valued at ~£5 million per annum). Although the value of fishing is low compared with agriculture, it represents 0·9 % of GDP, which is the highest proportion of

*P. F. Fox*

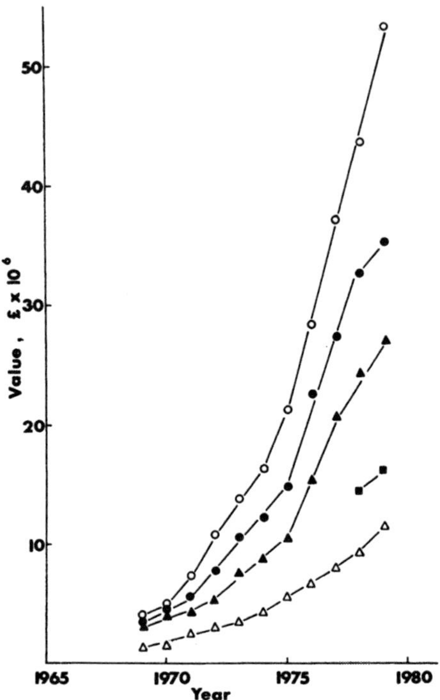



FIG. 14. Investment in fish-handling facilities and the value of trade in fish. Investments in fishing fleet, ○. Investments in on-shore processing facilities, △. Value of sea fish landings, ▲. Value of fish exports, ●. Value of fish imports, ■.

TABLE 10

LANDINGS OF PRINCIPAL FISH SPECIES EXCLUDING SALMON[a]

| *Species or group of species* | *Landings (metric tons)* | | |
|---|---|---|---|
| | *1977* | *1978* | *1979* |
| Total demersal | 18 887 | 17 940 | 21 092 |
| Herring | 23 129 | 27 717 | 27 383 |
| Mackerel | 22 695 | 27 505 | 24 217 |
| Sprats | 6 055 | 9 119 | 1 892 |
| Total shell fish | 11 722 | 11 403 | 11 114 |
| At foreign ports (total) | 5 849 | 6 699 | 5 233 |
| Grand total | 88 337 | 100 388 | 90 931 |

[a] Calculated from BIM Annual Report, 1979.

## TABLE 11
TRADE IN FISH AND FISHERY PRODUCTS, 1979[a]

| | Exports | | | Imports | | |
|---|---|---|---|---|---|---|
| | *Metric tons* | *Ir £million* | *% Total value* | *Metric tons* | *Ir £million* | *% Total value* |
| Fresh, chilled, frozen | 28 011 | 15·3 | 43·3 | 2 010 | 2·3 | 14·3 |
| Dried, salted, smoked | 10 671 | 7·1 | 20·1 | 1 591 | 1·7 | 10·6 |
| Prepared or preserved | 783 | 1·0 | 2·8 | 4 461 | 7·3 | 45·3 |
| Shell fish; fresh, chilled, frozen | 3 850 | 4·1 | 11·6 | 284 | 1·0 | 6·2 |
| Shell fish preparations | 2 242 | 4·9 | 13·9 | 94 | 0·3 | 1·9 |
| Direct exports ex vessels | 5 233 | 2·1 | 5·9 | — | — | — |
| Fish meal, oil, etc. | 4 764 | 0·8 | 2·3 | 14 443 | 3·5 | 21·7 |
| | | 35·3 | 99·9 | | 16·1 | 100 |

[a] Calculated from BIM Annual Report, 1979.

GDP represented by fish of any EEC country except Denmark. The principal destinations of exports are the UK (£9·4 million), the Netherlands (£6·7 million), France (£5·4 million) and Germany (£4·7 million) (1979 values).

There is considerable interest in mariculture in Ireland with about forty fish farms producing $\sim$£1·5 million worth of fish in 1980. The principal species farmed are salmon, trout ($\sim$450 tons), mussels, oysters (900 tons) and scallops. It is believed that there are many very suitable locations for fish farming around the Irish coast and that the industry has considerable potential for development.

Consumption of fish in Ireland is low: $\sim$5·3 kg per caput per annum in 1979, which shows a slight increase from 4·6 kg per caput in 1970.

**Dairying**

Ireland has a long tradition in dairying due, to a major extent, to its climatic and soil conditions. In addition to butter, many samples of which have been excavated from bogs, the ancient Irish diet included a variety of hard and soft cheeses and fermented milks; unfortunately, manufacture of these was discontinued 200–300 years ago. The export of butter from Ireland dates from at least 1550 when the trade was prohibited by the Lord Deputy, St. Ledger. Fairly reliable figures for the export of butter are available from 1641 onward and, with the exception of a period around 1940, there has been a more or less continuous increase (Fig. 15). At one time cheesemaking appears to have been of considerable importance but it died out during the 17th century and, with the exception of a brief period during World War I, cheese did not again assume a position of any significance until the 1960s.

Up to $\sim$1880 butter was manufactured on the farm from naturally ripened, gravity separated cream. It was purchased by merchants at local markets and exported via the Cork Butter Market, which was the largest and most successful butter exchange in the world during the 18th and 19th centuries. It set international prices and maintained its quality standards by introducing a butter grading scheme as early as 1770. Although its importance declined during the late 19th century, it continued to operate until the 1930s.

The development of the mechanical separator and pasteurisation of cream in the 1880s facilitated the conversion of dairying from a farm-yard to a factory-based industry. The first creamery in Ireland was established in Hospital, Co. Limerick, in 1884 and the first co-operatively owned creamery at Dromcollogher, Co. Limerick, in 1889. The Irish Agricultural

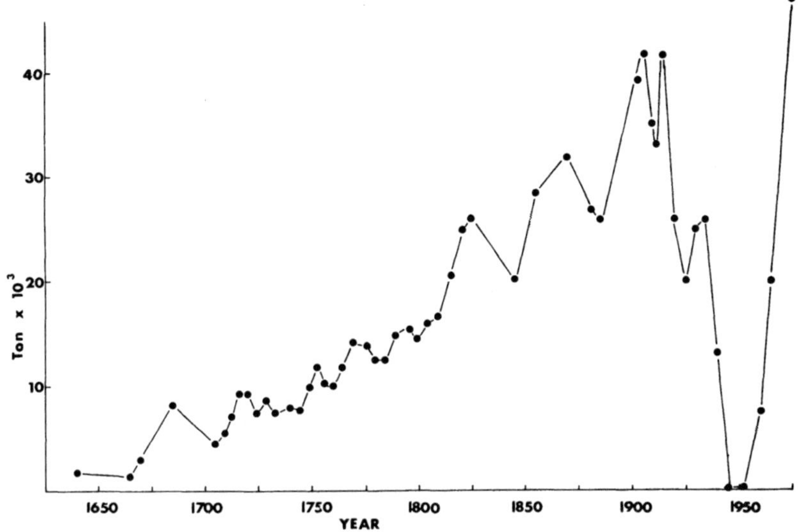


Fig. 15.    Exports of Irish butter, 1640–1975.

Organisation Society was established in 1894 as the national organising and advisory body to the Irish agricultural co-operative movement. Both privately and co-operatively owned creameries were rapidly established thereafter and by 1907 there were 780 creameries in operation. This number was too many for the milk supply available and, in an attempt to rationalise the dairy industry, the Irish Government established the Dairy Disposal Company Ltd. (DDC) in 1927 with a mandate to purchase proprietary creameries with the object of closing superfluous premises and transferring the remainder to farmer co-operatives. Rationalisation was rapid initially and by 1943 all creameries were owned by co-operatives or the DDC, which controlled 80% and 20% of manufacturing milk, respectively. However, it was not until 1975 that the DDC transferred its last creamery to co-operative ownership. Today, all milk is purchased by co-operatives although many of these are involved in joint ventures with private companies (usually British). In 1962 there were 156 central creameries in Ireland but considerable amalgamation of small creameries occurred between 1970 and 1975 and there are now about fifteen major processors with perhaps six smaller ones.

The Irish dairy industry has expanded rapidly and been drastically altered and modernised since 1960. It has reorganised into larger manufacturing groups; most of the branch creameries (collection points) have been closed with about 60% of the total milk supply now collected by road tankers. Refrigerated on-farm storage of milk is now the norm on larger farms, usually with alternate-day collection. An efficient marketing organisation, An Bord Bainne (Dairy Board), has been established. Some product diversification has occurred. A major factor militating against diversification, especially into the more consumer-oriented products, is the very marked seasonality of Irish milk production; there is a 16:1 ratio between peak (May) and trough (December) production of manufacturing milk. Only New Zealand has a seasonality pattern like Ireland, and for the same reason, i.e. in both countries dairy farming is very strongly dependent on grass. Perhaps, from the farmer's viewpoint, this is economically correct, and perhaps also from the national viewpoint, considering the increasing cost of energy, but it creates many problems for manufacturers: variations in the composition of raw material, continuity of supply of short-life products, surplus plant capacity, seasonality of employment. In spite of attempts by processors via incentive and bonus schemes to make the milk supply more uniform, there has been very little change to date.

With the exception of a few years, which can be explained by prices, animal disease eradication programmes and/or poor weather, there has

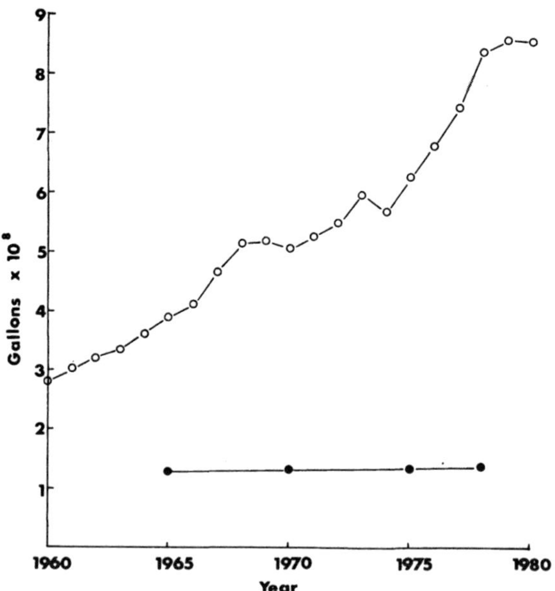



FIG. 16.   Output of manufacturing milk, ○, and liquid milk, ●, 1960–1980.

been a general and fairly rapid increase in milk output since 1960 (Fig. 16). This increase has occurred due to increased cow numbers and, more especially in latter years, to increased output per cow which is still somewhat low in comparison with other major countries.[10] Lower yields are due mainly to the grass-based farming system and do not necessarily reflect low profitability to the farmer.

Output of the principal dairy products is summarised in Figs 17 and 18. Butter and skim milk powder are the principal products and, in spite of the often-expressed view that greater diversification of milk utilisation is required, these two products have consistently represented ~70% of total manufacturing milk. The Irish dairy product mix is in fact remarkably

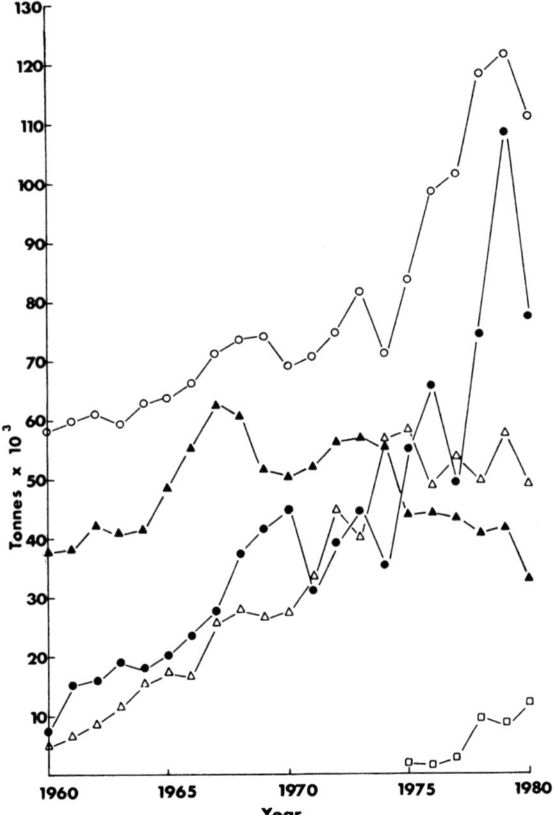


FIG. 17. Production of high-fat dairy products. Butter production, ○. Butter exports, ●. Butter oil, □. Cheese, △. Chocolate products, ▲.

*P. F. Fox*

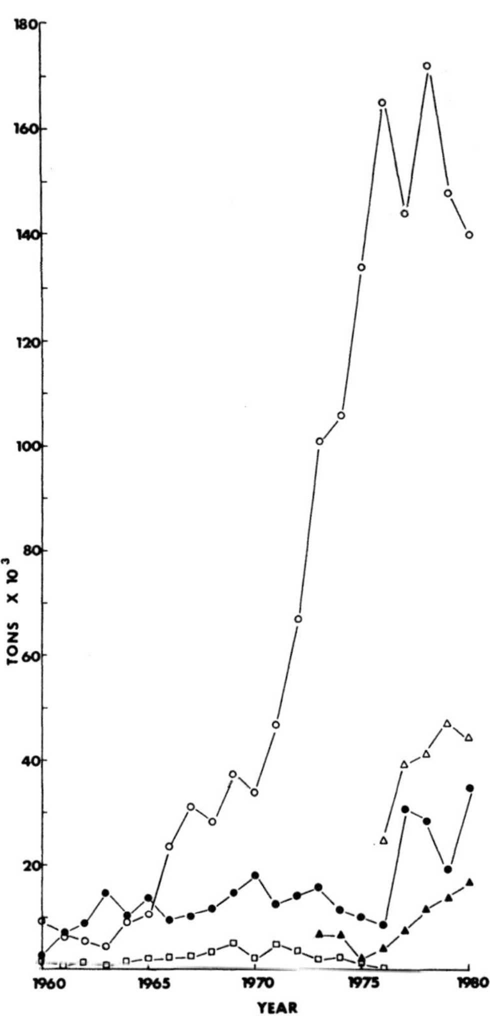



FIG. 18.   Production of concentrated and dehydrated dairy products. Skim milk powder, ○.
Whole milk powder, ●. Prepared milk powders, including infant foods, △. Caseins, ▲.
Condensed and evaporated milks, □.

similar to that of New Zealand, presumably reflecting the availability of major export markets for dairy products. Within the general areas of butter and skim milk powders there has been considerable diversification, especially with respect to a range of specification milk powders. There has been considerable interest in the manufacture of heat-stable powders for use in the preparation of reconstituted milk products and An Bord Bainne has formed joint-venture companies overseas in reconstitution/recombination plants. Casein (acid and rennet) has become a major product in recent years and, because of current EEC pricing policy, has recently been more profitable than skim milk powder. Some of the casein manufacturers have extended the use of their caseins downstream to the manufacture of cheese analogues. Manufacture of butter oil has had some impact on the proportions of milk fat used for butter manufacture.

Cheese production, mainly Cheddar, grew rapidly between 1965 and 1975 but has recently stabilised due to market prices *vis a vis* alternative outlets. However, production is tending to increase again with diversification of the range of cheese varieties produced, notably Gouda, Emmentaler, Romano-type and hard Greek varieties. There are at present thirteen cheese factories, all quite large and very highly mechanised. Much of the whey is still utilised as pig feed (Table 12), either in standard or concentrated form; some whey is spray-dried and some is incorporated into milk-based animal feeds. Two plants produce whey protein concentrate, by ultrafiltration, for use in human foods and three produce demineralised whey (by electrodialysis or ion exchange) for incorporation into baby food formulations. Heat-denatured lactalbumin is not produced on a regular

TABLE 12

UTILISATION OF WHEY (CHEESE AND CASEIN)[a]

| | *1976* | *1977* | *1978* | *1979* |
|---|---|---|---|---|
| *Production* | | | | |
| (Million gallons) | 120 | 153 | 177 | 200 |
| *Use* | | | | |
| Animal feed (%) | 52 | 58 | 50 | 47 |
| Whey powder (%) | 23 | } 37 | 3·5 | 3 |
| Other products (%) (Baby food, lactose WPC, etc.) | 17 | | 31 | 43 |
| Unaccounted for | 8 | 3·5 | 16 | 7 |

[a] From data supplied by Department of Agriculture.

basis. There has been research interest in the manufacture of 'texturised' casein for use as a meat extender but it has not been commercialised, probably for economic reasons. To my knowledge, milk protein hydrolysates from casein or whey proteins, by acids or enzymatic means, have not been commercialised here although there is research interest in such products.

The manufacture of 'humanised' milk-based baby food formulations is an important and growing sector of the Irish dairy industry. It is likely that even more highly modified preparations will be developed in this sector.

## CONCLUDING REMARKS

In this presentation I have, hopefully, demonstrated that the production and processing of 'food proteins' is a major aspect of the Irish economy. The average Irish diet contains at least twice the recommended daily intake of protein which is mainly of animal origin and of high biological value. Irish agriculture produces and exports 5–6 times as much beef and protein-rich dairy products (skim milk powder, caseinates, cheese) as are consumed domestically. Although some of us may be unable to afford it, it is unlikely that Ireland, at least from the supply viewpoint, will suffer from a shortage of food proteins.

Economically, the production of protein-rich foods, especially meat and dairy products, is of major national significance. That the principal agricultural products produced in Ireland are protein-rich is, however, completely accidental. I think that it is fair to say that we do not have a food protein industry, or perhaps not even a food industry. We have a large agricultural industry which, although it is declining in its absolute importance, is still the major sector of the Irish economy. The food industry exists only because it is necessary, for marketing, technical and sometimes economic reasons, to process agricultural produce. If such constraints did not exist, the output of Irish farms would probably be exported unprocessed, as is still the case with a significant proportion of cattle. Although it is probably too idealistic and totally unrealistic to suggest that the food industry should exist to supply people with food rather than to process agricultural produce for economic reasons, it appears to me that it would be more desirable if farmers reacted to the needs of the food industry, and ultimately of the consumer, rather than *vice versa*. The seasonality of the supply of milk and cattle is a particular case in point. Unlike many other countries, we do not, in Ireland, have a Minister of Food, not even a nominated section within a government department. Several people have recommended the establishment of such and I would support them. Although increased bureaucracy is undesirable, it is unlikely to cause a deterioration of the present situation in which at least four government departments are concerned with food-related matters.

However, even with all the present imperfections, the Irish food industry, or, more specifically, the food protein industry, has very considerable scope for improvement. Although it is far from perfect, the dairy sector is, with the exception of sugar processing, the best organised sector of the Irish food industry. One could be cynical and say that this is of necessity, since milk is a perishable commodity. While perishability may have been a significant factor in the establishment of the industry, the present industry is genuinely concerned and progressive. Current pricing policies are partially responsible for progress, or lack of it, toward diversification and the principal products in the mix are determined largely by these policies, but there is a consistent trend toward more consumer-oriented products. This trend is very likely to continue and accelerate, as indicated by the increasing involvement of dairy companies in research and development. Better utilisation of milk proteins is almost certain to acrue from these developments.

The meat industry is less well organised and developed. There is a weak trend toward the more extensive processing of carcasses and at least the majority of cattle are now slaughtered in Ireland. While recognising that there are major economic constraints, there would appear to be tremendous opportunities for developments in meat processing, both in the production of consumer-packed cuts and more extensively processed, downstream products. The better utilisation of blood for the production of a range of functional proteins would appear to hold potential, although there are problems of product acceptability and of marketing.

The most underdeveloped aspect of the Irish food protein industry is in the area of tertiary products. Examples that come to mind include: consumer ready meat-based meals, special dietary protein-based products, protein hydrolysates, functional proteins (other than milk proteins), non-conventional proteins, perhaps based on imported raw materials. One must be impressed by the success of the Industrial Development Authority in attracting chemical, pharmaceutical and, more recently, electronic engineering companies, to Ireland. With the exception of the two big British and some other dairy companies, one must be equally impressed by the absence, or at least the presence on a very small scale, of the multi-national food companies from the Irish scene. Not that we should have to depend on multi-national companies for progress, but their presence does facilitate development, especially in marketing.

To end on an optimistic note, the raw material for a highly developed food protein industry exists in Ireland; the technical expertise is also available, so why not marry the two and produce the products?

## ACKNOWLEDGEMENTS

Unless otherwise stated, the data for the Figures and Tables in this paper were taken from various issues of *Irish Agriculture in Figures*, compiled by An Foras Taluntais, 19 Sandymount Avenue, Dublin, with some supplementation by data taken directly from publications by the Central Statistics Office, Dublin.

Data for the period 1850–1930 in Fig. 8 was taken from reference 7. Data for Fig. 10 was supplied by Coras Beostoic & Feola (Irish Meat Board). Data for the period 1750–1930 in Fig. 12 was taken from reference 7. Figure 13 was constructed from data in various Annual Reports of the Pigs and Bacon Commission and Fig. 14 from data in various Annual Reports of An Bord Iascaigh Mhara (Irish Sea Fisheries Board). Data for the period 1650–1950 in Fig. 15 was taken from references 7 and 9. Figures 17 and 18 were constructed in part from data from *Irish Agriculture in Figures* and in part from data in Annual Reports of An Bord Bainne (Irish Dairy Board).

## REFERENCES

1. *Irish Agriculture in Figures*, An Foras Taluntais, 1980.
2. *Irish Agriculture in Figures*, An Foras Taluntais, 1981.
3. CREMIN, F. M. and MORRISSEY, P. A. *Irish J. Medical Sci.* 1976, **145**, 18.
4. DWYER, E., WALSH, T. and GORMLEY, T. R. *Farm & Food Research* 1981, **12**(1), 17.
5. FROHBERG, R. C. *Wheat Protein Conference, U.S.D.A., Peoria, Ill.*, 1979.
6. LINDLEY, J. A. and REDMAN, D. G. *F.M.B.R.A. Bulletin* No. 2, 1981, 80.
7. O'DONOVAN, J. *The Economic History of Livestock in Ireland*, Cork University Press, 1940.
8. *Annual Report*, Bord Iascaigh Mhara, 1979.
9. LYONS, J. *Agricultural Ireland.* Various issues between June, 1959 and December, 1961.
10. DUNKLEY, W. L. and PELISSIER, C. L. *J. Dairy Sci.* 1981, **64**, 975.

# 2

# World Supply and Demand of Food Proteins*


A. SIMANTOV

*Food, Agriculture and Fisheries Division,
Organisation for Economic Co-operation and Development (OECD),
Paris, France*


## INTRODUCTION

World food supplies in general and protein supplies in particular seem to be growing at a slower rate than requirements and the situation in many developing countries has reached a most critical point: present inadequate food consumption levels are made even more dramatic by rapidly growing populations. Major international conferences have tried to mobilise public opinion and government policies in order to step-up food production and to find a workable solution to what some people rightly call the most intricate problem of our times and the one which revolts humanity more than anything else.

Talking about the food problem several years ago I said: 'Some people attribute the problem to poverty and they are right: but poverty is the consequence of insufficient growth in productivity. Some others attribute it to the rapid demographic explosion and they are right: but demography becomes a problem only when sufficient productive employment cannot be provided. Others attribute it to lack of technological breakthrough in production and they are right: but all known technology is not applied yet because of insufficient investment, both physical and intellectual. Others attribute it to a growing disparity between the production and consumption capacity of the rich and that of the have-nots and they are right: but this is the cumulative effect of generations of strenuous efforts to produce wealth coupled with our inability to make a more egalitarian society compatible with the necessity of increasing overall wealth.'[1]

* The views expressed in this paper are the author's and do not necessarily correspond to those of the Organisation to which he belongs.

I have practically nothing to add to this statement. What I shall try to do here is to relate these general propositions to proteins.

## A PRELIMINARY COMMENT: PROTEINS AND CALORIES

When discussing world supply and demand for proteins there is usually an underlying and implicit belief that the problems of hunger and malnutrition in the world are due primarily to protein deficiencies in the diet. Until a few years ago this belief was so strong as to practically equate lack of nitrogen with malnutrition in the developing countries. Even more recently one still reads statements like 'The food deficit in the world is primarily a deficit of proteins ... It is the lack of proteins that causes most of the malnutrition in the world. ... The protein crisis has reached today a planetary dimension...'[2] and 'If this situation should continue, the physical, economic and social development of the future generations may become completely arrested.'[3]

This view is being modified as a result of a simultaneous examination of diets for protein and energy. It has been found that protein malnutrition is the indirect result of inadequate energy intake. The protein problem has therefore tended to be considered essentially a socio-economic problem, as low-income people do not have the means to purchase the amount of food which will provide them with the minimum nutrient requirements.[4]

Because protein may be used either for energy or for tissue formation (but not always in the same proportions) and because its efficiency in tissue formation depends upon the kind of protein it is, protein needs cannot be stated as so many grams per day for a person of a given age, sex and body weight. Protein requirements are lower than what many people thought they were and have therefore not received the same degree of acceptance as energy requirements; the principal reservation about them is that they are too low compared to what people eat in the rich countries where it is difficult not to eat more tasty animal foods when it is possible to afford to do so.

One of the reasons which motivated the former belief that protein deficiency was at the root of the food problem is to be found in the fact that a protein can usefully and economically be used by the human body (and by animals, too) only when it contains the essential amino acids (those which cannot be synthesised within the body) in appropriate proportions. It is known that if some essential amino acids are present in less than the required amounts, the amount of fully utilisable protein that can be formed is determined by the most limiting of them. As whole (hen's) egg protein is in this respect the best protein, it has been taken as a reference and, together with other animal products, has tended to be considered as an indispensable food.[5,6]

It is at present thought that a person consuming adequate quantities of basic foodstuffs (with a diversified intake of various products) can obtain adequate protein if the caloric requirements are met. Thus the satisfaction of protein requirements cannot be dissociated from the caloric availabilities, and the problem of hunger and malnutrition becomes therefore a problem of adequate production and distribution of calories obtained from foods containing a wide spectrum of complementary amino acids so as to provide the best possible nutritional result from limited available resources.

## PRESENT SITUATION

Various studies have been made to assess the present nutritional situation in the world and the situation likely to develop in the next 10 to 30 years, based on different assumptions.[7-9] All these studies indicate that malnutrition is at present the dominant feature; in many parts of the world people simply do not get enough food and they are under-nourished; in others, the diet lacks one or more essential nutrients giving rise to serious deficiency diseases; in others, people take in too many calories and consume an excess of one or more component of a reasonable diet, i.e. they are over-nourished. A recent assessment of the present situation indicates that 20 % of world population have an excessive food intake, 5 % have a satisfactory diet, 15 % are in an intermediate position, 20 % have a Chinese-type diet, 30 % are undernourished and 10 % are affected by famine.[10] This indicates that about three-quarters of mankind have food deficiencies and that two-fifths find themselves in a serious situation, with several hundred million at the edge of famine.

To provide a map of the state of nutrition in the world is not very easy. It is true that the state of nutrition in Africa, in some countries in South and South East Asia, or in parts of Latin America is particularly deficient, and there are indications that the situation has worsened recently in some of these countries. But using average figures for continents, as those in Table 1, is extremely misleading, as the country differences within the same region are often very great and the difference within a country between low- and high-income groups is much greater than the difference between rich and poor countries. The examples in Tables 2 and 3 illustrate these differences.

## TABLE 1
### LEVELS OF FOOD CONSUMPTION IN 1970[11]

| | Consumption of principal foodstuffs (kg per head) | | | | | | | Calories per day | | Basic calories per day[a] |
| --- | --- | --- | --- | --- | --- | --- | --- | --- | --- | --- |
| | Cereals for direct consumption | Starchy roots | Sugar and sugar products | Meat | Eggs | Fish | Whole milk | Total | Of which: of animal origin | |
| North America | 90·3 | 55·1 | 56·0 | 109·5 | 17·8 | 10·7 | 157·3 | 3 318 | 1 324 | 11 300 |
| Western Europe | 123·6 | 88·3 | 39·5 | 68·0 | 12·7 | 17·6 | 90·8 | 3 133 | 1 102 | 9 700 |
| Oceania | 106·9 | 58·8 | 52·8 | 114·1 | 13·3 | 6·8 | 167·4 | 3 261 | 1 498 | 12 200 |
| Others | 178·5 | 32·8 | 30·0 | 25·6 | 14·5 | 51·6 | 43·2 | 2 554 | 449 | 5 200 |
| Total developed countries with market economies | 122·6 | 67·4 | 43·3 | 74·4 | 14·6 | 21·3 | 104·7 | 3 091 | 1 064 | 9 500 |
| Africa | 138·4 | 177·4 | 9·2 | 12·8 | 1·1 | 6·9 | 18·5 | 2 188 | 141 | 3 000 |
| Latin America | 126·9 | 100·8 | 43·8 | 36·8 | 4·7 | 6·9 | 59·2 | 2 528 | 443 | 5 200 |
| Near East | 185·7 | 23·4 | 19·1 | 14·9 | 1·5 | 2·2 | 25·3 | 2 495 | 236 | 3 900 |
| Asia and Far East | 193·7 | 25·5 | 28·5 | 4·5 | 0·8 | 7·8 | 16·4 | 2 082 | 124 | 2 800 |
| Total developing countries with market economies | 173·3 | 61·6 | 27·0 | 12·1 | 1·6 | 7·0 | 24·6 | 2 211 | 189 | 3 300 |
| Asian countries with centrally planned economies | 206·7 | 101·7 | 5·6 | 18·7 | 3·6 | 9·2 | 2·9 | 2 174 | 203 | 3 400 |
| USSR—Eastern Europe | 207·2 | 120·3 | 40·9 | 48·8 | 9·3 | 18·6 | 163·1 | 3 265 | 792 | 8 000 |
| World | 173·8 | 77·2 | 26·9 | 29·5 | 5·4 | 11·5 | 49·2 | 2 480 | 425 | 5 000 |

[a] These figures were calculated on the basis of seven vegetable calories being required for the production of one animal calorie. This method being too rough-and-ready, the figures in this Table should be used with caution. Furthermore, there are considerable differences in the composition of vegetable calories from one region to another; for example, in North America or Japan, cereals occupy a very important place, whereas in Oceania pastures constitute the basis of animal feed.

## TABLE 2
### INEQUALITIES IN FOOD CONSUMPTION (CALORIES/DAY) WITHIN GEOGRAPHIC REGION[12]

|  | *Average 1964–1966* | |
|---|---|---|
|  | *Total calories* | *Animal calories* |
| Africa (average)[a] | 2 154 | 121 |
| Rwanda | 1 908 | 53 |
| Senegal | 2 348 | 224 |
| Latin America | 2 470 | 427 |
| Haiti | 1 904 | 94 |
| Uruguay | 3 039 | 1 262 |
| Asia (average)[b] | 1 984 | 129 |
| Indonesia | 1 798 | 45 |
| Singapore | 2 454 | 385 |

[a] Excluding South Africa and Rhodesia.

[b] Excluding Japan and China.

## TABLE 3
### INEQUALITIES IN FOOD CONSUMPTION WITHIN COUNTRIES[13]

| *Country* | *Consumption in calories/day* |
|---|---|
| India | |
| (Maharastra 1971) | |
| *Income per head:* | |
| Under 25 Rs/month | 1 540 |
| Over 75 Rs/month | 2 990 |
| Tunisia | |
| (Rural area 1965–1968) | |
| *Total expenditure per head:* | |
| Under 20 Dinars/year | 1 780 |
| Over 150 Dinars/year | 3 200 |
| Brazil | |
| (North-east, Urban area, 1966) | |
| *Income per family:* | |
| Under 100 Cruz./year | 1 240 |
| Over 1 200 Cruz./year | 4 100 |

They provide a serious argument in favour of those who maintain that food deficiencies in the world result primarily from poverty and inadequate distribution of incomes and also from inadequate education and information.

The question which usually comes to mind when considering the food supply and demand situation in the world is whether the world as a whole has the capacity to produce all the food—both in terms of calories and of proteins—which is required to provide a satisfactory diet for the entire population of the world.

At this stage of the discussion it could be stated, without a great risk of error, that the world has today the ability to produce all the food which is at present required. In fact, there are some estimates which put the present shortfall in availabilities to only the equivalent of about 40 million tons of cereals.* My belief is that the shortfall is substantially greater than this, but it is a quantity which technically could be easily obtained, either by increased production or by preventing part of the enormous losses of food after it is produced. Available technology, as evidenced by the techniques used by the upper 20 % of farmers and by the leading food industries, indicates that the present deficiencies in food production and utilisation are primarily a matter of inadequate policy and management decisions, i.e. are man-made.

It is, however, rather unrealistic to expect that these policy and management decisions can be easily solved and that human attitudes can be changed in the short run. The immediate future will continue to be characterised by the co-existence of food deficits in many regions of the developing world—both in terms of calories and of proteins—and of production surpluses in industrialised countries where food consumption is already high and where food is frequently destroyed or given to animals.


## THE FUTURE OUTLOOK


What about the future, i.e. between now and the end of the century? What are the requirements likely to be? Will the food and agricultural system be capable of meeting these requirements? I do not propose to project or to forecast future supply and demand: many authoritative and detailed

---

* A USDA report[14] indicates that about 39 million tons of wheat, less than 3 % of world cereal production, would be enough to raise the calorie intake of the acutely malnourished to the recommended minimum.

studies have been conducted on this subject by several national and international institutions.[7-9] What I prefer to do here is, first, to recall briefly some of the basic factors, such as population, incomes and technology, which are likely to affect future trends and to provide some very rough quantitative indications of these trends and, secondly, to discuss some features of the food and agricultural economy, which are difficult to quantify but which could have a profound impact on the future situation, in particular with respect to proteins. I have in mind such questions as the pattern of consumption in the world, the use of non-conventional foodstuffs and the production of energy from biomass.

A preliminary question needs to be answered: is it possible to assess future food demand independently of supply? In fact, the level and pattern of demand, in particular in many developing countries, corresponds broadly to the level and pattern of production. With international trade acquiring a larger part in the satisfaction of food demand in various countries and with the flow of information about food habits spreading rapidly all over the world, there will probably be a tendency for demand in each country or region to acquire a force of its own, with the consequence of gradually reducing the link between consumption and production in a given country. The same development will result from the reduction of the farm population and, therefore, of the farm subsistence sector. These developments, however, are likely to occur at a slow pace, so that, for a long time to come in many developing countries, local food production will determine, by-and-large, food consumption levels.

## Three Basic Factors Influencing Demand and Supply
### *The Growth in Population*

World population is increasing at 1·9 % per year, but at substantially higher rates in the food-deficient developing countries. There is reason to believe that growth rates have reached their historic peak and will decelerate in the next 10–20 years, with the possible exception of Africa; world population is expected to increase between now and the end of the century by 1·7 % a year.* On the assumption that consumption per head remains at its present level, a 1·7 % growth in population alone involves a 40 % growth in requirements by the end of the century. It is only after the

---

* The annual population growth rate for Africa, Far East, Latin America, Near East and developed countries between 1963 and 1975 has been 2·6 %, 2·5 %, 2·8 %, 2·7 % and 1·0 % respectively and is expected to be 3·0 %, 2·0 %, 2·5 %, 2·6 % and 0·6 % respectively between 1980 and 2000.

end of the century that present policies to restrain population growth, if conducted with determination, will make their impact felt on food demand.

### The Growth in Incomes

Any increase in per capita incomes results in an increase in food demand, and at lower levels of income the increase in demand is proportionately higher. This means that a 2 % per year improvement in average per capita incomes can result in an increase in demand of some 25–30 % by the end of the century. Yet, a 2 % per year improvement in per capita incomes is remarkably small, as, at this rate, it would take 14 years in Africa and 24 years in Asia to reach the present world average consumption of calories and 42 and 52 years, respectively for animal proteins!

If the improvement of incomes at the lower end of the scale is higher as a result of much-needed more egalitarian income distribution policies in many developing countries, the increase in food demand can be substantially higher than the above-mentioned figures. To illustrate the size of this potential demand for both calories and proteins, it has been estimated[7] that, in a country with 400–500$ average per capita income and where 20 % of the population receive 50 % of the income and 60 % receive 25 % a growth of 4 % in average income would give the following results, depending on the distribution of the newly created wealth:

> If this wealth goes entirely to the richest 20 % of the population the total demand for calories will increase by only 0·26 % and demand for calories of animal origin by 0·77 %.
> If it goes entirely to the bottom 60 % of the population total demand for calories will increase by 3·2 % and demand for calories of animal origin by 5·75 %.
> If it is distributed equally throughout the population, the total demand for calories will increase by 1·4 % and demand for calories of animal origin by 2·9 %.

### The Growth in Urbanisation

Although population is expected to increase by some 1·7 % a year, the growth of cities is expected to be two-and-a-half times faster.* The most populous cities—or agglomerations—will be in the developing countries: some of them will have about 30 million people by the end of the century. This development runs parallel with a relative reduction of the ·farm

* According to UN statistics the urban population increased by 3·3 % per year during the decade 1960–1970 and the rural population by 1·3 %.

subsistence sector. It involves a profound change in the pattern of food demand and a dramatic increase in the quantities of food which will need to be moved from producer to consumer, and also a dramatic increase in the degree of processing which is likely to be required.

In spite of this rapid urbanisation, the fact remains that the bulk of the increase in population will remain in agriculture and in rural areas. If agricultural productivity does not increase rapidly enough, the increased population pressure in rural areas may result in lower food availabilities for urban centres which will have to resort, more than at present, to supplies from abroad.

These three elements put together should give an *approximate view of the food requirements* by the end of the century. That the availability of food should double is a conservative estimate, especially if those who are at present particularly malnourished should obtain a minimum decent food intake and nothing more. This means that availability should increase by a minimum of 4% per annum: this is not technically impossible to reach, as 2–2·5% could come from increased production and 1·5–2% from prevention of losses and from better utilisation, especially in livestock feeding. The main problem in achieving this result will still be in the policy and management field.

## The Development of Technology

Increased production, prevention of losses, and better utilisation of available supplies indicate the lines along which technology needs to be developed. It is evident that future production increases will need to be obtained more than in the past through higher yields, as the reserves of new land are shrinking: these higher yields will, however, need to be obtained with inputs different from those used so far when fossil energy and energy-based products were relatively cheap, and when the more accessible sources of water have been fully exploited. This indicates that much innovative thinking will be needed to step up productivity of both farm production and food technology.

It is worth recalling here that the development and application of new technology in food and agriculture depends to a very large extent upon the technological environment of the national economies. It has been shown that new technologies are applied in agriculture at approximately the same rate as the national economy grows:[15] this means that in periods of slow economic growth, like the one in which we are at present, it becomes more difficult to accelerate the growth of food production.

Yet agricultural production is primarily a biological process and some of the fundamental research taking place at present, like the use of solar energy through a better photosynthetic process and the possibility for plants to biologically fix the nitrogen of the air, could have a profound impact on the ability of the system to provide the food which is required. The first agricultural revolution was based on proper fertilisation; the second on mechanisation and automation; the third is likely to be in the areas of genetic engineering and biotechnology.

## Major Issues Likely to Affect the Future Food Economy
### The Pattern of Food Consumption

As incomes grow, the share of animal products in the diet increases. This means that the higher the degree of economic and social development of a country, the larger is the increase in aggregate food and feed requirements resulting from a unit increase in final food demand. For example, in a country of Western Europe, an increase by 1 % of final per caput food demand will probably result in a decline by 0·6 % of the calorie intake not originating from animal food and an increase by 3 % of the calories of animal origin: this aggregate will necessitate an increase of about 2·6 % in total food and feed utilisation (even without an increase in final food demand, the replacement of non-animal calories by animal calories will result in an increase in total food and feed utilisation).

On the other hand, in a country like India an increase by 1 % of final per caput food demand will probably result in an increase by 0·9 % in the calorie intake not originating from animal food and an increase by 3 % of the calories of animal origin: this aggregate will necessitate some 1·4 % increase in total food and feed utilisation.*

* This calculation is based on the assumption that to produce one animal calorie it is necessary to convert, through the animal, seven calories of vegetable origin. In fact, these ratios vary according to the type of animal, the type of final product obtained, the type of the farm enterprise, the availability of all necessary amino acids and oligo-elements. For example, in countries of North Western Europe with highly advanced systems of production, the following conversion rates are commonly observed. In other countries, the feed requirements can be substantially higher.

|  | Consumption of energy (calories for calories) | Consumption of proteins (grams for grams) |
|---|---|---|
| Broilers | 4·7 | 3·0 |
| Eggs | 5·5 | 4·0 |
| Pig-meat | 2·9 | 4·5 |
| Pig-meat for processing | 4·4 | 6·5 |
| Beef | 10·0 | 8·0 |
| Veal | 16·0 | 12·5 |
| Milk | 4·7 | 4·2 |

The question which arises is, whether food consumption patterns should necessarily continue to be similar to those we have today in high income countries, especially after the realisation, as mentioned earlier in this paper, that adequate protein can be supplied to the body if the diet provides adequate calories of a varied origin. While animals, especially ruminants, can utilise products which otherwise, with the present state of technology, are not yet utilisable by human beings, livestock production requires increasing quantities of cereals and other feeds, such as cassava or soy beans, which can be used directly for human consumption: even grassland could be diverted to the production of cereals in certain circumstances.

To understand the size of the issue it is interesting to determine the food consumption patterns and levels in the world in terms of original calories, as illustrated in the last column of Table 1. Realisation of these differences has given rise to the belief that, by reducing consumption of livestock products in the rich countries, it would be possible to wipe out famine and malnutrition in the world. In fact, a reduction in consumption of 100 animal calories in the developed countries (the USSR and East Europe included), would allow the average food ration in the developing countries (China included) to be increased by about 285 calories per day. This would enable the world shortage of energy food to be *statistically* eliminated. A reduction of 200 animal calories in the industrialised countries would make it possible, on the assumption advanced above, to eliminate the calorie deficit in the poor countries and increase the consumption of animal products in Africa and Asia by more than half. This obviously assumes that food thus left unconsumed in the rich countries would be transferred *in toto*, and used in the poor countries exclusively for direct human consumption: this is a doubtful hypothesis requiring careful examination.

It is a fallacy to believe that this would happen and this issue has been adequately discussed in other papers.[16] But what we could say here is that it would be in the general interest of the populations of developing countries—the bulk of the world's population, especially in the future—not to adopt diets which have become a traditional pattern in the nowadays rich countries. An increase in production of a well-balanced commodity mix remains, however, a necessity in those countries and it should be a challenge to scientists to help produce, save and utilise the maximum of nutrients with the available, scarce, physical, economic and managerial resources. What should at any cost be avoided is that scarce food be used in a given country to feed livestock—or to be transformed into energy, as discussed later on—at the expense of badly needed direct human consumption.

Another point which is worth stressing here is that the overall food situation in the world could be improved if products like tapioca which are rich in carbohydrates but poor in proteins and which are at present fed to people in developing countries, could be used as livestock feed in rich countries *in lieu* of cereals which could then be shipped to developing countries for direct human consumption, thus providing the much needed protein intake. Products like soy beans could complement the feed ration in the rich countries. Theoretically, it will be possible to avoid this exchange of products by adding to the diet of the people in poor countries soy protein in an acceptable form. This may become possible in the future, but it is probably not a practical course to follow at present. This leads us to the discussion of the next issue: the growing importance of non-conventional products.

*The Use of Unconventional Products*

Technology has always allowed us to obtain new products or to develop new utilisations for existing agricultural raw materials. Good examples of such developments are the appearance of margarine and of cereal-derived sugars. But it was not until the last quarter of a century that foodstuffs of exclusively industrial origin made their appearance, and the indications are that more of such products will come on the market in the years to come. Since many of these products concern new utilisations of existing proteins or the production of new proteins, it is worth analysing their possible impact on the world supply and demand balance.

Unconventional protein products can make an important contribution towards a better world food balance by reducing animal production costs. Recent OECD Reports,[17,18] illustrate the various pathways now being followed for that purpose. Moreover, by improving yields and by changing the land use pattern, an attempt is being made to expand long-term production of oilcake protein faster than the production of cereals and pulses. The development of high protein cereals; for example in the case of wheat, of hybrid maize rich in particular amino acids, and of a rye–wheat hybrid is promising; it has been estimated that an increase by 1 % of the average protein content of cereals would correspond to 11 million tons of protein, or the equivalent of protein contained in 30 million tons of soy beans.

The use of non-protein nitrogen in the feed of ruminants can save vegetable proteins for direct human use. In 1971, the FAO estimated that a 12 million tons deficit of animal meal and oilcake could be turned into a 3 million tons surplus, on the assumption that 2 million tons of urea were used in animal feed. Other possibilities are offered by substitutes for milk proteins in milk replacers, the use of whey in human nutrition, the recycling of organic waste, the production of single-cell proteins and the improvement of animal feed conversion ratios.

The technology for producing unconventional protein for direct human use is not part of my brief. However, I would suggest that rather than run the risk of offering imperfect imitations of conventional foods, researchers and manufacturers should aim to develop new dishes which completely revolutionise traditional cooking practices. The future market for unconventional proteins will probably be found in a whole range of dishes as yet unknown and undreamed of by former generations who compiled cookery books. Such dishes will enable the functional properties of these proteins to be used to the best advantage. For example, the incorporation of whey proteins in 'pasta' has been studied as a means of dispensing with meat in some school meals in the United States. There is also a recipe for high-protein fruit pie. Similarly, proteins are being added to acidic effervescent drinks. Such innovations will gradually gain acceptance as young consumers become accustomed to assimilating new kinds of food and drink. Relative price levels, dietetic qualities, physical properties, taste properties and ease of use are the main reasons for using the new proteins.

The researchers working on these proteins have drawn up a number of utilisation formulae which seem well suited to the diverse food situations prevailing in the developing countries. Besides the use of concentrates such as fish protein concentrate, algae and leaf proteins, or yeasts grown on alkanes, recourse is being made to the blending of foods as a means of balancing the amino acids required, and to the enrichment of cereals and soy proteins with industrially-produced amino acids.

What will the overall impact of such developments be on the world supply and demand of proteins? It is difficult to answer this question because of the uncertainties about the rate of commercial application of technological progress in food science, i.e. the cost/price environment for both raw materials and end products. What can be said with a reasonable degree of assurance is that the avoidance of crop losses after harvest and the increased use of unconventional food and feedstuffs can make a greater contribution to a better world food balance in the shorter run—and possibly in the longer run—than the increase in farm production. This does not mean that increased farm production does not remain by far the primary objective, but it means that in certain circumstances an increase in financial and human resources devoted to the prevention of losses and to food science can bring bigger benefits than if devoted to raising agricultural production.

*Agriculture as a Producer of Energy*

Much has been said and written about the implications of the rise in the costs of energy for agriculture and food.[19] In the context of this discussion there are two aspects of the energy problem which need to be particularly addressed: the first deals with the technological developments which are likely to occur in the production of food in general and of proteins in particular; the second deals with the use of agricultural products for non-food uses.

The first of these aspects, i.e. developments in food technology, has already been touched upon earlier in this paper. To the extent that the rise in energy prices is not confined to agriculture but is a concern for the entire economy, there is no reason to expect that the adjustments needed in agriculture will be more difficult to achieve than those which will be required in other sectors of economic and social activity. To the extent that agriculture and food production are essentially biological processes and that biological engineering and controlled fermentations are the directions in which new technology is moving, food production is well placed to take advantage of these developments. Even if this new technology is developed in industrialised countries, there is no serious reason to believe that the advantages to be derived from it will not be available to the entire world.

The second aspect, i.e. the possible use of food for energy production, is, on the contrary, a source of concern for the world supply and demand balance of food in the longer run. While the number of non-food utilisations of agricultural products has been increasing over the years, on the whole the quantities involved have remained marginal and have never imperilled the supplies of food. But since the increases in the price of energy, and especially in anticipation of further increases, it is fair to wonder whether agricultural products will continue to have food as their almost exclusive use. It is known that ambitious programmes have been set up to produce ethanol, not only by using by-products of agriculture and forestry—which would have had only a limited impact on food and agriculture—but by using entire crops. The examples of sugar cane in Brazil and of maize in the USA are illustrative of this trend. It is also known that plans are under way for the use of vegetable oils for the production of energy.

Such an outlook poses at least two basic questions which are difficult to answer given the present state of technology and especially of the economics of biomass utilisation. The first question deals with the value which will be attributed to the by-products resulting from the use of agricultural commodities for energy purposes: they consist of vegetable proteins of a high nutritive value for livestock, as a by-product of maize or of oilseeds used for the production of energy. For the time being these 'derived' vegetable proteins will obtain the same price as those already available; but if the production of energy were to expand, the prices of these protein by-products and the trade flows would be affected. Proteins available for food or feed consumption per unit of agricultural commodities produced may thus become larger and their prices relatively lower.

The second question deals with the danger that an expanded use of farm products for energy purposes would cause a real and important rise in the price of food. Food demand being very inelastic and its satisfaction being a priority requirement, food prices will have to rise substantially in order to discourage the use of basic foodstuffs for energy purposes. The price of food is likely to be, in the future, indirectly fixed by the price of energy, since, in the supply/demand equation, demand could considerably increase whereas supply is not likely to follow in the same proportions in spite of the cultivation of new marginal and low productivity soils.

Although it is difficult to quantify these possible developments, the question could, however, be asked whether such a technological and economic trend will not induce the high income groups (countries and individuals) to use for their own purposes an even greater part of basic agricultural production than is already the case today. The market is likely to respond more easily to the demand for non-food products derived from agricultural commodities than to the requirements of food by low-income groups. Will government policies aid such a development or will they oppose it? This is a serious question which deserves a better understanding and strict scrutiny.


## CONCLUDING REMARKS

The preceding discussion has shown in which direction and at what pace food demand is likely to develop in the remainder of the century. This level can vary greatly according to the policies which governments will implement in the years to come in such crucial areas as demography, economic growth and employment, income distribution and education. Whether the world food system will allow demand (and requirements) to be met will certainly depend upon the rapidity of technological progress, but even more importantly, upon management and policy decisions. These decisions are not confined to developments in individual countries but relate also to the ability of the world community to allow for a better economic integration of the various regions of the world. It is the combination of adequate policies with respect to economic and social development, to higher food production and better utilisation of available food and feed, and to the integration of the various national economies which will enable the food situation in the future to be kept under control instead of developing into a nightmare.

## REFERENCES


1. SIMANTOV, A., Food and feed: Future needs, in *Agriculture in the Whirlpool of Change*, University of Guelph, 1974, 145–68.
2. CEUZIN, P., La crise des proteins, *Sciences et Avenir*, April, 1973, 331–4.
3. United Nations 1968 quoted by SUKHATME, P. V. The protein problem, its size and nature, *Journal of the Royal Statistical Society*, 137, Part 2, 1974, 166.
4. SUKHATME, P. V., Nitrogen in malnutrition, *Ambio*, 6, 1977 (2–3) 137–40.
5. Japan Essential Amino Acids Association, *Significance of Fortification of Cereals with Amino Acids in Japan and South-East Asia*, Tokyo. October, 1968.
6. SCRIMSHAW, S. and YOUNG, V. *Scientific American*, September, 1976, 57.
7. OECD, *Study of Trends in World Supply and Demand of Major Agricultural Commodities*, Paris, 1976.
8. FAO, *Agriculture: Toward 2000*, Rome, 1979
9. *The Global 2000 Report to the President*, Washington DC, 1980.
10. KLATZMANN, J., Une typologie de l'alimentation dans le monde, *Economie Rurale*, No. 129, January–February, 1979, 3–10.
11. SIMANTOV, A., Feeding the World, *Grindlays Bank Review*, March, 1975.
12. FAO. *Agricultural Commodity Projections, 1970–1980* Vol. II, Rome, 1971.
13. *World Food Conference. Assessment of the World Food Situation*, 1974.
14. USDA, *Report Assessing Global Food Production and Needs as of April 15, 1979*, ESCS Report No. 61, Washington DC, 1979.
15. SIMANTOV, A., The dynamics of growth and agriculture, *Zeitschrift für Nationalökonomie*. **XXVII** (3) 1967, 328–51.
16. SIMANTOV, A., World food consumption—can we achieve a balance? *Food Policy*, 1(3), May, 1976, 232–8.
17. OECD, *Unconventional Foodstuffs for Human Consumption*, Paris, 1975.
18. OECD, *Nitrogenous Products of Industrial Origin for Animal Fooding*, Paris 1974.
19. SIMANTOV, A., Agriculture and the energy challenge, *Journal of Agricultural Economics*. **XXXI** (3) September, 1980.

# 3

# Relationships Between Structure and Functional Properties of Food Proteins

J. E. Kinsella

*Institute of Food Science, Cornell University, New York, USA*

## INTRODUCTION

Foods are consumed, ultimately, to satiate hunger and obtain essential nutrients. However, when foods are abundant and available people tend to select foods for hedonistic reasons, i.e. their sensory qualities (flavour, odour, colour, texture), and nutritional criteria become secondary. Organoleptic attributes reflect the physical and chemical properties of food constituents. While a minor component may impart the principal organoleptic characteristic in some foods, e.g. amyl acetate in bananas, in most common foods quality reflects the interactions among the major constituents. Thus, water, carbohydrates, lipids and proteins affect quality.[1]

Proteins contribute significantly to the functional behaviour and quality of several major foods: milk, meat, cheese, eggs and bread. The organoleptic qualities of these primary foods depend upon the particular physical properties and interactions of the protein components. In addition, isolated proteins are being used to an increasing extent to formulate or fabricate new protein-rich foods, e.g. extruded foods, textured meat analogues, whipped toppings, protein-rich beverages. These applications depend upon the physical and chemical properties of the proteins.

## FUNCTIONAL PROPERTIES

In the jargon of the food scientist, functional properties of proteins are those physical and chemical properties which affect the behaviour of proteins in food systems during processing, storage, preparation and

                              *J. E. Kinsella*

consumption. It is these characteristics which influence the 'quality' and organoleptic attributes of foods.[2,3] Several typical classes of functional properties are shown in Table 1.

There are several examples of functionality in traditional foods, e.g. the viscoelastic properties of wheat gluten; the textural properties, succulence and colour of meat; the curd-forming properties of caseins and the whipping properties of egg white albumin.

TABLE 1
FUNCTIONAL PROPERTIES OF PROTEINS IN FOOD APPLICATIONS

| *General property* | *Functional criteria* |
|---|---|
| Organoleptic | Colour, Flavour, Odour. |
| Kinesthetic | Texture, Mouthfeel, Smoothness, Grittiness, Turbidity. |
| Hydration | Solubility, Wettability, Water absorption, Swelling, Thickening, Gelling, Syneresis, Viscosity. |
| Surface | Emulsification, Foaming (aeration, whipping), Film formation. |
| Binding | Lipid-binding, Flavour-binding. |
| Structural | Elasticity, Cohesiveness, Chewiness, Adhesion, Network crossbinding, Aggregation, Dough formation, Texturisability, Fibre formation, Extrudability. |
| Rheological | Viscosity, Gelation. |
| Enzymatic | Coagulation (rennet), Tenderisation (papain), Mellowing ('proteinases'). |
| 'Blendability' | Complementarity (wheat–soy, gluten–casein). |
| Antioxidant | Off-flavour prevention (fluid emulsions). |

Enzymatic activity may be classified as a functional property although, in the narrower context in which functional properties are usually discussed, enzymatic functions are not included. Enzymes can be used to alter the chemical and physical characteristics of food components and thereby affect quality attributes and processing characteristics; hence, they should be included as functional ingredients.[4]

Functional properties are most important in determining the usefulness of proteins in food systems. As the number of processed and fabricated foods increases, greater reliance will be placed on the consistent behaviour of ingredients in specific formulations. Proteins (novel or traditional) *per se*, as dry powders, have very limited appeal to potential users or to consumers. To facilitate their use in foods and their conversion to a range of desirable ingredients, they must possess or impart appropriate characteristics, following interactions with other components of foods (e.g. water, lipids) or following processing.

The type of functional property required in a protein or a protein-mix varies with the particular food system in question (Table 2). Thus, in a meat system, water binding, solubility, swelling, viscosity, surface activity and

TABLE 2

TYPICAL FUNCTIONAL PROPERTIES PERFORMED BY FUNCTIONAL PROTEINS IN FOOD SYSTEMS

| *Functional property* | *Mode of action* | *Food system* |
| --- | --- | --- |
| Solubility | Protein solvation | Beverages |
| Water absorption and binding | Hydrogen bonding of water; Entrapment of water (no drip) | Meat, Sausages, Breads, Cakes |
| Viscosity | Thickening; Water binding | Soups, Gravies |
| Gelation | Protein matrix formation and setting | Meats, Curds, Cheese |
| Cohesion–adhesion | Protein acts as adhesive material | Meat, Sausages, Baked goods, Pasta products |
| Elasticity | Hydrophobic bonding in gluten; Disulphide links in gels | Meats, Bakery |
| Emulsification | Formation and stabilisation of fat emulsions | Sausages, Bologna, Soup, Cakes |
| Fat absorption | Binding of free fat | Meats, Sausages, Doughnuts |
| Flavour-binding | Adsorption, entrapment, release | Simulated meats, Bakery, etc. |
| Foaming | Form stable films to entrap gas | Whipped toppings, Chiffon desserts, Angel cakes |

gelation are the most important properties in determining the usefulness of a protein and its impact on the final product quality. Different food applications require different characteristics, e.g. in a beverage, the protein should be soluble; in a comminuted meat, it should absorb moisture and gel upon heating; in a whipped topping it should have thermostable foaming properties (Table 3).

Ideally, protein ingredients should possess a broad array of functional

### TABLE 3
FUNCTIONS OF INGREDIENT PROTEIN IN MEAT-BASED PRODUCTS

Improves uniform emulsion formation and stabilisation
Gelation improves firmness, pliability and texture
Facilitates cleaner, smoother slicing
Reduces cooking shrinkage and drip by entrapping/binding fats and water
Prevents fat separation
Enhances binding of meat particles without stickiness
Improves moisture-holding and mouthfeel
May impart antioxidant effects

properties and food technologists continually seek proteins with versatility and broad spectrum functionality. However, only a few types of protein meet these criteria. Thus, milk proteins, by virtue of their heterogeneity, their range of physical properties and amphipolar characteristics, are the premier proteins used in food processing for multiple functions—emulsion stabilisation, water holding, coagulation, adhesion, etc. Egg white proteins are also versatile, being the premier foaming proteins and having excellent adhesion and gelation properties. However, many commercial proteins have limited functionality, e.g. gelatin. In addition, several preparations with reasonably good properties have other limitations; e.g. soy proteins have good functional properties[3] but bound off-flavours[5] and the marked thermal stability of glycinin[6] limit their use in many foods.

Although the importance of functional properties has been recognised by commodity specialists, i.e. meat, dairy and cereal scientists, for many years, the widespread awareness of their importance in food technology is relatively recent. This has been accentuated by the increased emphasis on novel food manufacturing and formulation. Thus, attempts to utilise less expensive proteins, to extend traditional foods (textured soy-bits), to simulate traditional foods (milk, meats, whipped toppings), to fabricate food analogues and to develop new functional ingredients (gelling, foaming, emulsifying, acid soluble proteins) have emphasised the need for more information on the functional properties of specific proteins. The availability of a large number of functional ingredients has significantly broadened the development of new food products (Tables 4 and 5).

A significant functional attribute required in food proteins is their thermal stability during processing. This is exemplified by the relative heat stability of the casein system in milk which facilitates processing, i.e. pasteurisation and UHT treatment. In contrast, the whey proteins are quite heat labile, a factor which presents a challenge in the preparation of whey

### TABLE 4

PRINCIPAL TRADITIONAL FOOD PRODUCTS USED AS SOURCES OF FUNCTIONAL PROTEIN

| Commodity source | Functional protein |
|---|---|
| Milk | Milk powder, Caseins, Whey proteins |
| Egg | Egg white, Lipoprotein of egg yolk |
| Meat | Collagen, Muscle, Blood |
| Fish | Collagen, Muscle |
| Cereals | Wheat gluten, Maize zein |
| Oilseeds | Flours, Concentrates, Isolates from Soybean, Cottonseed, Peanut, Sesame |
| Yeast | Extracts, Dried yeast |

proteins with good functional properties. The major soy protein component, glycinin, is extremely heat stable (transition temperature $> 90\,°C$). Thus, to realise its gelling and water-holding properties, very high processing temperatures are necessary. These are not compatible with several conventional processing conditions because high temperatures have deleterious effects on other quality attributes.

Several factors affect the functional properties of proteins (Table 6): the inherent properties of the protein *per se*, whether native or denatured; the methods and conditions of isolation; the degree of purification; processing alterations, e.g. refining, drying; other components of the system and environmental conditions (pH, temperature, salt type and concentration), all interact to provide the final functional properties.[2,3,7] Despite their importance, functional properties of proteins have not received systematic

### TABLE 5

PRINCIPAL FUNCTIONAL PROTEINS USED IN FOOD FORMULATIONS

| Protein | Functional property | Mechanism |
|---|---|---|
| Gluten | Viscoelasticity, Foaming, Adhesion | H, SS, Hp, E |
| Myosin | Adhesion, Emulsifying, Gelling | E, Hp, SS |
| Casein | Adhesion, Emulsifying, Coagulation, Viscoelasticity | Hp, E |
| Collagen | Gelation, Foaming | Hp, H |
| Egg white | Foaming, Coagulation/gelation, Binding | E, Hp |

H = hydrogen bonding; Hp = hydrophobic association; E = electrostatic and SS = disulphide.

*J. E. Kinsella*

## TABLE 6
SOME OF THE FACTORS INFLUENCING THE FUNCTIONAL PROPERTIES OF PROTEINS IN FOODS

| *Intrinsic* | *Environmental factors* | *Process treatments* |
|---|---|---|
| Composition of protein(s) | pH | Heating |
| Conformation of protein(s) | Redox status | pH |
| Mono- or Multi-component | Salts, Ions | Ionic strength |
| Homogeneity–heterogeneity | Water | Reducing agents |
| | Carbohydrates | Storage conditions |
| | Lipids | Drying |
| | Surfactants | Physical modification |
| | | Chemical–enzymatic |
| | Flavours | modification |

treatment in terms of definition, classification and standardisation of methods for measuring these properties.[7] Currently, there is much ambiguity and inconsistency in the terms applied by various workers. There are tremendous variations in the methods being used to measure functional properties, yet the methods employed and the equipment (geometry, configuration, energy input, temperature controls of the system) used, all affect the final assessment.

The literature is replete with papers on the functional properties of proteins as measured by a variety of methods under differing conditions. The tendency has been for each investigator to devise methods and/or conditions to suit a particular situation with limited concern for an experimental design which would facilitate explanation of the physical property being studied. Much of the published data are of very limited value for practical or comparative purposes.[7] This situation reflects the heterogeneous nature of the proteins involved, the very complex series of interactions, which are not amenable to easy measurement, involved in a particular functional property, the multiple interacting factors which impact on the resultant reactions and the lack of appropriate measuring devices accurately quantifying the results of these interactions in terms of the inherent properties of the protein *per se* and environmental factors. Ideally, in measuring the functional properties of proteins, equilibrium conditions should be attained so that the thermodynamic criteria are met and the particular measurement reflects the physical properties of the protein in question.[8]

Because of the complexity, variability and range of food systems, the dissimilarity of their proteins and their variable history, it has been difficult to standardise tests for measuring specific functional properties. Thus, it is difficult to generalise from the available information and it may be misleading to extrapolate data from one system to predict the behaviour of a particular protein in other systems. Proteins from different sources vary markedly in their physical properties and it may not be possible to design standardised methods that are suitable for measuring a specific functional property of all proteins. Thus, initially, it may be necessary to develop methods that can be used to compare the functional properties of proteins from the same general commodity source. Until suitable methods are developed it may be desirable to use actual food systems or model systems that mimic the food system and are designed to highlight the particular function in question.

In examining functional properties of food proteins, there is a major need to understand the basic information relating functional properties to particular conformational or structural features of the protein. This is limited by the lack of appropriate methods for measuring functional properties consistently and for relating functional properties to a particular physical property or conformational state of a protein. To facilitate the development of proteins with particular functional properties it is first necessary to define which physical properties are pre-eminently necessary for a specific function. To achieve this it is necessary to develop good methods for measuring functional properties and to devise methods which elucidate relationships between structure and function. Obviously, in approaching this, one should select a well characterised protein and use it in a model system which simulates foods wherein functional proteins are the major active components. When relationships between structure and function are understood, then it may be quite feasible to modify inexpensive proteins to impart required functional properties. Thus, to gain a clear understanding of functionality, the structural features of the protein, its molecular properties and, particularly, the secondary interactions that are required for particular functions, need to be clearly understood.

In order to have more general applicability, measurements should be based upon fundamental physical/chemical properties of the major protein component. In order to predict the behaviour of a protein it is necessary to determine its composition and physical/chemical properties under a specified set of conditions; however, to date there are no methods that can predict the behaviour of proteins in food systems.

The food scientist is faced with a situation wherein the proteins being used possess the tertiary structures and conformations that presumably have evolved to perform specific biological functions and are not necessarily designed for specific functional applications. In several instances, the physical properties of particular food proteins have dictated the manner in which they can be used and the consumer has generally taken this for granted. Thus, the peculiar nature of milk proteins under appropriate conditions allows the formation of a curd from whence the cheese foods evolved. Similarly, in the case of soy proteins, their ability to coagulate in the presence of calcium facilitated the manufacture of tofu. The particular composition and conformation of egg white protein made it the premier whipping protein and led to the evolution of foam-based foods. The consumer has been conditioned to these and hence much effort has been made in endeavouring to simulate the texture of meats using other proteins. In order to manipulate these proteins and/or simulate their functional properties, it is important to understand their structure and physical properties and the relationships between structure and specific functional properties.


## PROTEIN STRUCTURE AND FUNCTION

While the nutritional value of a food protein is related to its essential amino acid content and digestibility, the appeal and universal acceptance of conventional food proteins (meat, eggs, milk) are attributed to their attractive and highly palatable nature, reflecting their physical and chemical properties. One challenge to the modern food scientist is to define the physical base responsible for these desirable properties and apply this knowledge to the fabrication of analogues of these foods.

Understanding the functional properties of proteins should be facilitated by a detailed knowledge of the intrinsic properties of the protein, i.e. amino acid composition, primary sequence, conformation, molecular size and shape, charge distribution, intra- and intermolecular bonding, as these are affected by environmental conditions (Table 7).


### Amino Acids

The amino acid composition or, more particularly, the content of different classes of amino acids and their disposition in the polypeptide chain, affect functional properties. A high proportion of apolar residues affects interpeptide interactions, solubility and surface activity. Hydrophobic regions directly influence surfactant properties, e.g. emulsifying, foaming and

TABLE 7

STRUCTURE–FUNCTION RELATIONSHIPS IN FOOD PROTEINS

1. Amino acid composition (major groups)
2. Amino acid sequence (segments/polypeptides)
3. Secondary/tertiary conformation (compact/coil)
4. Surface charge, hydrophobicity/polarity
5. Size, shape (topography)
6. Quaternary structures
7. Secondary interactions (intra- and inter-peptide)
8. Disulphide/sulphydryl content
9. Environmental conditions (pH, Redox, salts, temperature)

flavour binding.[9] Hydrophobic interactions are major forces accounting for the compact tertiary folding of proteins, e.g. soy glycinin.

The presence of charged amino acids enhances electrostatic interactions which may be involved in stabilising globular proteins and in water binding, e.g. Asp and Glu can bind 6–7 molecules of water per charged residue.[10] This is important for hydration, solubility, gelation and surface activity. The contents of cysteine and cystine significantly affect the structure and function of proteins. Sulphydryl groups may be oxidised to form disulphide bonds (intra- and intermolecular); thiols and disulphides can undergo interchange reactions which markedly affect structure and function, e.g. gluten formation. Except in cases where a particular type of amino acid predominates, however, knowledge of amino acid composition is of limited value in predicting tertiary structure or physical properties. Thus, α-lactalbumin and lysozyme have very similar amino acid compositions but they differ markedly in physical properties, the mere replacement of glutamine with valine changes the conformation and functional properties of haemoglobin,[11] and casein variants show significant differences in properties.[12]

**Protein Structure**

The structure of protein is categorised as primary, secondary or tertiary, depending on the progressive state of spatial arrangement of polypeptide chains (Table 8). The sequence of amino acids represents the primary structure of proteins. Polypeptides containing in excess of 100 amino acids may assume numerous conformational arrangements in space but, without any stabilising effects, would be in continuous conformational flux or random coil. However, although the only information transmitted from the gene is the sequence of amino acids, each protein species in the native state exhibits a unique three-dimensional structure that determines its biological and functional properties.

The polypeptides of a nascent protein tend to coil in a characteristic fashion to form localised secondary structures, e.g. $\alpha$-helix or $\beta$-pleated sheet. The driving force for this folding is of hydrophobic origin due to the favourable change in free energy of the solvent upon folding of the polypeptide. The secondary structures are stabilised by hydrogen bonding.

TABLE 8

ORGANISATION OF PROTEIN STRUCTURE

| *Structure* | *Description* |
|---|---|
| Primary | Sequence of amino acids in the polypeptide |
| Secondary | Folding of segments of primary structure ($\alpha$-helix); stabilised by hydrogen bonds; hydrophobically driven |
| Tertiary | Pattern of folding of secondary structure into a compact molecule stabilised by hydrophobic forces, hydrogen, van der Waals and disulphide bonds |
| Quaternary | Association of secondary or tertiary structures into an oligomeric unit. Stabilised by hydrophobic forces, hydrogen and disulphide bonds |

In the $\alpha$-helix the successive amino acids are coiled and the resultant polypeptide forms a helical backbone containing the peptide linkages (3·6 amino acid residues per turn of the helix) from which the side chains of the amino acids extend outward. The multiplicity of H-bonds imparts stability to the helix, especially in 'anhydrous' domains of the polypeptide coils. In aqueous media (where the quantitative importance of H-bonds may be weakened) van der Waals' interactions (1–3 kcal/mole over 3–5 Å) between adjacent residues and hydrophobic interactions between long or bulky apolar side chains contribute to stabilisation of the $\alpha$-helical structure.[13]

Because of certain structural constraints imposed by particular sequences of amino acids, $\alpha$-helix formation is not always possible. Thus, polypeptides with sequences containing significant amounts of amino acids with bulky side chains or highly charged groups (at particular pH values) tend to destabilise the $\alpha$-helix. Polypeptides with proline are prevented from forming $\alpha$-helices. In the case of proline, the nitrogen atom, being part of the rigid pyrrolidone ring, prevents free rotation; furthermore, the lack of substitute hydrogens on the nitrogen limits hydrogen bonding. Thus, whenever proline or hydroxyproline occur in a polypeptide they cause a bending or kinking of the polypeptide chain and break or interrupt the α-helix structure.

β-Pleated sheet regions are formed where inter-polypeptide chain interaction is possible because β-turns or folds permit adjacent polypeptide chains to associate, mostly via hydrogen bonding and, to a lesser extent, via hydrophobic and electrostatic interactions. The parallel polypeptide chains in the β-pleated sheet are almost fully extended without a helical arrangement. Only in rare cases (e.g. myosin, fibrinogen, collagen, keratin) is the polypeptide of a protein totally in an α-helix or β-pleated sheet. In most proteins polypeptide segments may possess α-helical segments which may be separated by areas of non-structured sequences, called random coil.

The tertiary structure refers to the preferred three-dimensional arrangement of the folded polypeptide (α-helix, β-pleated sheet, random coil) chains. In a typical tertiary structure (e.g. myoglobin), the polypeptides are tightly folded to give a compact molecule in which most of the polar amino acid residues are located on the outer surface and are hydrated; most of the apolar groups are internal in the hydrophobic region from which water is essentially excluded; proline residues are located at the bends or folds in the polypeptide chain where iso-leucine, serine and charged amino acid residues (α-helix disrupting molecules) also tend to be located. These fold regions lack helical structure and tend to be random. Approximately 30 % of the amino acids are located in the folded regions. The gross conformation, i.e. tertiary structure, of globular proteins tends to be similar in general organisation but the degree of folding and relative proportions of α-helix and random coil vary immensely among proteins. Some proteins are loosely folded and the tertiary structure is stabilised by disulphide bonds (e.g. insulin); others, although compactly folded, also have S—S bonds, (e.g. lysozyme, glycinin). Lysozyme (125 amino acids) has 25 % α-helix and a very convoluted, tightly folded structure with four interpeptide S—S bonds.[14−16]

In an aqueous environment, proteins are usually folded in a compact state with the polar groups exposed to the aqueous phase and the hydrophobic groups located internally. In solution, globular proteins exist in the same closely packed state with individual amino acids occupying the same volume as they do in the crystalline state. In some cases water molecules occur internally but are completely isolated from the surrounding water and are engaged in hydrogen bonding with protein side chain groups or with other structured water molecules. These water molecules play a structural rôle in stabilising the native protein conformation. In solution, proteins are highly hydrated, being surrounded by a multilayer of water of solvation. This is generally called unfreezable water. It covers the immediate surface of the macromolecule but extra clusters occur about ionic and polar side-chains of the protein.[13-16]

Many globular proteins with molecular weights exceeding 50 000 daltons, are oligomeric, consisting of two or more individual polypeptides (protomers) associated. This association is referred to as quaternary structure. Haemoglobin, a tetramer composed of two pairs of similar subunits, is perhaps the best characterised 'quaternary' protein. The $\alpha$- and $\beta$-subunits possess about 70 % helical structure and are quite similar to myoglobin. The subunits associate spontaneously, presumably resulting from hydrophobic interactions.

Van Hold[17] proposed that proteins with more than 28 mole per cent of certain non-polar amino acid (valine, proline, leucine, *iso*-leucine and phenylalanine) tend to self-associate. Thus, the $\alpha$- and $\beta$-chains of haemoglobin associate, whereas myoglobin, which has <28 % apolar amino acids, does not self-associate. An explanation for this phenomenon is that, below a critical fraction of hydrophobic residues, it is possible to bury all of these completely within the molecule. When this fraction is exceeded the folding of molecules cannot accommodate all hydrophobic residues internally and some are exposed on the surface. When enough such residues appear on the surface, hydrophobic patches are created and these can be protected from the polar medium only by a self-association, thereby resulting in subunit association, e.g. casein, actin, glutenin.


**Protein Folding**

An unfolded linear polypeptide spontaneously folds to the same structure *in vivo* and *in vitro*. Therefore, the conformation of proteins is largely under thermodynamic control although some kinetic constraints may apply. The conformation adopted by a polypeptide is such as to ensure a minimum free energy of the system, i.e. the native protein assumes its lowest energy state. This hypothesis is supported by instances of reversible denaturation wherein a globular protein can be transformed to a random coil form and then be capable of reassuming its native globular structure because it is thermodynamically favourable.[16-18]

The thermodynamic motive force for protein folding is the sum of the common (attractive and repulsive) forces acting on separate components and segments of the polypeptide. Both entropic and enthalpic factors contribute to the folding but it is generally felt that the dominant driving force is the negative entropy effect caused by the exposure of apolar polypeptides to the aqueous solvent phase. This results in the unfavourable

$(-\Delta S)$ structuring of the water in the immediate vicinity of the hydrophobic peptide moieties. Consequently, folding of hydrophobic regions out of the water into the protein interior increases the entropy $(+\Delta S)$ of the previously structured surrounding water. Although the enthalpy $(\Delta H)$ of this process is usually slightly positive—and therefore unfavourable—the large positive $\Delta S$, i.e. entropy, greatly favours folding. Hydrophobic interactions are non-specific because they result from solvent properties rather than attractive bonding between specific groups. Once apolar groups are forced within a critical distance (3Å), van der Waals and electrostatic interactions, hydrogen bonding and disulphide bonds may stabilise the final structure.[15-19]

## Non-covalent Forces Stabilising Protein Structure

Functional properties are related to structure. Because changes in the structure of food proteins are frequently important in specific functions, e.g. surface denaturation of egg white during interfacial film formation, or the unravelling of gluten proteins during dough development, it is important to understand the forces responsible for the native structure of proteins and the factors which affect these in relation to their function in the native or denatured states. The forces involved in stabilising the tertiary (and quaternary) structure of proteins may involve hydrogen bonding, van der Waals, dipole and electrostatic interactions and hydrophobic associations. Covalent disulphide bonds are also important in several proteins (Table 9).

TABLE 9

BOND ENERGIES OF SOME BONDS IN PROTEINS

| Bond | Bond energy[a] (kcal/mole) |
|---|---|
| Covalent | |
|   C—C | 83 |
|   S—S | 50 |
| Non-covalent | |
|   Hydrogen bond | 3–7 |
|   Ionic/electrostatic bond | 3–7 |
|   Hydrophobic association[b] | 3–5 |
|   Van der Waals bonds | 1–2 |

[a] Free energy required to break the bonds.
[b] Free energy required to unfold a non-polar side-chain from the hydrophobic protein interior into the aqueous environment at 25 °C.

The hydrogen bond, which is ionic in character, refers to the sharing of a hydrogen atom attached to an electronegative atom (i.e. nitrogen, oxygen or sulphur) with another electronegative atom (i.e. oxygen of carboxyl or carbonyl groups). The free energy favouring the formation of hydrogen bonds ranges from 2 to 10 kcal/mole. Proteins contain several groups capable of hydrogen bonding. In proteins, the $\alpha$-helical and the $\beta$-pleated sheet structures are stabilised by hydrogen bonding within and between the peptides. Although the existence of such hydrogen bonds in proteins is well established, there has been considerable disagreement about their contribution to protein stabilisation because, in aqueous media, H-bonds are thermodynamically unstable. Water is, itself, a strong hydrogen bonding agent; both the donor and acceptor groups of the solute (e.g. N—H and C=O groups of peptide bond) are capable of hydrogen bonding with water as well as with each other. Schachman[20] estimated that the free energy contributed by hydrogen bonds to the stability of proteins is only about $-0.15$ kcal/mole of residue, i.e. negligible. However, in the internal non-aqueous, hydrophobic regions of certain segments of proteins, H-bonds may contribute to the stability of particular structures. Hydrogen bonding is particularly important in gluten development.

Electrostatic interactions may occur between the negatively charged amino acids (aspartic and glutamic acids) and the positively charged (arginine, lysine and histidine) residues. Terminal carboxyl and protonated amino acids may also participate in ionic interactions. These interactions are pH-dependent and the $pK_a$'s of these groups vary depending on the local environment. The extent to which ionic interactions contribute to the stability of protein structures is not clearly understood. Only marginal stabilisation can be expected because, even in a relatively low ionic strength environment, sufficient concentration of counter-ions may minimise charge/charge interaction. Furthermore, in most proteins charged groups tend to be exposed on the surface in contact with solvent. However, ionic interactions may be important in some cases. For example, in the case of chymotrypsin the $\alpha$-amino group of the terminal *iso*-leucine interacts with the carboxyl anion of aspartic acid and stabilises the specific conformation of the protein. These two charged groups are actually buried in a hydrophobic environment and the ionic bond has been shown to have a stabilisation energy of $\sim 3$ kcal/mole.[21] Perutz[22] adduced evidence for the importance of electrostatic forces in stabilising protein structure and function with particular emphasis on the rôle of electrostatic interactions in haemoglobin and the thermal stability of enzymes from thermophilic organisms. Ionic interactions with substrates can enhance the catalytic conformation of enzymes, e.g. ribonuclease and trypsin.[23]

Williams[24] discussed the importance of electrostatic interactions, particularly those involving cations, in many biological functions, especially regulatory actions. In comparison with hydrophobic interactions, these have low energy barriers and provide a mechanism for rapid reaction and alteration of protein conformation, thereby being important in biological regulation.

Electrostatic and ionic interactions/attractions are very important in the physical properties of several food proteins. The assembly of myofibrillar proteins and the interaction between myosin and actin depend upon electrostatic effects. Electrostatic factors influence the formation of interfacial films and affect the intermolecular interactions between contiguous molecules which contribute to the rheological properties of these interfacial films in emulsions and foams.[9,25,26] Furthermore, net surface charge and electrostatic repulsion are important in the stability of emulsions and foams.[27]

Electrostatic interactions between the phosphorylated caseins, via calcium crossbridging, are important in stabilising the micelle structure formed by self-association of casein polymers. Dephosphorylated $\alpha$- and $\beta$-caseins can both form micelles,[28,29] in fact, self-aggregation is enhanced in the case of dephosphorylated $\beta$-casein, indicating that some electrostatic repulsion occurs in normal $\beta$-casein. However, micelles formed from dephosphorylated $\alpha_s$-casein are loose, highly hydrated and relatively unstable,[28] indicating that calcium crossbridging is important for the formation and stabilisation of normal micelles.

Green and Marshall[30] demonstrated the importance of electrostatic interactions in the stabilisation of casein micelles and how this knowledge may be exploited for practical purposes, i.e. by enhancing the action of rennet in accelerating coagulation of milk proteins. Thus, cationic compounds, e.g. lysozyme, polylysine and spermine, by neutralising the anionic charge on $\kappa$-casein, enhance coagulation.[30] Further research in this area may lead to the development of acceptable non-rennet, non-enzymatic clotting agents or agents that amplify the effect of rennet.

Modification of food proteins to alter net charge and improve functional properties has received much attention in recent years.[31] A classical example of protein charge modification is the removal of the anionic glycomacropeptide from $\kappa$-casein by rennets to permit coagulation and curd formation in cheese manufacture.

The succinylation of certain novel proteins from soy, leaf, yeast, fish and peanut increases net charge and markedly enhances their solubility and emulsifying and foaming properties.[9,26,31–33] Reversible modification using citraconylation has provided a method for the preparation of functional protein from microbial sources.[34]

Van der Waals forces are generally non-specific, short-range forces whose magnitude is inversely proportional to the sixth power of the distance between the molecules. They have an energy of 1 to 3 kcal/mole and are most effective at 3 to 5Å. They are attributed to polar interactions between induced dipoles and adjacent atoms and therefore are related to the polarisability of the molecules or atoms involved. Contribution of van der Waals interactions to the stability of proteins is indicated by the packing density of amino acid residues in the interior of globular proteins. These interactions are individually very weak but collectively they may be important in stabilising domains of proteins although their involvement in conformation and folding is questionable.

Hydrophobic interactions are the aggregate forces due to repulsion of apolar groups by water (entropy driven). These interactions actually cause folding, and facilitate other stabilising interactions of proteins and are among the principal forces responsible for protein conformation. Proteins contain significant amounts of apolar amino acids (alanine, valine, leucine, *iso*-leucine, proline, phenylalanine, tryptophan, tyrosine, methionine). Polypeptides containing a high proportion of these are poorly soluble, i.e. thermodynamically they interact unfavourably with water. Thus, in the immediate vicinity of these residues, water is forced into a more ordered H-bonded cage-like arrangement and the negative entropy effect due to the structuring of water is very significant. The effect is minimised if the polypeptides containing the apolar residues fold in such a manner as to minimise contact with water. Thus, proteins usually fold into globular structures where the polar groups are exposed and the apolar groups associate in the anhydrous hydrophobic interior.[19]

An estimation of the magnitude of the hydrophobic effect was gained from thermodynamic data on the transfer of apolar compounds from organic solvents into water.[19] The approximate free energy change ($\Delta G_t$), i.e. gain, for the transfer of apolar amino acids from water to ethanol is approximately $-3400$, $-2970$, $-2650$, $-2600$, $-2800$, $-2420$, $-1680$, $-1300$, $-730$ cal/mole for tryptophan, *iso*-leucine, phenylalanine, proline, tyrosine, leucine, valine, methionine and alanine, respectively.[19] The negative free energy change for each residue is significant and increases with size of the group, with aromatic residues being most hydrophobic. When an apolar residue or hydrocarbon is transferred from an aqueous to an apolar medium, the following changes in thermodynamic parameters occur: $\Delta G^\circ < 0$; $\Delta H^\circ > 0$; $\Delta S^\circ > 0$. The negative $\Delta G$ indicates a favourable reaction and this is accounted for by the large entropy ($\Delta S^\circ$) increase which is due to the extensive disordering of the water structure upon removal of the apolar groups.

If similar thermodynamic phenomena prevail in proteins, the free energy contributed by all apolar residues (averaging $-2\,\text{kcal/mole}$ residue) is a major force responsible for conformation.

The inherent properties of water are responsible for the hydrophobic effect,[35] and the magnitude of the hydrophobic interactions depends very much on the structure of the dispersing water. Agents or conditions which alter water structure change hydrophobic interactions,[36] e.g. fluorides or sulphates, which enhance water structure, stabilise protein structure, whereas perchlorate, which destructures water, weakens hydrophobic-dependent structure and destabilises protein structure.[36-38]

Decreases in temperature reportedly weaken hydrophobic interactions whereas increasing temperatures up to $60\,^\circ\text{C}$ enhances them. As temperature is lowered the water becomes more extensively hydrogen-bonded and more structured; therefore, it is generally felt that the entropy advantage of folding non-polar groups into the hydrophobic interior of proteins is lessened at lower temperatures and structuring and stabilisation by hydrophobic interactions are expected to decrease.[15] Because hydrophobic interactions are weakened as temperature is lowered, some proteins have a high propensity towards cold-dissociation, e.g. oligomeric proteins, casein, and soy protein.

Using the transfer data of Tanford,[35] Bigelow[39] calculated the average hydrophobicity of a large number of proteins (Table 10). These data are useful in relating hydrophobicity to physical properties. For example, data are being accumulated showing a relationship between the surface hydrophobicity of food proteins, their surface activity and emulsifying properties[40] (Table 11). The hydrophobic polypeptide loops of proteins facilitate orientation at the interface and enhance film formation.[27]

Hydrophobic interactions are of great practical significance in food products in relation to structure and function. Thus, they are critically involved in the assembly and structure of myofibrillar proteins (actin) in muscle, the casein micelles in milk and gluten in dough. There is strong evidence relating the elastic properties of gluten to the hydrophobic effect, i.e. the spontaneous refolding of extended gluten films following relief of stress.

The major proteins of milk exist in colloidal suspension as spherical

                 *J. E. Kinsella*

## TABLE 10
THE AVERAGE HYDROPHOBICITY AND POLARITY OF SOME FOOD PROTEINS[39]

|  | $MW \times 10^3$ | Q | NPS | P | Charge |
|---|---|---|---|---|---|
| Tropomyosin | — | 870 | 0·20 | 1·94 | 0·47 |
| Ovomucoid | 27 | 920 | 0·25 | 1·50 | 0·30 |
| Collagen | 80 | 960 | 0·19 | 0·89 | 0·16 |
| Lysozyme | 15 | 970 | 0·26 | 1·18 | 0·16 |
| Myosin | 500 | 1 020 | 0·28 | 1·43 | 0·35 |
| Actin | 57 | 1 050 | 0·31 | 1·15 | 0·27 |
| Gliadin | 120 | 1 080 | — | — | — |
| Conalbumin | 87 | 1 080 | 0·31 | 1·30 | 0·32 |
| Myoglobin | 17 | 1 090 | 0·32 | 1·12 | 0·34 |
| Ovalbumin | 46 | 1 110 | 0·34 | 0·92 | 0·24 |
| BSA | 65 | 1 120 | 0·32 | 1·22 | 0·33 |
| α-Lactalbumin | 16 | 1 150 | 0·34 | 1·11 | 0·28 |
| α-Casein | — | 1 200 | 0·38 | 1·27 | 0·24 |
| β-Lactoglobulin | 17 | 1 230 | 0·37 | 0·96 | 0·28 |
| Zein | 50 | 1 310 | — | — | — |

MW = mol. weight; Q = total hydrophobicity divided by total number of residues; NPS = sum of Trp, Ile, Tyr, Phe, Pro, Leu and Val residues as a fraction of total residues; P = polarity ratio roughly all polar side chains/all non-polar side chains; charge is sum of Asp, Glu, His, Lys, Arg as fraction of total amino acids.

## TABLE 11
RELATIVE HYDROPHOBICITY (FLUORESCENCE INTENSITY, $S_0$), SURFACE TENSION, INTERFACIAL TENSION AND EMULSIFYING ACTIVITY INDEX VALUES OF VARIOUS PROTEINS[40]

|  | So | Surface tension (Dynes/cm) | Interfacial tension (Dynes/cm) | Emulsifying Activity Index ($m^2/g$) |
|---|---|---|---|---|
| Bovine serum albumin | 1 400 | 57·9 | 10·3 | 166 |
| κ-Casein | 1 300 | 54·1 | 9·5 | 185 |
| β-Lactoglobulin | 750 | 59·8 | 11·0 | 151 |
| Trypsin | 90 | 64·1 | 12·0 | 93 |
| Ovalbumin | 60 | 61·1 | 11·6 | 57 |
| Conalbumin | 70 | 63·7 | 12·1 | 105 |
| Lysozyme | 100 | 64·0 | 11·2 | 55 |

aggregates ($10^8$–$10^9$ daltons) composed of subunits consisting of $\alpha_{s1}$-, $\alpha_{s2}$-, $\beta$- and $\kappa$-caseins in the ratio 4:1:4:1.[41] The individual caseins contain a relatively high concentration of apolar amino acids that occur in clusters along the polypeptide, giving rise to several markedly hydrophobic regions. The average hydrophobicities of $\alpha_{s1}$-, $\beta$- and $\kappa$-caseins are 1170, 1335 and 1205 cals/mole residue, respectively.[42] Because of their hydrophobicity, these proteins aggregate spontaneously into mixed polymers by an entropy-driven process.[43] The micelles thus formed are stabilised in the presence of calcium by the $\kappa$-casein component.[42–45] The importance of hydrophobic interactions in the functional behaviour of milk proteins is demonstrated by the effects of low temperature and chaotropic agents (urea) on the stability of casein micelles.[45]

Micelle formation represents a mechanism by which relatively high concentrations of insoluble polypeptides can be maintained in suspension and rendered stable to a range of heat treatments. The coagulation of casein following rennet action (i.e. removal of the anionic glycomacropeptide moiety) requires the formation of a critical amount of para-$\kappa$-casein (average hydrophobicity, 1300 cal/mole residue) to maximise hydrophobic interactions which result in coagulation.[46]

The hydrophobicity of proteins affects flavour. While most intact proteins have little flavour, they influence flavour because they may contain bound flavour compounds; also, they can modify flavour perception and, upon partial proteolysis, certain proteins yield bitter peptides which are commonly encountered in casein, soy and gluten hydrolysates. The bitterness of these peptides is related to their net hydrophobicity or apolar character. The average hydrophobicity (HQ) is the sum of the hydrophobicity of the peptide[39] divided by the number of amino acid residues.[47] HQ (cal/mole residue), reflects the amino acid composition and peptides with high amounts of apolar amino acids have high HQ values. Generally, peptides with HQ values above 1400 cal/mole residue taste bitter, irrespective of the sequence of amino acids, while peptides with HQ < 1300 do not taste bitter[47,48] (Table 12).

Thus, hydrolysates of proteins with high HQ values, e.g. casein (HQ = 1605), zein (HQ = 1480) or soy (HQ = 1540), tend to yield more bitter peptides upon proteolysis than meat proteins (HQ = 1300).[48] The amount of bitter peptides formed reflects the extent of hydrolysis of these proteins and the type of enzyme used. Guigoz and Solms[48] listed 26 bitter peptides of varying molecular sizes from hydrolysates of casein and identified them with sequences in the original molecule. The HQ values ranged from 1493 to 2525 and the flavour from slightly bitter to extremely

### TABLE 12
EXAMPLES OF BITTER PEPTIDES OBTAINED FOLLOWING THE
PROTEOLYSIS OF CASEIN[48]

| *Peptide* | *Hydrophobicity* *(HQ, cal/residue)* |
|---|---|
| Ala-Pro-Phe-Pro-Glu-Val-Phe-OH | 1 984 |
| Leu-Phe | 2 535 |
| Phe-Tyr-Pro-Glu | 2 167 |
| Ala-Pro-Lys | 1 610 |

bitter. They also listed bitter peptides from soy, zein and other proteins. Most of the peptide sequences could be obtained from any food protein but incomplete proteolysis of highly hydrophobic proteins produces more of them. This phenomenon is of significance in relation to enzyme technology and protein utilisation and is a limitation in the preparation of protein hydrolysates by proteinases for food applications, i.e. for beverages, foaming agents, etc. Bitterness in cheese is frequently associated with the increased proteolysis due, for example, to the use of microbial proteinase as rennet substitutes.

The binding of off-flavours to proteins, especially soy proteins, involves hydrophobic interactions.[10] Agents that alter hydrophobic relationships in proteins may be exploited to remove off-flavours.[49]

### Disulphide Bonds

Covalent disulphide (SS) bonds (50 kcal/mole) are important in stabilising the tertiary structure of many extracellular and secretory proteins. Generally, SS bonds are formed following attainment of the most thermodynamically stable conformation.[16] Proteins vary in the number of SS bonds present, e.g. ribonuclease (13·6 kdaltons) has 4, lysozyme (14 kdaltons) has 4, soybean trypsin inhibitor (8 kdaltons) has 7, bovine serum albumin (BSA) (66 kdaltons) has 17 and soybean 11S (360 kdaltons) has 21, SS bonds. In these proteins, the SS bonds enhance tertiary structure and confer molecular stability, particularly where non-covalent interactions are weak. Where hydrophobic interactions are minimum, the formation of SS bonds may be necessary for the stabilisation of some tertiary structure. Disulphide linkages may add rigidity to specific domains of a protein for particular purposes. They may permit deformation for functional reasons and ensure restoration of the original tertiary structure.[24] Thus, in certain proteins with physiological functions, e.g. BSA, where a range of variable environmental conditions may be encountered and because binding of apolar compounds (fatty acids, prostaglandins) may tend to alter its conformation, the presence of SS bonds may stabilise the tertiary structure. It is also conceivable that the presence of SS crosslinks impairs proteolysis and thereby minimises degradation of functional proteins (BSA, trypsin inhibitors) or storage proteins (glycinin 11S, glutenin).

Disulphide bonds are very much involved in determining the digestibility of proteins; thus, the 11S fraction of soy and other legumes is highly resistant to proteolysis, and reduction of the disulphide crosslinks enhances the rate of digestion *in vitro*.[50] Proteins containing SS bonds tend to be quite heat stable and show higher temperatures and enthalpies of denaturation compared with non-SS-bonded tertiary structures. The marked thermal stability of soy 11S protein is a major factor limiting its application in foods.[6,51] The trypsin inhibitors of soybean have several disulphide crosslinks which help stabilise these inhibitors against thermal denaturation.

The formation and reduction of SS bonds are integral events which affect functional properties of protein in foods. This is epitomised by thiol oxidation and thiol–disulphide interchange interactions which occur during development of the viscoelastic properties of gluten.[52] The stability of milk proteins is affected most by the thermal instability of $\beta$-lactoglobulin.[53] This is the most heat-labile of the major milk proteins and, when milk is heated, $\beta$-lactoglobulin denatures and aggregates. These aggregates can destabilise the micellar system, possibly by binding calcium or altering pH.[44,53] However, $\beta$-lactoglobulin contains a free sulphydryl group and during heating of milk this interacts with the disulphide bond of $\kappa$-casein and, via thiol–disulphide interchange, forms a disulphide linked $\beta$-lactoglobulin–$\kappa$-casein complex.[54,55] This eliminates destabilisation effects unless $\kappa$-casein is limiting for complex formation and micelle stabilisation. This reaction is exploited by forewarming milks intended for the manufacture of some milk powders to improve their functionality in bread and bakery uses. Conceivably, improvement is due to elimination of the free thiol group of the $\beta$-lactoglobulin which otherwise might catalyse excessive thiol–disulphide interchange in the dough.[56] Incidentally, the formation of the $\beta$-lactoglobulin–$\kappa$-casein complex in milk impairs rennet action.[45]

## Other Stabilising Factors

Finally, cations and/or prosthetic groups play an important rôle in enhancing the stability of a number of proteins (especially enzymes), e.g. in the case of alkaline phosphatase and α-amylase, divalent metal ions enhance thermal stability; in myoglobin and haemoglobin, the haem group confers stability. These are significant in food processing; for example, the activity of alkaline phosphatase, which is used as an index of thermal processing of milk, is significantly affected by magnesium concentrations.[57]

The quaternary structure of proteins is governed by the same general classes of stabilising interactions, particularly hydrophobic interactions, crossbridging (e.g. Ca–caseinate) and disulphide crosslinking (e.g. the acidic and basic subunits of glycinin and glutenin). Manipulation of the quaternary structure to alter the physical properties of these proteins in many food applications is feasible.

## Protein–Water Interactions

Perhaps the most critical criterion governing the usefulness and functional application of food proteins is the manner in which they interact with water (Table 13). The nature of these interactions affects applications depending on water sorption, i.e. wettability, dispersability, viscosity, thickening, dough formation, gelation, solubility and surfactant properties. The water-holding capacity of protein, e.g. in meat, affects juiciness and flavour release. The various types of protein–water interactions are important in different applications.[58]

The hygroscopic properties of protein powders, i.e. water vapour sorption capacity, can markedly affect their functional properties and their storage stability. This has been observed in casein systems, soy and fish protein isolates. The adsorption of moisture during storage can facilitate chemical reactions, e.g. browning reactions, and result in polymerisation of proteins, thereby impairing solubility and hydration properties.

Wettability and dispersability are important in food manufacture where large amounts of protein are being dispersed to make aqueous systems (beverages, batters). The ease and rapidity of hydration of protein particles determines whether an intractable, pasty, clumpy mass or a smooth, uniform dispersion is obtained. The wettability of a protein powder depends on surface activity of the particle, the proportion of hydrophilic residues and the size and microtopography of the particles. Generally, proteins are more difficult to disperse when denatured.

As certain proteins absorb water they display a tendency to swell without disintegrating, e.g. soy protein, myofibrillar proteins, caseinates. Swelling reflects the uptake of water by the protein matrix with loosening of the polypeptides. The degree of swelling is influenced by the intermolecular

TABLE 13

CLASSIFICATION OF WATER THERMODYNAMICALLY ASSOCIATED WITH PROTEINS AT INCREASING WATER ACTIVITIES[58]

| | |
|---|---|
| Structural Water: | Water H-bonded to specific groups, participates in stabilisation of structure, unavailable for chemical reaction. |
| Hydrophobic Hydration Water: | Structured cage-like water surrounding apolar residues; like structural water, is very much involved in stabilising protein structure. |
| Monolayer Water: | The first adsorbed water monolayer, H-bonded; unavailable as solvent, may be available for chemical reactions; ranges from 4–9 g/100 g protein. |
| Unfreezable Water: | This roughly includes all water (structural monolayer, and perhaps some adsorbed multilayer water) that does not freeze at normal temperature—amounts to $0.3$–$0.5$ g/g protein and corresponds to water up to $A_w = 0.9$. The amount varies with polar amino acid content and includes some water available for chemical reactions. |
| Capillary Water: | Water held physically in clefts or by surface forces in the protein molecule (e.g. water entrapped in gels, cheese curd); physical properties similar to those of bulk water. |
| Hydrodynamic Hydration Water: | This represents the water 'loosely' surrounding the protein that is transported with the protein during diffusion (centrifugation); it has properties typical of normal water. |

forces or bonding (hydrogen bonding, electrostatic interactions) between adjacent polypeptides and the facility with which water can disrupt these and replace protein–protein with protein–water bonding. Swelling markedly affects the amount of water held by a food protein and thereby affects viscosity, body, juiciness and texture of the food products.[59]

The viscosity of a protein dispersion is important in beverages, fluids and soups. Viscosity is a function of protein type (molecular size, shape, surface charge, type, concentration), pH, temperature, heat treatment, shear rate, ion type and ion concentration. Edsall[60] studied relationships between molecular shape, size, orientation and concentration of protein particles, the shear rate and apparent viscosity, and noted significant effects depending upon the orientation of these molecules. The various physical factors affecting the hydrodynamic properties and intrinsic viscosity of protein molecules of different shapes and sizes and the theoretical basis for calculating viscosity have been reviewed.[59-64]

Gelation and coagulation are important events in thermosetting and matrix formation. Gelation usually results from the unravelling of protein by heat which, upon subsequent cooling, sets to a structured matrix, entrapping water. This is an important feature of proteins in meat systems, custards and gels. Coagulation is an irreversible protein aggregation reaction and usually results from destabilisation of soluble systems or denaturation of the protein with concomitant protein–protein association and loss of solubility.

Water-holding capacity is an index of the amount of water retained within a protein matrix under certain conditions. This usually includes bound water, entrapped water and hydrodynamic water and the index very much depends upon the conditions of its determination.

Solubility is a very important functional property of food proteins because many other functional properties, e.g. foaming, emulsifying and gelation, depend upon the initial solubility of the particular protein. In addition, the degree of solubility is a very good index of prior treatment of that protein. Protein solubility depends upon the protein, its conformation, surface charge, polarity, the pH, ion type and concentration and temperature of the solvent.[2]

*Protein Hydration*

Water associates with proteins in a number of different ways (Table 13). Thermodynamically, a significant amount of water binds to protein as part of the structural matrix (constitutional water) and as hydrogen-bonded, sorbed water layers (interfacial water). Physically, water is entrapped and, hydrodynamically, layers of water are associated with a protein in solution.[41,58,65]

Crystalline proteins may possess a few molecules of 'structural' water as part of the tertiary structure. This water engages in hydrogen bonding between polar groups of the protein and is actually involved in stabilising the native structure of protein. This is the truly bound water (about 6 mg/g protein) and is unavailable for chemical reactions; it affects the structure and conformation of the protein molecule.

When a protein is exposed to an environment of progressively increasing relative humidity or water activity ($A_w$), a classical sigmoidal sorption isotherm is obtained, reflecting the progressive increase in the amount of water associated with the protein. At low $A_w$, $<0.3$, water is selectively H-bonded to charged and polar residues to form a monolayer of bound water (vicinal water) (3 g/100 g protein). This also includes the 'iceberg' or clatharate water structured around the hydrophobic residues (hydrophobic hydration). Most proteins have 5–9 % monolayer water (BET monolayer) which represents structured water that is unavailable as solvent but may be available for certain reactions. Subsequently (between $A_w$ 0·3 and 0·9), multilayers of water build up beyond the vicinal layer. This interfacial layered water is strongly influenced by the nature of the protein; generally its mobility is less than that of normal water and, in the region of hydrophobic groups, it is perhaps more structured and of different configuration than in the neighbourhood of hydrophilic groups where hydrogen bonding occurs. Vicinal water amounts to 0·25–0·35 g $H_2O$/g protein. The multilayer water is the water adjacent to the free water; it exhibits reduced vapour pressure and molecular mobility and much of it is unfreezable at 0 °C.

Finally, at $A_w$ > 0·9, liquid water–bulk phase water may associate loosely with the protein. This may cause swelling, or may be trapped within a matrix of polypeptides (gel) or exist as capillary water. It represents the major portion of water held in foods such as cheese curd, meats, meat emulsions and gels. It is freely available for chemical reactions and solvent functions. While its structure resembles that of bulk water, considerable force is required to move it from the protein mass.[58]

As the proportion of water is increased, the protein may actually go into solution. Proteins in solution may be surrounded by a water layer, hydrodynamic hydration, which is transported along with the protein during diffusion. This is independent of entrapped and surrounding water. Hydrodynamic hydration water may or may not be included in the water-binding capacity of proteins as determined by different methods.[58] Usually, water-binding capacity includes all the types of hydration water plus some water remaining loosely associated with the protein following centrifugation. The operational term 'bound' water is widely used in the literature, although it is not universally defined. It is often referred to as that water which does not freeze at a specific temperature; as water which is not available as solvent; as water which shows different physical properties or as water that moves with the macromolecule in a sedimentation field. Bound water is really that water in the vicinity of a macromolecule whose properties differ detectably from those of the bulk phase water in the same system.[58]

The amount of bound water in proteins ranges from 0·3 to 0·5 g/g-protein. Proteins of widely different composition and structure show roughly similar water-binding properties (Table 14). For example, fibrous

TABLE 14

HYDRATION (g/100 g PROTEIN) OF DIFFERENT PROTEINS[58]

| Protein | BET monolayer | $A_w$ 0·92 | NMR | Calorimetry |
|---|---|---|---|---|
| Collagen | 9·52 | 45 | 50 | 35 |
| Casein | 5·50 | 40 | 39 | 43 |
| $\beta$-Lactoglobulin | 6·67 | 32 | — | 55 |
| Ovalbumin | 5·65 | 30 | 33 | 32 |
| Serum albumin | 6·73 | 32 | 40 | 32 |
| Myoglobin | — | 42 | 42 | — |
| Soy protein | 5·80 | 33 | 35 | — |

and globular proteins both bind 0·3–0·5 g $H_2O$/g protein; however, the former is relatively insoluble compared with the latter group. Thus, apparently there is no relationship between water binding and solubility when expressed as g $H_2O$/g protein. This may reflect the manner in which bound water is expressed. Fennema[65] recommended that, rather than expressing hydration as g $H_2O$/g protein, one should express it on the basis of surface area. For example, $\alpha$-chymotrypsin binds 0·37 g and 0·4 g $H_2O$/g protein in the native and denatured state, but the surface areas of chymotrypsin in native and denatured states are 10 000 and 36 000 Å$^2$, respectively. The surface area of globular proteins is significantly less than that of fibrous proteins, hence hydration per unit area of the latter is much less. However, surface area alone is not a full explanation since differences in the nature of the protein, surface polarity, charge, hydrophobicity and topography all markedly affect the degree of hydration.

**Factors Affecting Protein Hydration**

Several factors affect water binding by food proteins (Table 15). The extent of protein hydration correlates strongly with the polar residues (carboxyls and basic groups); amides generally inhibit water binding. Interaction between water molecules and hydrophilic groups (hydroxyl, amino, carbonyl and imino groups, etc.) occurs via hydrogen bonding. H-bonded water bridges between polar groups in different regions of the protein may also occur. Co-operative hydrogen bonding may occur in macromolecules wherein the binding of one water molecule to a polar group facilitates the hydrogen bonding or bridging of another molecule of water to an adjacent polar group in the protein.

Polar amino acids with ionised side chains bind the greatest amount of

## TABLE 15
### FACTORS AFFECTING WATER BINDING BY FOOD PROTEINS

Amino acid composition
Protein conformation (shape, size)
Surface topography
Surface polarity/hydrophobicity
Ionic concentration
Ion species, anion/cation
pH
Temperature

water; non-ionised amino acids bind an intermediate amount (2–3 molecules) and the hydrophobic groups bind little or no water (Table 16). Kuntz[66] derived a formula: $a = f_c + 0.4f_p + 0.2f_n$; where $a = g$ bound water/g protein, $f_c$ = fraction of charged amino acid side chains, $f_p$ = fraction of polar amino acid side chains, $f_n$ = fraction of apolar amino acid side chains. Basically, this indicates that the more polar the protein, the greater the amount of water it can bind.

At low $A_w$ ($<0.3$), each polar group in a protein binds one molecule of water up to the monolayer level. Thus, ovalbumin, which has 313 moles polar groups/$10^5$ g, binds 329 moles $H_2O/10^5$ g protein. However, as the relative humidity is increased the polar groups progressively bind more

## TABLE 16
### WATER-BINDING CAPACITY OF AMINO ACIDS IN SYNTHETIC POLYPEPTIDES (MOLE $H_2O$/RESIDUE)[41]

| *Polar and ionised* | | *Polar but unionised* | |
|---|---|---|---|
| Asp⁻ | 6 | Asp (pH 4·0) | 2 |
| Glu⁻ | 7·5 | Glu (pH 4·0) | 2 |
| Tyr⁻ (pH 12) | 7·5 | Tyr | 3 |
| Lys⁺ | 4·5 | Arg (pH 10·0) | 3 |
| Hist⁺ | 4 | Lys (pH 10–11) | 4·5 |
| *Hydrophobic* | | *Unionisable* | |
| Ala | 1·5 | Asn | 2 |
| Val | 1 | Gln | 2 |
| Leu | 1 | Pro | 3 |
| Phe | 0 | Ser | 2 |
| Ile | 1 | Trp | 2 |
| Met | 1 | | |

water (Table 17). Thus, at 10–15 % relative humidity they bind one mole whereas, at 80 % relative humidity, on average, they bind at least two moles of water. The increased binding is due to the formation of multilayers of $H_2O$ around these particular groups. Bull and Breese[67] found that six water molecules are bound per polar group at $A_w = 0.92$.

TABLE 17

THE AMOUNT OF WATER ASSOCIATED WITH HYDROPHILIC GROUPS OF PROTEINS AT INCREASING RELATIVE HUMIDITIES[68]

| Sorption site | Moles water per sorption site at relative humidity (%) | | |
|---|---|---|---|
| | 10% | 50% | 80% |
| —COOH | 0·92 | 2·0 | 2·5 |
| —NH$_2$ | 0·83 | 2·1 | 2·7 |
| —OH Aliphatic | 0·09 | 0·3 | 0·6 |
| —OH Phenolic | 0·25 | 1·0 | 1·8 |
| —CO—NH$_2$ | 0·06 | 0·2 | 0·6 |

In food systems, the amount of water bound by polar groups depends upon their steric availability. Polar groups are generally exposed to the aqueous environment and are available for binding. Predictions based on the amino acid composition of several proteins, including ovalbumin, BSA and myoglobin, indicate that the calculated water of hydration, i.e. 0·45, 0·42 and 0·45 g/g protein, was very close to observed values. In most instances, it slightly exceeded the experimental value, indicating that some groups were unavailable for water binding.

Protein conformation can affect the thermodynamics of water binding by affecting the availability of polar or hydration sites. Transition of a protein molecule from a compact globular conformation to a random coil results in exposure of previously buried peptide bonds and amino acid side chains that can interact with water. Denaturation, i.e. a change in tertiary structure, has little effect on the amount of water bound by proteins when expressed on a weight basis. Usually, there is an increase of 0·02–0·1 g $H_2O$/g protein upon denaturation (e.g. egg albumin, serum albumin and tropocollagen in the native state bind 0·32, 0·33 and 0·31, respectively and 0·33, 0·46 and 0·52 g $H_2O$/g, respectively in the denatured state). The extra water reflects the availability of additional binding sites and an increase in hydrophobic hydration resulting from the unfolding of apolar polypeptide sequences.

In many systems, denaturation causes aggregation which may actually decrease water binding by replacing water–protein interactions with protein–protein interactions. Processes causing aggregation, e.g. extrusion texturisation, result in decreased water binding. However, modest heating, by unravelling tightly folded tertiary and quaternary structures, may enhance water binding by forming newly structured networks, e.g. heat gelation of soy, muscle (myosin) proteins and gelatin.

Although few data are available, it is obvious that surface topography affects water binding and Fennema[65] reported that water binding is related to available surface area. Water binding is also affected by pH.[66,67] Ionised amino acid groups bind considerably more water than non-ionised groups; reducing pH below 4·0 converts carboxy groups to the protonated, non-ionised forms, thereby reducing the amount of bound water. Thus, glutamic acid in the ionised and protonated states binds 6–7 and 2 moles of water/mole residue, respectively. Similarly, BSA at pH 7·0 and 3·0 binds 0·45 and 0·32 g $H_2O$/g protein at $A_w = 0·9$.[65] pH also affects the magnitude of the net charge on proteins which directly affects ionic interactions. At the isoelectric pH, where the net charge is zero, the electrostatic interaction is maximum and water binding is minimum. Thus, for example, an homogenate of beef muscle has relative water-binding capacities of 101 %, 42 % and 135 % that of normal muscle at pH 3·5, 5·0 (isoelectric) and 10, respectively.[65]

Salts, via their effects on electrostatic interactions, markedly affect water binding by proteins. At low concentrations of neutral electrolytes, e.g. NaCl, the amount of water bound to proteins increases with increasing salt concentration. This generally continues up to ionic strengths of 0·1–0·15 M. With monovalent electrolytes, there is a decrease in bound water around 2 M.[67,69] At this concentration, ions compete for water, suppress the electrical double layer surrounding the macromolecule, change protein conformation and hydration and may result in precipitation. Cations have different capacities to affect water binding, e.g. when dispersed in chloride salts (2 M) of a caesium, rubidium, potassium, sodium and lithium, egg albumin binds 0·2, 0·16, 0·15, 0·08, 0·08 g $H_2O$/g protein compared with 0·3 g in pure water.[67] The effect of the ions is related to the size of their hydrated radius; the smaller the unhydrated radius, the larger the hydrated radius because of surface charge density. The larger the hydrated radius, the greater is the 'dehydration' of the protein, i.e. Li > Na > K > Cs > Ca. With anions, the larger the hydrated radius, the less is its ability to reduce water binding ($ClO_4^- > NO_3^- > I^- > Br^- > Cl^-$). In relation to the relative effectiveness of cations and anions, it is probable that their effects are markedly influenced by the intensity of their surface charge which is influenced, of course, by their atomic radii.[67,69] At high salt concentrations competition of ions and proteins for existing water is such that electrostatic protein–ion interactions are apparently of little importance.

The increased water binding of meat proteins upon the addition of salt may be due to the preferential anion binding, e.g. chloride, by the protein molecules.[65,69] The binding of chloride by protein enhances hydration, possibly because the $Cl^-$ increases protein negative charge and results in repulsion between adjacent polypeptides, thus enhancing water imbibition.[70] At pH values below the isoelectric point, the positive charge of the protein is neutralised by $Cl^-$, thus decreasing its water-holding capacity.[65–67,70]

Modification of proteins by succinylation, etc., significantly improves hydration and water-binding capacity due to the increased number of charged groups and possibly to the expansion of the molecule as a result of increased electrostatic repulsion.[32]

Solubility is a very important property of a functional protein because it provides an index of native structure and because it is a critical prerequisite for using a protein in beverages, fluid foods and for emulsion and foam formation. Ideally, solubility at saturation represents the equilibrium between the solid and soluble phases.[8] However, protein solubility depends on the operational conditions and the methods used for its determination. The solubility of food protein depends on the inherent properties of the protein (composition, size, surface charge) and the prevailing environmental conditions. Most food proteins are heterogeneous populations of different proteins with different properties, each of which may respond differently to different processing conditions. Some fractions may be very susceptible to denaturation, others may be highly stable.

In solution, most globular proteins are compactly folded, with the polar and charged groups mostly exposed to the aqueous environment and the non-polar residues folded inwards, away from the aqueous phase. Globular proteins tend to be spherical to give maximum surface area for hydration of polar groups. However, as molecular size increases there is a tendency for the ratio of hydrophilic to hydrophobic residues to decrease. Because volume and surface area increase in proportion to the cube and square of the radius, respectively, the surface area to the volume ratio of a sphere decreases with increasing size and thus fewer polar groups can be accommodated. However, the surface area can be increased with constant volume by making the shape more asymmetrical, i.e. certain high molecular weight proteins assume ellipsoidal shapes to accommodate more polar residues on the surface.[71] If these constraints cannot be met, then subunit oligomeric structures may result.

The solubility of many proteins depends on tertiary and quaternary interactions as exemplified by the casein of milk and the 11S soy protein. Individual proteins of milk tend to have low solubility, but by associating to form a micelle and by having a surface rich in $\kappa$-casein moieties, the milk protein system is quite 'soluble' (colloidal suspension). The 11S soy protein is made up of acidic and basic subunits; association of the basic subunit with the acidic subunit tends to enhance the solubility of the former, particularly at high temperatures.[6]

Several factors affect the solubility of proteins, e.g. composition and ratio of polar to apolar amino acids, their sequence in the polypeptides, molecular size, conformation, pH, temperature, salts (species and concentration). Generally, other things being equal, solubility decreases with molecular weight and increases with polarity of the surface. Most proteins show typical pH-solubility profiles with minimum solubility in the isoelectric region, where electrostatic interaction is maximal. This behaviour can be altered by salts, which may counteract these attractive coulombic forces and cause solubilisation, even in the isoelectric range.

Ions markedly affect solubility via electrostatic, hydration and water-structuring effects.[72-74] The $NH_3^+$ and $COO^-$ groups in polypeptides tend to interact electrostatically with each other or with counterions in the solution. The strength of the interaction ($F = e^+ \cdot e^- / Dr^2$) reflects the strength of the charge ($e$) and varies inversely with the distance ($r$) apart and the dielectric constant (D) of the medium. In media with high dielectric constants the interaction between charged groups tends to be repressed while interaction with water is enhanced. For example, in water, globulin molecules interact with each other via charged groups and are insoluble but, when salt is introduced, the intermolecular charges are masked by counterions and globulins become more hydrated and soluble.

The solubilising effect of a salt depends upon its concentration, the charge of the ions and their effect on the structure of water. Ions affect solubility because the charged groups on proteins interact with inorganic ions and solubility is enhanced to a certain degree (salting-in effect) whereas at higher ion concentrations, water activity is reduced and protein–protein interactions ensue with concomitant insolubilisation (salting-out).

Thus, for most globular proteins, solubility first increases and then decreases with increasing salt concentration. However, in the case of soy proteins, salts initially decrease and then increase solubility.[8,37] Thus, oligomeric proteins may behave differently from globular proteins. In the case of the former, low salt levels may enhance hydrophobic interactions by suppressing electrostatic repulsion within the quaternary structure while, at higher salt concentrations, water is destructured and the proteins dissociate and unfold. The polymerisation of actin follows this pattern: G-actin polymerises to F-actin in $0.1$ M NaI but depolymerises in $0.5$ M NaI.[72]

The solubility of a protein reflects its conformation which, in turn, is governed, to a large extent, by the structure and physical state of water.[72,73] The ability of a protein to fold and locate most of its apolar residues internally enables it to go into solution. Because hydrophobic forces play a significant rôle in protein solubility, factors which alter water structure, e.g. salts and urea, alter protein conformation and solubility.[37,73−75]

Several authors have systematically determined the effects of various anions and cations on protein solubility (salting-in and salting-out capacity). With chloride anion, the following cations: $Ca^{++} > Mg^{++} > NH_4^+ > Li^+ > Cs^+ > Na^+ > K^+$ tend to enhance solubility (salting-in) of proteins while with Na cation, the following anions: $SCN^- > ClO_4^- > I^- > NO_3^- > Br^- > Cl^- > F^- > HPO^{2-} > SO_4^{2-}$ have a similar effect.[38,74] The effects of these ions can be related to their effects on the structure and physical properties of water.[74] The anions, $SCN^-$, $ClO_4^-$, $I^-$, $Br^-$ and $Cl^-$ are associated with large positive entropy changes, most probably due to their structure-breaking effects on water.[38] Alkali and $F^-$ ions appear to have net structure-breaking effects. These ions increase entropy, i.e. beyond the first saturated region of water molecules they create a region in which the water structure is broken down or depolymerised as compared with ordinary hydrogen-bonded water. Thus, the transfer of an apolar group to solutions of such ions involves a lower negative entropy than transfer to water. Hatefi and Hanstein[38] tabulated the relative effects of various anions (Table 18). The partial molar entropy of the aqueous ion (cal/mole/deg at 25°C) was highest for $SCN^-$ and lowest for $F^-$. The proton chemical shift was greater (less constrained) and the change in surface tension was lower in the order of $SCN^-$, $ClO_4^-$, $I^-$, $Br^-$, $Cl^-$ and $F^-$, indicating greater water structure-breaking and lowering of water polarity.

Anions attract the H atoms of water and the net effect is a weakening of the hydrogen bonds between water molecules. This results in an increase in surface potential and a progressive decrease in surface tension rise ($\Delta y$) with the increase in partial molar entropy effects of these anions which

TABLE 18

EFFECTS OF DIFFERENT ANIONS ON WATER STRUCTURE AND PROPERTIES[38]

| Anion | $S_{io}$ (*m*) | $S_i$ (hyd) | $\Delta\gamma$ (*k*) | $\delta$ (*Na*) |
|---|---|---|---|---|
| $SCN^-$ | 36·0 | −8·1 | 0·42 | — |
| $ClO_4^-$ | 43·2 | −8·3 | 0·64 | 1·32 |
| $I^-$ | 25·3 | −10·1 | 1·30 | 0·92 |
| $Br^-$ | 19·7 | −14·4 | — | 0·78 |
| $Cl^-$ | 13·5 | −18·2 | 1·65 | 0·58 |
| $F^-$ | −2·3 | −31·2 | 1·80 | — |
| $SO_4^{2-}$ | 4·4 | — | — | — |

$S_{io}$ (m) = partial molar entropy of the aqueous ion (cal/mole degree at 25 °C). $S_i$ (hyd) = partial molar entropy of hydration (cal/mole degree at 25 °C). $\Delta\gamma$(k) = surface tension increase of potassium salt (dyne/cm). $\delta$ (Na) = proton chemical shift of Na anion in 1 M solution of electrolytes compared with water as standard (ppm).

make water more lipophilic and weaken the hydrophobic interactions in proteins. The anions are called chaotropic anions because they reduce the free energy required to transfer apolar groups into water[38] and their effects follow the Hofmeister series of lyotropic anions. Thus, they increase the tendency of proteins to unfold and to assume a 'more' random coil structure, and reduce the temperature (and enthalpy) of protein unfolding and denaturation.[74] The effects of these anions on the solubility and association of some proteins are summarised in Table 19. While it is generally held that the lyotropic effect of salts is observed only at relatively high concentrations (> 1 M), recent data show that they can be effective at much lower concentrations.[37]

Dissociating agents such as urea and its isoelectronic analogue, guanidinium chloride, effectively disrupt the H-bonded structure of water and thereby weaken hydrophobic interactions and facilitate protein unfolding.[75]

It should be of interest to study the effects of chaotropic agents on the functional properties of proteins. It has been observed that chaotropic agents, e.g. salts, urea and alcohols, facilitate autoxidation in mitochondrial systems by disrupting membranes, i.e. lipoprotein structures.[38] Similar effects may occur in food systems; for instance, the addition of nitrate to meat may actually loosen membrane structures and facilitate autoxidation. Similarly, in the case of milk, the presence of urea may alter its thermal stability by modifying the polarity of water. Chaotropic salts

## TABLE 19

THE RELATIVE POTENCIES OF AQUEOUS ANIONS ON DENATURATION, DEPOLYMERISATION OR SOLUBILITY OF SOME PROTEIN SYSTEMS[38]

| Protein system | Anion species | | | | | | | | |
|---|---|---|---|---|---|---|---|---|---|
| | $SO_4^{2-}$ | $F^-$ | $AcO^-$ | $Cl^-$ | $Br^-$ | $NO_3^-$ | $ClO_4^-$ | $I^-$ | $SCN^-$ |
| | | | | | *Relative Potency* | | | | |
| *Denaturation* | | | | | | | | | |
| Collagen/Gelatin | 1 | — | 3 | 4 | 5 | 6 | 7 | 8 | 9 |
| Ribonuclease | 1 | — | 3 | 4 | 5 | — | 7 | — | 9 |
| Ovalbumin | — | — | — | 4 | 5 | — | — | 8 | 9 |
| *Depolymerisation* | | | | | | | | | |
| Actomyosin | 1 | — | — | 4 | 6 | 5 | 7 | 8 | 9 |
| F-Actin | 1 | — | — | — | 6 | 5 | 7 | 8 | — |
| Antigen–antibody complex | — | — | — | — | — | — | 8 | 7 | 9 |
| *Solubility* | | | | | | | | | |
| Electron-transport proteins (M) | — | 2 | — | 4 | — | — | 7 | — | 9 |

and urea are effective at reasonably low concentrations and usually have little chemical reactivity. They affect dissociation, depolymerisation of protein aggregates, solubilisation of protein/protein complexes, denaturation of proteins and thermal stability. These compounds, about which much information is available concerning their effects on protein structure, could provide an excellent tool for studying the importance of hydrophobic interactions in protein structure and function.

In the remainder of this paper the functional properties of proteins in two specific types of food, cereals and meats, are examined in detail.

# STRUCTURAL–FUNCTIONAL RELATIONSHIPS IN FOOD PROTEINS

## Cereal Proteins

Bread is one of the few universal food staples. The physical properties of bread reflect the unique properties of the major protein component, gluten. The ability of gluten from wheat to form an extensible, viscoelastic film upon hydration enables it to entrap gas and form a solid foam which becomes stabilised during baking, thereby accounting for its unique functional status.

Addition of water to flour results in hydration of the glutenin and gliadin which, following kneading for 3–5 min, gradually form a continuous film throughout the hydrated dough. As yeast fermentation proceeds the evolved carbon dioxide becomes entrapped in vacuoles surrounded by this impervious gluten film. A good gluten enables dough to accommodate many gas bubbles and to expand under pressure without rupture or coalescence of vacuoles. With the accumulation of more gas during proofing, and the expansion during baking, the dough mass expands as a foam. During baking, proteinaceous films containing starch granules encapsulate the gas bubbles. The hydration and partial gelatinisation of starch granules enable them to become flexible, stretch and distort during the oven baking. During baking the gluten film is denatured but stabilised and, together with the partially gelatinised starch, forms the structural matrix of baked bread[76,77]

Breadmaking potential (i.e. loaf volume) is closely related to the protein content and composition of wheat flour. There is a linear relationship between loaf volume and protein (particularly glutenin) content of flour from 8–18 %, provided adequate oxidant is added.[76] Breadmaking flours are divided into hard or strong and soft or weak types. Hard wheat flours contain a high percentage of protein (glutenin) which forms viscoelastic gluten with good gas-retaining properties capable of being made into a well-risen loaf possessing good crumb grain and texture. These flours require more water, have excellent handling properties and are not critical in their mixing and fermentation requirements so that good breads are obtained over a range of manufacturing conditions. In contrast, soft flours give relatively soft, weak, non-elastic gluten with poor gas-retaining properties; they yield doughs of inferior handling qualities and have critical manufacturing requirements. They require less work (kneading/mixing) than strong flours. In the case of chemically leavened products, however, a low protein flour is preferred. Thus, for the mechanical mixing used in commercial bread preparation, a tough, hard flour is required whereas, for home breadmaking, a much softer flour is desirable.[76]

Proteins are the major functional components of flour and are usually classified according to their solubility characteristics (Table 20). They affect

TABLE 20
OSBORNE PROTEIN COMPONENTS OF WHEAT

| *Extracted by* | *Protein* | *Amount* <br> *( % of total protein)* |
|---|---|---|
| Water | Albumins | 16·0 |
| NaCl (0·5 M) | Globulins | 3·5 |
| Ethanol (70 %) | Gliadin | 33·5 |
| Acetic acid (0·05 N) | Glutenin | 15·0 |
|  | Insoluble glutenins | 32·0 |

water absorption, oxidation requirements, work input, fermentation tolerance and, of course, the quality of the final product. The major functional proteins in flour are gliadin and glutenin which possess a similar amino acid composition (Table 21) with an unusually high content of glutamine and proline. Few of the carboxyl groups of glutamic and aspartic acids are free to ionise and the low content of lysine, histidine and arginine, capable of acquiring positive charges in solution, results in gluten proteins having low ionic character. The relatively large number of non-polar amino acids facilitates hydrophobic interactions, while the unusually high glutamine content facilitates hydrogen bonding; hence gluten proteins have strong aggregation tendencies.[76]

Gliadin is composed of discrete, slightly folded, single-chain polypeptides (average molecular weight: 40 000 daltons) each containing two intramolecular disulphide bonds. Its structure is mostly ( ∼80 %) random coil but is compactly folded and stabilised by the two disulphide crosslinks,

TABLE 21
AMINO ACIDS OF WHEAT PROTEINS (g/16 g N)

|            | *Flour* | *Gliadin* | *Glutenin* |
|------------|---------|-----------|------------|
| Try        | 1·5     | 0·7       | 2·2        |
| Lys        | 1·9     | 0·5       | 1·5        |
| His        | 1·9     | 1·6       | 1·6        |
| Arg        | 3·1     | 1·9       | 3·0        |
| Asp        | 3·7     | 1·9       | 2·7        |
| Thr        | 2·4     | 1·5       | 2·4        |
| Ser        | 4·4     | 3·8       | 4·7        |
| Glu ($NH_2$) | 35·0  | 41·0      | 34·2       |
| Pro        | 11·8    | 14·3      | 10·7       |
| Gly        | 3·4     | 1·5       | 4·2        |
| Ala        | 2·6     | 1·5       | 2·3        |
| Cys 1/2    | 2·8     | 2·7       | 2·2        |
| Val        | 3·4     | 2·7       | 3·2        |
| Met        | 1·3     | 1·0       | 1·3        |
| Ile        | 3·1     | 3·2       | 2·7        |
| Leu        | 6·6     | 6·1       | 6·2        |
| Tyr        | 2·8     | 2·2       | 3·4        |
| Phe        | 4·8     | 6·0       | 4·1        |
| $NH_3$     | 4·0     | 4·7       | 3·8        |

reduction of which greatly increases viscosity. Hydrated gliadin readily forms long fibrils that are very adhesive, cohesive and extensible. In gluten, gliadin modifies the toughness, elasticity and inextensibility of glutenin.[76–79]

Glutenin, which possesses strong elastic and cohesive properties, is a polymeric protein composed of aggregates of heterogeneous polypeptides ranging in molecular weight from $10^5$ to $10^7$ daltons. The subunits are disulphide crosslinked and associate mostly via hydrophobic interactions, with 1-2 disulphide interpeptide crosslinks. Hydrogen bonding is also involved, especially in dry, anhydrous gluten. Most, but not all, of the polypeptide components are disulphide linked in a linear fashion.[52,80,81] Reduction of the disulphide bonds of glutenin results in a substantial decrease in molecular weight and yields a heterogeneous mixture of polypeptides ranging in molecular weight from 40 000 to >100 000 daltons.[79,80] Therefore, glutenin consists of relatively low molecular weight components intermolecularly crosslinked by disulphide bonds. Both intra- and intermolecular disulphide bonds occur in such a way that each glutenin component has an average of only one or two intermolecular disulphide bonds, thus favouring a linear polymeric arrangement.[52] Hydrophobic interactions are also involved in the association of glutenin subunits, as indicated by their dissociation by urea and sodium dodecylsulphate.[82] The large asymmetric glutenin molecules form a tough, rubbery, cohesive mass when hydrated and are of fundamental importance in cohesiveness and elasticity in dough. Insoluble glutenin has several intermolecular disulphide bonds, making it into a large aggregate that contributes substantially to the viscoelasticity and strength of doughs. This insoluble component is composed of subunits similar to those of glutenin but there is some branching of polymers.[78,80,82]

The proteins of wheat have different cohesive strengths, as reflected in the respective tensile strengths of 3·38, 1·42 and 1·75 psi for glutenin, gliadin and gluten.[83] Glutenin forms films with greater tensile strength than gliadin films but the latter stretch further before breaking because of their weaker intermolecular associations. Gluten, a mixture of gliadin and glutenin, has intermediate film properties which are flexible but cohesive enough to retain gas during breadmaking.

**Secondary Interactions**
In addition to disulphide bonds, molecular interactions via secondary forces are extremely important in the functionality of gluten. Gliadin and glutenin exist in an anhydrous state in flour. Based on the high amounts of glutamine and asparagine, it is probable that the component polypeptides associate via hydrogen bonds, and perhaps some ionic interactions, in addition to the intermolecular disulphide bonds. In the absence of water, hydrophobic interactions are not involved. The interactions between polypeptides, following hydration and some thiol–disulphide interchange, are critical for the development of the viscoelastic properties (cohesiveness, plasticity, tensile strength and elasticity) of gluten in dough. These interactions may involve hydrogen bonds, electrostatic interactions, van der Waals forces and hydrophobic interactions. The strength of interpeptide hydrogen bonds in aqueous solution is extremely small and consequently there has been a tendency to minimise their importance in stabilising proteins in solution. However, water may be excluded from certain segments or domains of gluten and, therefore, hydrogen bonds may be involved in its functional properties. Hydrogen bonding between neighbouring polypeptides may contribute to the viscous cohesive properties of gluten and impart plastic behaviour. Furthermore, in gluten, the absence of significant amounts of charged residues permits maximum hydrogen bonding between adjacent residues in the anhydrous, hydrophobic interior of the gluten mass.

Modification of the H-bonding groups alters the rheological properties of gluten. Acetylation of glutamine reduces the ability of the gluten proteins to enter into hydrogen bonding and weakens their cohesiveness. When $D_2O$ is used instead of water, extensibility is decreased, elasticity increased and the gluten is strengthened because deuterium bonds have somewhat higher bond energies than hydrogen bonds. It has been stated that extensive H-bonding may account for the aqueous insolubility of glutenin but this is unlikely because water effectively competes for protein hydrogen bonds and facilitates gluten development in the absence of kneading, indicating that breaking of H-bonds in the anhydrous gluten permits rearrangement.[84]

Because gluten proteins have very few charged groups they are poorly soluble in neutral solvents but, at extremes of pH, unfolding occurs because of the excess of similarly charged residues. The low ionic character of gluten proteins facilitates the formation of cohesive films at normal pH values because the lack of strong electrostatic repulsions permits extensive H-bonding and hydrophobic interactions. However, some ionic interactions occur since the addition of NaCl (0·1 M) increases rigidity and reduces extensibility of dough.[76,85]

Although van der Waals and dipole-induced interactions are weak (0·5 kcal/mole), collectively they are effective when molecules come closer than 5 Å and are involved in stabilising tightly folded groups of molecules and may account for part of the work required for gluten development. The secondary forces engaged in stabilising interactions are dependent on the propinquity of adjacent molecules ($<4$ Å) and may be visualised as static forces, whereas the forces for the dynamic properties of these molecules are hydrophobic in origin. The net hydrophobicity of gluten is $\sim 1100$ cal/ mole. In flour the hydrophobic segments of the polypeptides associate by hydrogen bonds and van der Waals forces but when water is added to gluten the apolar polypeptides are forced to fold in such a way as to minimise their interactions with water; the molecules thus folded may then be stabilised by other secondary interactions. In dough, much of the water ($\sim 50\%$) is free and is a very significant factor in dough development and gluten rheology. The fact that water is essential for gluten development strongly indicates the importance of hydrophobic interactions for its cohesiveness and elastic properties and this has been further demonstrated by the effects of water destructuring agents (urea, dimethylformamide, acetamide) in dissociating gluten components and weakening viscoelastic properties.

Rheologically, gluten displays plastic and elastic characteristics. Plastic

                           *J. E. Kinsella*

deformation in response to shear stress may reflect limited rupture of hydrogen and stressed disulphide bonds or slippage along lipid films. Elasticity is a dynamic phenomenon and conceivably involves hydrophobic effects. Thus, during active extension, hydrophobically associated polypeptide folds of gluten are pulled apart, free water flows into the vacated areas and large segments of apolar polypeptides become exposed to the water. Water molecules adjacent to the polypeptides are forced into a more structured state (lowered entropy). However, upon release of the stress force, the extended polypeptides tend to refold spontaneously with a concomitant increase in entropy.

### Disulphides and Thiol–Disulphide Interchange

Approximately 1·5 % of the amino acids in gluten are cystine and cysteine in an approximate ratio of 15:1 but this ratio varies immensely among wheat flours.[76,81] In wheat, the disulphide bonds of gliadin are intramolecular whereas glutenin components have, on average, two intramolecular disulphide bonds and at least one intermolecular disulphide link which permits the formation of a predominantly linear polymer. In contrast, the glutens of other cereals have more numerous intermolecular disulphide bonds which cause extensive crosslinking and on hydration the resulting doughs are very tough, with limited elasticity.[76,85,86]

Wheat flour contains from 4–19 and 83–130 microequivalents of thiol and SS groups per gram of protein, respectively;[76,87–89] 10 to 30 % of the thiol groups occur in non-protein fractions (5 % in gluthathione, 10 % in highly reactive small peptides) and differences in the rheological properties of flours may be related to the distribution of thiol groups between the protein and non-protein compounds.[85] Thiol groups play an important rôle in dough development by breaking SS bonds and engaging in thiol–disulphide interchange reactions.[52,81,84] The stiffness of a dough increases and elasticity decreases with increasing disulphide and decreasing thiol content; hence the ratio of disulphide to thiol is a parameter related to dough strength.[76,78,80,87] The quantity of reactive thiol groups in a flour markedly affect the rheological and mechanical properties of gluten by catalysing the rupture of stressed disulphide bonds.[52] If excessive reactive thiol groups are present, extensive rupturing of disulphide bonds ensues, resulting in a weak, extensible gluten with limited resistance or elasticity. It is generally felt that the thiol groups of protein have limited reactivity whereas the low molecular weight thiol compounds (e.g. gluthathione) are quite active. Some interchange can occur between gliadin and glutenin polypeptides during mixing to form disulphide-linked gliadin–glutenin polymers.

Low molecular weight thiol groups can catalyse the interchange reaction and subsequently be regenerated. To prevent excessive thiol–disulphide interchange, oxidants are added to flour to reduce the amount of reactive thiols. Oxidants may selectively oxidise low molecular weight thiols because both have high mobility in flour whereas protein thiol groups may be sterically hindered. Oxidising agents, by promoting the formation of disulphide crosslinks, increase the toughness and decrease the extensibility of dough. The presence of thiol groups allows for a viscous (permanent) deformation of dough. Thus, the modulus of elasticity and viscosity increase with increasing disulphide content and decrease with increasing thiol content.[52,85] Hence, oxidation of the thiol groups increases both viscosity and the modulus of elasticity of gluten. Experimental results show that the amount of oxidation required is very small, e.g. oxidation of $\sim 4\%$ of the thiol groups can increase the modulus of elasticity by $>60\%$, indicating the sensitivity of this physical property to the number of disulphide bonds.

During kneading, stress points in the gluten may be relieved by disulphide reduction or interchange, thereby facilitating development. Thiol–disulphide interchange reactions facilitate network deformation by rupture of disulphide bonds in regions of high stress and reformation of the disulphide bonds in regions of lower stress. These reactions, in which thiol reagents act as catalysts, have been demonstrated in flour and it is assumed that the formation of the protein network in dough during mixing is very much dependent on them, although only a limited number of disulphide bonds may be involved.[52,76,80,85] The interchange reaction is most rapid in wheat flours and much slower in rye dough.[52]

Thus, thiol–disulphide interchange is important for facilitating the unravelling of glutenin, for the formation of a limited number of heteropolymers (gliadin–glutenin), for permitting viscous flow, plastic deformation and relief of stress, thereby avoiding breakage and shortening of the gluten polypeptides.[52,76]

*Hydration and Kneading*

Addition of water (65 g water/100 g flour) is essential for optimum gluten development, i.e. for hydration, for unravelling glutenin and gliadin, for development of the physical properties of the gluten, for solubilisation and mobilisation of thiols, oxidants, and enzymes and for starch gelatinisation.

About 33 g water/100 g flour is unfreezable, i.e. bound, most of which is associated with the protein.[90,91] Free water is apparently essential for lubrication and is also necessary for the gradual development of the viscoelastic matrix.

Working and kneading are necessary for developing the dough matrix by facilitating the chemical reactions, dispersing the protein and unravelling the coiled polypeptides. In addition, air is incorporated, the oxygen of which is effective in oxidising lipids and excess thiol groups.[76,84,87] This conceivably strengthens the matrix via the formation of disulphide bonds and facilitates hydrophobic interactions between the unravelled polypeptides. Since proteins amount to only 10–12 % of flour, kneading is necessary to develop the viscoelastic matrix uniformly throughout the dough. Generally, a minimum of protein, i.e. $5 \times 10^{-5}$ gram of protein per square centimetre of starch surface ($\sim 7.5 \%$ protein), is necessary for dough formation.[85] When glutenin alone is mixed with starch it exhibits a very long dough development time which is reduced by the addition of gliadin, indicating that gliadin is important in the formation of dough. Proteinases are sometimes added to flour to decrease the toughness of high-glutenin flours.[76,80,85]

Upon mixing with water and kneading, a smooth homogeneous dough is produced which stiffens as mixing proceeds until a point of maximum consistency is reached, usually in 3–5 min, but this depends upon the strength of the flour. Development time reflects the gradual unravelling of the gluten proteins, their rearrangement and reorganisation into parallel fibres that facilitate cohesiveness and extensibility. Mixing unravels the polypeptides, i.e. this exposes functional groups and permits strong association by non-covalent forces. Association is greatly facilitated by thiol–disulphide interchange and the replacement of protein–protein hydrogen bonds by protein–water H-bonds. The coiled polypeptides allow extensibility along the long axis but permit recoil upon removal of the stress, i.e. elasticity. Interaction between adjacent polypeptides contributes lateral cohesion and resistance to flow. The smaller gliadin molecules, being less tightly bound up in the matrix, facilitate hydration, fluidity and expansion of the dough. If dough is too tough or too cohesive, it will not rise upon the evolution of gas.

Flours that take excessively long to develop are classified as too strong. There is a high correlation between mixing time requirement (and tolerance to mixing) and the content of residual protein, i.e. protein insoluble in dilute acetic acid. Generally, stronger flours contain high levels of glutenin and residual protein. During mixing, these insoluble proteins initially are part of a gelatinous mass which decreases with mixing time as solubilisation of the glutenins occurs on unravelling and disaggregation. Glutenins from other cereals do not show this increased solubility with mixing. Thiol reagents, e.g. cysteine or gluthathione, accelerate the rate of solubilisation of the highly polymerised glutenin.

### 'Isolated' Gluten

Gluten is also an important protein in many fabricated foods because of its functional properties.[76,77,92] Of the ~20 million kilograms produced in the US annually, 70 % is used in bakery products, 12 % in breakfast foods, 9 % in pet foods and 4 % in meat analogues. In baked goods, gluten is used to supplement wheat flours, to provide additional mixing strength and enhance mixing tolerance. It is used in high fibre breads to permit expansion of loaf volume and in hamburger buns to impart structural strength to the hinge.

In the development of fabricated foods and in the manufacture of many textured foods containing different components, binding (i.e. adhesiveness and cohesiveness) of the components is very important. Binding agents are important in several foods, e.g. processed meats, baked goods, cereals, snacks, to impart the final characteristic texture. The adhesive power of a protein connotes its ability to glue other components together and cohesiveness relates to the ability of that material to interact with similar molecules to form a continuous matrix and impart texture.[77] Thus, in the fibre spinning process, egg white is used as an effective adhesive to glue the fibres together into bundles. Wheat gluten is also widely used as an adhesive agent in food systems. In breakfast cereals, it is used to bond the ingredients prior to cooking. Its viscoelasticity, cohesive, gelling, binding and fat emulsification properties are extremely important in simulated and processed meat products.[93] Gluten is a particularly useful functional ingredient for binding meat pieces together,[93] as in the manufacture of simulated sausage, bologna and formed meat products. Gluten may also be used in whipped and aerated foods where its foaming properties are exploited. Finally, because of the range of its physical properties, gluten is an excellent functional protein for blending with other ingredients for specific applications, e.g. gluten–soy or gluten–casein blends.

### Muscle Proteins and Meat Quality

The composition of muscle is 55–78 % water, 15–22 % protein, 1–15 % lipid and 4 % carbohydrate, minerals and small organic compounds. Proteins, the most significant component with respect to the properties and quality of meat, are classified as sarcoplasmic, stromal and myofibrillar proteins. Sarcoplasmic proteins amount to $\sim 30\%$ of muscle protein and are composed of enzymes, myoglobin and cytoplasmic proteins. The stromal proteins, which constitute 10–15% of the muscle proteins, are composed mostly of collagen (40–60%) and elastin (10–20%). These, especially collagen, affect the toughness of meat (particularly from old animals) and, where levels are high, they reduce water-holding capacity and impair emulsifying properties of comminuted meats.[94] However, in sausage-type meats, they impart structure upon gelation following thermal processing. The myofibrillar proteins, which amount to $\sim 55\%$ of muscle protein, constitute the myofilaments which are the assembled myofibrillar proteins and exist as such at the ionic concentration (0·17 M) in the muscle cell but are disaggregated and solubilised at higher ionic strength (0·5 M). The myofibrils, which account for $\sim 85\%$ of muscle volume, are composed of myosin, actin, tropomyosin, troponins and actinin (see Bailey, Ch. 13) and are the most important functional components in meat.[94–97]

Myofibrillar proteins, their structure, conformation and interactions, affect the eating quality of meat, i.e. juiciness and tenderness.[96,98] Tenderness reflects the ease with which meat can be masticated and juiciness depends upon the amount of liquid released from the meat during biting and mastication. Juiciness is a unique and most desirable feature of good quality meat and reflects the water-holding capacity of meat (WHC). According to Goll *et al.*,[94] 90% of the WHC of meat is ascribed to the fluid that is trapped in the lattice spacings between the protein filaments of the myofibrils, the remainder being bound water. About 10% of the latter is tightly bound as a monolayer around the filaments and the remainder is loosely bound layered water. Therefore, most of the water in muscle is trapped in the lattice spacings and hence the WHC of meat is high if these lattice spacings are large, as in relaxed muscle.[96]

The WHC is influenced greatly by processing, handling conditions, rigor mortis, storage, heating, drying, freezing and by the immediate environmental conditions.[70] Swelling of the filaments of meat enhances WHC whereas shrinkage decreases it. Minimum WHC occurs at pH 5, the isoelectric point of actomyosin. As pH is increased, WHC increases. At the isoelectric pH, the proteins are closely attracted and squeeze out the water whereas, above this pH, there is repulsion which increases spacing and allows entrapment of water with swelling.[96] With rigor mortis, there is a decrease in pH to about 5·5 and a concurrent decrease in WHC. The decrease in WHC is attributed to a decrease in pH towards the isoelectric range but 60% of the decrease is due to degradation of ATP because polyphosphates have a large water-holding capacity.[65,70] Subsequent ageing of the meat post rigor results in a moderate increase in WHC. This is attributed to an increase in pH, a loosening of the protein matrix during ageing, changes in ion–protein relationships (i.e. adsorption of potassium, release of calcium) and the action of proteolytic enzymes at the Z line which may loosen the matrix[96,99,100] (see also Dutson, Ch. 17).

At pH values above the isoelectric point, NaCl increases the WHC of protein whereas, below it, the opposite effect occurs. This is explained by the stronger interaction of the chloride anion with the muscle proteins where net negativity is actually increased[70,96] causing the matrix to swell and increase the WHC. Ionic strength influences WHC; maximum WHC occurs at an ionic strength of 0·8 M NaCl. At ionic strengths above this, WHC decreases apparently by a salting-out effect due to water binding by the salt and concurrent dehydration of the protein. Anions and cations affect water bound to protein in accordance with the Hofmeister and lyotropic series; however, their effects on bound water are often opposite to their effects on the WHC.[65] Several salts are routinely used in processed meats to enhance WHC, e.g. polyphosphates, which have high water-binding capacity.[96]

Myofibrillar proteins contribute greatly to the tenderness and juiciness of meat, particularly in meats in which connective tissue is low. In such tissues, the phenomenon of rigor, cold shortening and resultant toughness is due to the contraction state of the myofibrillar protein.[101] (See Tarrant, Ch. 14, and Dutson, Ch. 17.)

**Functional Properties in Meat Products**
With the increased interest in processed and formed meat products, greater reliance is being placed on the functional properties of the protein components. Processed meat products such as emulsions (hot dogs, sausages, etc.) and sectioned and formed meat products, e.g. turkey hams, depend very much on the formation of a continuous matrix throughout the product. This property is contributed mostly by the protein components. The formation of a matrix determines the texture, juiciness and general mouthfeel of these products. In different products, they reflect the emulsifying properties and binding and gelling properties of the muscle proteins following processing and cooking.[102]

*Fat Emulsification*
The nature and properties of meat emulsions were reviewed by Saffle.[103] During manufacture of sausages and bologna, fat should be uniformly

dispersed throughout the product in emulsified droplets and the emulsion should be stable to heat and set upon cooking. During mixing, the water and salt-soluble proteins in the meat coat the fat droplets to form a membrane matrix surrounding the fat droplet. Myosin is the most important functional protein in the formation of this emulsion membrane; actin and actomyosin are apparently of lesser importance.[104] However, in meat emulsions, all the proteins may be important in enhancing the viscosity of the continuous phase, thereby minimising coalescence of the emulsified fat droplets, particularly during the early stages of heating. Myosin is the most surface active protein of meat and apparently has the ability to migrate to, and locate at, the interface and unfold to form a cohesive matrix around the fat droplet. This is accompanied by some reorientation, unfolding and possibly some denaturation of the myosin. During subsequent heating, the membrane material thermally sets and the accompanying gelation of the matrix protein serves to stabilise the emulsion.

### Adhesion

Formed meats, i.e. products obtained by sectioning meat into smaller pieces, followed by mechanical mixing, i.e. tumbling, in the presence of salt and phosphates, are becoming popular because they can be used to simulate boneless, high quality cuts. They can be prepared uniformly, are amenable to portion control and they eliminate waste. During their preparation the mechanical working frays the muscle surfaces so that they become soft, pliable and develop a creamy, tacky exudate. This exudate, composed of soluble muscle proteins, provides adhesion and structural integrity following heating. Numerous workers have shown that the binding of these pieces depends upon the salt extraction of myofibrillar proteins.[105-108] Myosin is an essential binding and gelling component and actomyosin is very important for the viscoelastic properties of the gel system. Myosin confers superior adhesive properties compared with actomyosin at salt concentrations up to 1 M.[105-108]

During manufacture of formed meats, the extraction of myosin as an effective binder requires the addition of at least 2% salt and 0·3% phosphate. However, when crude myosin or a mixture of crude myosin and sarcoplasmic proteins are added, little or no added salt is required to attain optimum binding for the restructured meat products.[105] Myosin levels from 0 to 8% progressively increase the binding ability. The heavy chains of myosin are apparently involved in a three-dimensional network which is necessary to bind the pieces of meat.[106,107]

Because salt plays a major rôle in this process, some workers concluded that ionic interactions are very important in the binding phenomenon.[107,108] However, the principal rôle of salt is to disaggregate, solubilise and extract the myofibrillar proteins. It is conceivable that hydrophobic interactions and disulphide crosslinking are important in the adhesion of sectioned meats, particularly since the heavy meromyosin is rich in hydrophobic peptides and myosin contains 42 sulphydryl groups.[109]

Because of the importance of binding chunks of meat in comminuted formed meat and emulsified sausage products, and because other ingredient functional proteins are being used increasingly to extend meat proteins, the binding properties (i.e. relative capacity to bind meat pieces upon heating) of various proteins were compared by Siegel and Schmidt.[107] In the presence of 8 % salt and 2 % phosphate, the order of binding of meat pieces was: wheat gluten, 175; egg white, 120; control (i.e. myosin), 107; skim milk, 74; bovine blood plasma, 71; isolated soy protein, 66; sodium caseinate, 0. Salt was very important in developing the binding properties of all these proteins.


*Gelatin*

Thermal gelation is a critical function of proteins in formed meats. The gel imparts the appropriate viscoelastic properties, juiciness and texture and stabilises the emulsion in these particular meat preparations. Myosin appears to be the most important protein for gelation. The viscosity of myosin solutions (0·5–1·0 %) progressively increases upon heating from 30 to 60 °C and gelation occurs above 60 °C.[110] Gelation of myosin, which is irreversible, occurs following unfolding of the rigid helical coil rod structures to randomly coiled polypeptides which interact to form a gel matrix. Scanning electron microscopy confirms the existence of this network gel structure in myosin gels.[111] Myosin gels exhibit a high water-binding capacity, strong elastic properties and little syneresis. Gelation of myosin occurs optimally at temperatures between 60 and 70 °C, pH 6·0, at about 0·5 molar (2–3 %) KCl. Apparently, the heavy chains of myosin are important in thermal coagulation, even though other studies indicate that the intact myosin molecules form the best gels upon heating.

Information concerning the mechanism of gelation of myosin and the actual conformational changes and the binding forces involved have not been reported.[102] Heat denaturation involves dissociation of the light and heavy chains and conformational changes of the individual molecules. The effects of ionic strength on gelation indicate the participation of electrostatic interactions although, conceivably, hydrophobic interactions are the most important forces leading to the formation of a gel following thermal unfolding of the polypeptides. Furthermore, formation of disulphide bonds between myosin molecules may occur via oxidation.

The extent of involvement of other myofibrillar proteins in the formation of gels in heated meats has not been determined. Schmidt *et al.*[102] reported that, in the presence of pyrophosphate, actin and tropomyosin contribute to the increased viscosity of heated meat proteins.

Thus, the biochemical and conformational properties of the proteins in meat, which are essential for physiological functions of muscle, are also important in determining the quality attributes of meat and meat products. More information is needed detailing the molecular events occurring in post-mortem meat because these are obviously extremely important in governing the quality of the edible product. Thus, factors affecting the association of the actomyosin complex have implications in tenderness, water-holding capacity and solubility of muscle components, all of which, in turn, affect other desirable functional attributes, particularly in cooked and processed meats.


## RESEARCH NEEDED

Although the unique functional properties of several food proteins have been exploited and enjoyed for years, only relatively recently has research begun to focus on the relationship between the physical properties of these proteins and their functional behaviour in food systems. Ironically, much of the impetus for this originated with attempts to simulate traditional products (meat, cheese) from less expensive oilseed proteins or to use these proteins as functional ingredients to replace established sources, i.e. casein, egg white, gluten.

Food technologists in the industrial sector and basic scientists are realising the critical nature of, and need for, comprehensive and organised information on functional properties, i.e. the physico-chemical characteristics that affect the applications of all proteins (both conventional and novel) in food systems. Thus, much long-term and systematic research is needed to elucidate the fundamental relationship between the structure and physical properties of proteins (Table 22). Such research requires the development and adoption of standardised methods for measuring specific functional

TABLE 22
RESEARCH NEEDS

1. Definition of functional properties
2. Development of standardised methods based on physical properties
3. Relating functional properties to structural features and secondary interactions
4. Manipulation of protein structure by chemical or physical alterations
5. Assessment in food systems

properties. However, development of the best methodology depends upon an understanding of the physico-chemical basis of functionality.

Research on functional properties should initially focus on the basic structure of the major food proteins, their conformation and the forces contributing to their structure. Their physical properties (solubility, thermal stability, viscosity, ligand binding, film forming ability and surfactant properties), particularly those known to be related to functional properties, should then be studied. The effects of pH, salts and other common food components on these properties should be systematically determined.

The physico-chemical basis underlying the functional properties of conventional proteins should be examined, i.e. why do the proteins of egg white form excellent thermostable foams; why does gelatin form gels; what is the mechanism by which myosin acts as a binder and thermosetting agent in meat systems; why does the coagulum formed by caseins develop into the unique texture of cheese? There is copious information in the literature which partly explains the relationships between the physical properties and the functional properties of these proteins; however, only in a few instances (perhaps gelatin) is the molecular basis understood. In many cases, this is because the function in question represents the combined properties of a heterogeneous group of proteins, each performing a specific action which, in the aggregate, represents the overall functional attribute. This must be recognised when a traditional ingredient protein is being replaced by new functional proteins.

When a clear understanding of the relationship between structure and function is obtained for a particular system (egg white foam, muscle, cheese), it should be possible to simulate that system more successfully by judiciously replacing some or all of the necessary protein components. Knowledge of the relationship between structure and function should facilitate the development of acceptable novel foods by using combinations of different proteins, including modified proteins and, of course, other functional ingredients. Furthermore, the fabrication of new foods that meet consumer acceptance should be more feasible and successful. Ultimately, it should be possible to catalogue the physico-chemical properties of all food proteins and to categorise proteins according to their capacity to perform specific functions. This would minimise the inefficient trial and error system currently used in research and development efforts involving functional proteins.

## REFERENCES


1. SOLMS, J. and HALL, R., *Criteria of Food Acceptance*, 1981, Forster-Verlag, AG, Zurich, Switzerland.
2. KINSELLA, J. E., *Crit. Rev. Food Sci. and Nutr.*, 1976, **7**, 219.
3. KINSELLA, J. E., *J. Am. Oil Chem. Soc.*, 1979, **56**, 242.
4. FEENEY, R. and WHITAKER, J., *Food Proteins*, Advances in Chemistry, Series No. 160, Am. Chem. Soc., Washington, DC, 1977.
5. KINSELLA, J. E. and DAMODARAN, S., in *Analysis and Control of Less Desirable Flavours in Foods and Beverages*, G. Charalambous (ed.), Academic Press, NY, 1980.
6. GERMAN, B., DAMODARAN, S. and KINSELLA, J. E., *J. Ag. Food Chem.*, 1982, **29** (in press).
7. POUR-EL, A., in *Protein Functionality in Foods*, J. Cherry, ed., ACS Symposium Series No. 147, Am. Chem. Soc., Washington, DC, 1981.
8. SHEN, J. L., in *Protein Functionality in Foods*, J. Cherry (ed.), ACS Symposium Series No. 147, Am. Chem. Soc., Washington, DC, 1981.
9. KINSELLA, J. E. in *Protein Structure and Deterioration*, J. Cherry (ed.), (in press), ACS Symposium Series, Am. Chem. Soc., Washington, DC, 1982.
10. BULL, H. B. and BREESE, K., *Arch. Biochem. Biophys.*, 1968, **128**, 488.
11. POILLON, W. and BERTLES, J., *J. Biol. Chem.*, 1979, **254**, 3462.
12. THOMPSON, M. P., in *Milk Proteins: Chemistry and Molecular Biology*, Vol. II, H. A. McKenzie (ed.), Academic Press, NY, 1971.
13. FRANKS, F., in *Water—A Comprehensive Treatise*, Vol. 4, F. Franks (ed.), Plenum Press, NY, 1975.
14. KAUZMAN, W., *Adv. Protein Chem.*, 1959, **14**, 1.
15. BRANDTS, J. F., in *Structure and Stability of Macromolecules*, S. Timasheff and G. D. Fasman (eds.), Marcel Dekker, NY, 1969.
16. ANFINSEN, C. and SCHERAGA, H., *Adv. Protein Chem.*, 1975, **29**, 205.
17. VAN HOLD, N., in *Food Proteins*, J. Whitaker and S. Tannenbaum (eds.), Avi Publishing Company, Westport, CT, 1977.
18. LEVENTHAL, C., *J. Chem. Phys.*, 1968, **65**, 44.
19. TANFORD, C., *The Hydrophobic Effect*, John Wiley and Sons, NY, 1973.
20. SCHACHMAN, H. K., *Cold Spring Harbor Symposium Quant. Biol.*, 1963, **28**, 409.
21. FERSHT, A. and REGUENA, Y., *J. Mol. Biol.*, 1971, **60**, 279.
22. PERUTZ, M., *Science*, 1976, **201**, 1187.

23. FLAGEL, H., ALBERT, A. and BILTONEN, R., *Biochemistry*, 1975, **14**, 2616.
24. WILLIAMS, R. J., *Biol. Rev.*, 1979, **54**, 389.
25. KINSELLA, J. E., *Food Chemistry*, 1981, **7**, 273.
26. WANISKA, R. and KINSELLA, J. E., *J. Ag. Fd Chem.*, 1981, **29**, 826.
27. GRAHAM, D. E. and PHILLIPS, M., in *Theory and Practice of Emulsion Technology*, A. L. Smith (ed.), Academic Press, NY, 1976.
28. BINGHAM, E., FARRELL, H. and CARROLL, R., *Biochemistry*, 1972, **11**, 2450.
29. YOSHIKAWA, T., *Agr. Biol. Chem.*, 1974, **38**, 2051.
30. GREEN, M. L. and MARSHALL, R. J., *J. Dairy Res.*, 1979, **46**, 365.
31. KINSELLA, J. E. and SHETTY, J. in *Functionality and Protein Structure.* A. Pour-El (ed.), ACS Symposium Series No. 92, Am. Chem. Soc., Washington, DC, 1979.
32. FRANZEN, K. and KINSELLA, J. E., *J. Ag. and Fd Chem.*, 1976, **24**, 788.
33. PEARCE, K. P. and KINSELLA, J. E., *J. Ag. and Fd Chem.*, 1976, **26**, 716.
34. SHETTY, J. and KINSELLA, J. E., *Biochem. J.*, 1980, **191**, 269.
35. TANFORD, C., *J. Am. Chem. Soc.*, 1962, **84**, 4240.
36. DAMODARAN, S. and KINSELLA, J. E., *J. Biol. Chem.*, 1981, **256**, 3394.
37. DAMORDAN, S. and KINSELLA, J. E., in *Protein Structure Deterioration.* J. Cherry (ed.), ACS Publications, Am. Chem. Soc., Washington, DC, 1982.
38. HATEFI, Y. and HANSTEIN, W. G., *Proc. Natl. Acad. Sci.*, 1969, **62**, 1129.
39. BIGELOW, C., *J. Theor. Biol.*, 1967, **16**, 187.
40. KATO, A. and NAKAI, S., *Can. Inst. Food Sci. Technol. J.*, 1981 (in press).
41. KUNTZ, I. D. and KAUFMAN, W., *Adv. Protein Chem.*, 1974, **28**, 239.
42. BLOOMFIELD, V. and MEAD, R. J., *J. Dairy Sci.*, 1975, **58**, 592.
43. PAYENS, T., *Biophys. Chem.*, 1977, **6**, 263.
44. SCHMIDT, D. G., *Neth. Milk and Dairy J.*, 1980, **34**, 42.
45. WHITNEY, R. M. in *Food Colloids*, H. Graham (ed.), Avi Publishing Company, Westport, CT, 1977.
46. GREEN, M. L. and MARSHALL, R., *J. Dairy Res.*, 1977, **44**, 521.
47. NEY, K. H. *Lebensmitt, Unters-Forsch.*, 1971, **147**, 64.
48. GUIGOZ, Y. and SOLMS, J., *Chemical Senses and Flavour*, 1976, **2**, 71.
49. DAMODARAN, S. and KINSELLA, J. E., *J. Ag. and Food Chem.*, 1981, **29**, 1253.
50. ROTHENBUHLER, E. and KINSELLA, J. E. *J. Ag. and Food Chem.*, 1982, **30** (in press).
51. LILLFORD, P., in *Plant Proteins*, G. Norton (eds.), Butterworths, London, 1978.
52. EWART, J., *J. Sci. Food and Ag.*, 1972, **23**, 687.
53. DeWIT, J. N., *Neth. Milk and Dairy J.*, 1981, **35**, 47.
54. ROSE, D., *J. Dairy Sci.*, 1962, **45**, 1305.
55. MORR, C. V., *J. Dairy Sci.*, 1975, **58**, 977.
56. KINSELLA, J. E., *Adv. Food Res.*, 1971, **19**, 148.
57. WHITAKER, J. R., in *Food Proteins*, J. R. Whitaker and S. Tannenbaum (eds.), Avi Publishing Company, Westport, CT, 1977.
58. CHOU, D. H. and MORR, C. V., *J. Am. Oil Chem. Soc.*, 1979, **56**, 53A.
59. HERMANSSON, A., *J. Text. Studies*, 1975, **5**, 425.
60. EDSALL, J. T., in *Protein, Amino Acids and Peptides*, E. J. Cohn and J. T. Edsall (eds.), Hafner Publishing Company, NY, 1965.
61. DE LA TORRE, J. G. and BLOOMFIELD, V. A., *Biopolymers*, 1978, **17**, 1605.

62. McCammon, W. J. and Deutch, G., *Biopolymers*, 1976, **15**, 1397.
63. Nakajima, H. and Wada, Y., *Biopolymers*, 1977, **16**, 875.
64. Pradipasena, P. and Rha, C., *J. Texture Studies*, 1977, **8**, 311.
65. Fennema, O., in *Food Proteins*, J. R. Whitaker and S. Tannenbaum (eds.), Avi Publishing Company, Westport, CT, 1977.
66. Kuntz, I., *J. Am. Chem. Soc.*, 1971, **93**, 514.
67. Bull, H. and Breese, K., *Biopolymers*, 1976, **15**, 1573.
68. Leader, J. and Watt, L., *J. Coll. Interfac. Sci.*, 1974, **48**, 339.
69. Eagland, D. in *Water Relations of Foods*, R. B. Duckworth (ed.), Academic Press, NY, 1975.
70. Hamm, R., in *Meat*, D. Cole and R. Lawrie (eds.), Butterworths, London, 1975.
71. Mihalyi, E., *Application of Proteolytic Enzymes to Protein Structure Studies*, Vol. 1, CRC Press, FL, 1978.
72. Nagy, B. and Jencks, W. P., *J. Am. Chem. Soc.*, 1965, **87**, 2480.
73. Lewin, S., *Displacement of Water and Its Control of Chemical Reactions*, Academic Press, NY, 1974.
74. von Hippel, P. and Wong, K. Y., *Science*, 1964, **145**, 577.
75. Valasubraman, V. and Ramachandran, C., *Proc. Indian Acad. Sci.*, 1978, **87B**, 53.
76. Pomeranz, Y., in *Wheat Chemistry and Technology*, Am. Assoc. Cer. Chem., St. Paul, MN, 1971.
77. Wall, J. S. and Huebner, F. R., in *Protein Functionality in Foods*, J. Cherry (ed.), ACS Symposium Series No. 147, Am. Chem. Soc., Washington, DC, 1981.
78. Kasarda, D. D., Bernardin, J. E. and Nimmo, C., *Adv. Cer. Sci. and Technol.*, 1976, **1**, 158.
79. Bietz, J. A., *Cer. Foods World*, 1979, **24**, 199.
80. Wall, J. S., in *Recent Advances in Biochemistry of Cereals*, D. Laidman and R. Wynn-Jones (eds.), Academic Press, NY, 1979.
81. Ewart, J., *J. Sci. Food and Ag.*, 1968, **19**, 618.
82. Huebner, C. and Wall, J. S., *J. Ag. and Food Chem.*, 1980, **28**, 433.
83. Wall, J. S. and Beckwith, A. C., *Cer. Sci. Today*, 1969, **14**, 16.
84. Daniels, N. W. R. and Frazier, P. J., in *Plant Proteins*, G. Norton (ed.), Butterworth and Co. London, 1978.
85. Bloksma, A. H., in *Wheat Chemistry and Technology* (2nd edn), Y. Pomeranz (ed.), Am. Assoc. Cer. Chemists, St. Paul, MN, 1971.
86. Hoseney, R. C., *J. Am. Oil Chem. Soc.*, 1979, **56**, 78A.
87. Mecham, D., in *Wheat Chemistry and Technology* (2nd edn). Y. Pomeranz (ed.), Am. Assoc. Cer. Chem., St. Paul, MN, 1971.
88. Mecham, D. K., *Brot. Gebaeck*, 1967, **21**, 145.
89. Pomeranz, Y., *Adv. Food Res.*, 1968, **16**, 353.
90. Bushuk, W., *Cer. Chem.*, 1977, **54**, 320.
91. Leung, H. K., Magnuson, J. A. and Bruinsma, B., *J. Food Sci.*, 1979, **44**, 1608.
92. Kalin, F., *J. Am. Oil. Chem. Soc.*, 1979, **56**, 477.
93. Siegel, D. G., Church, K. E., and Schmidt, G. R., *J. Food Sci.*, 1979, **44**, 1276.

94. GOLL, D., ROBSON, R. M. and STOMER, M. H., in *Food Proteins*, J. R. Whitaker and S. R. Tannenbaum (eds.), Avi Publishing Co., Westport, CT, 1977.
95. OCKERMANN, M. W., in *Meat Proteins in Food Colloids*, H. Graham (ed.), Avi Publishing Company, Westport, CT, 1977.
96. ASHGHAR, A. and PEARSON, A. M., *Adv. Food Res.*, 1980, **26**, 24.
97. RICHARDSON, T., in *Food Proteins Improvement Through Chemical and Enzymatic Modification*, R. Feeney and J. R. Whitaker (eds.), ACS Adv. in Chem. Series No. 160, Am. Chem. Soc., Washington, DC, 1977.
98. CASSENS, R. G., *Food Tech.*, 1977, **31**, 76.
99. ROBBINS, F. M., WALKER, J. E., COHEN, S. H. and CHATTERJEE, S., *J. Food Sci.*, 1979, **44**, 1672.
100. PARISH, F. C., *J. Food Sci.*, 1979, **44**, 1668.
101. MARSH, B. B., *J. Food Sci.*, 1977, **42**, 295.
102. SCHMIDT, G. R., MOWSON, R. F. and SIEGEL, D. G., *Food Tech.*, 1981, **35**(5), 236.
103. SAFFLE, R. L., *Adv. Food Res.*, 1968, **16**, 105.
104. GALLUZZO, S. and REGENSTEIN, J., *J. Food Sci.*, 1978, **43**, 1761.
105. MACFARLANE, J., SCHMIDT, G. R. and TARNER, R. H., *J. Food Sci.*, 1977, **42**, 1603.
106. THENO, D. M., SIEGEL, D. and SCHMIDT, G. R., *J. Food Sci.*, 1978, **43**, 488.
107. SIEGEL, D. G. and SCHMIDT, G. R., *J. Food Sci.*, 1979, **44**, 1686.
108. SIEGEL, D. G., CHURCH, K. E. and SCHMIDT, G. R., *J. Food Sci.*, 1979, **44**, 1276.
109. HOFFMANN, K. and HAMM, R., *Adv. Food Res.*, 1979, **24**, 2.
110. ISHIOROSHI, M., SANEJINA, K. and YASUI, T., *J. Food Sci.*, 1979, **44**, 1280.
111. YASUI, T., ISHIOROSHI, M., NAKANO, H. and SANEJINA, K., *J. Food Sci.*, 1979, **44**, 1201.

# 4

# Optimisation of Human Protein Requirements

Sanford A. Miller and Geraldine V. Mitchell
*Bureau of Foods, Food and Drug Administration,
Washington, DC, USA*

Following World War II, the high moral purpose that led to the foundation of the United Nations also moved that organisation to consider the ways in which the more affluent parts of the world could help the development of the newly emerging nations with which they shared this planet. Early in the deliberations of the august and distinguished groups of experts convened over several years by the world organisation to consider this problem, it became clear that the food supply and its consequence, nutritional status, were of prime importance in assuring the well-being of all of the peoples of the world. Further, the relationship between health and productivity led these groups to strongly advocate planning for improvement of the food supply as a principal component of national policy. At first, major effort was directed towards the amelioration of the results of the worldwide famine that followed World War II. Thus, international organisations, such as FAO, directed their efforts towards increasing the yield of indigenous crops without particular concern for the nutritional quality of these products. With time and the luxury of increasing knowledge and sophistication, more attention was paid to the quality of the food product, in particular, the protein component of the diet. Largely as a result of the work of researchers in Africa and Latin America, a picture began to develop from which it was argued that the principal world nutrition crisis was the lack of protein and that a major international effort was required to avert an impending disaster in food requirements if the world supply of protein was not substantially increased. Through the years, the original arguments, based largely on the observations of kwashiorkor and marasmus as protein deficiency syndromes in children and the development of data arguing the high incidence of these diseases in the developing world, have been expanded and refined. Concern was expressed not only for the amount but also for the quality of protein being consumed. Schemes evolved to produce low cost amino acids to supplement and improve low quality indigenous protein sources. Proposals were made to explore a variety of exotic protein sources, ranging from protein extracted from single cell organisms grown by fermentation techniques on a variety of substrates through development of higher quality strains of existing plant sources of protein to the harvesting of Antarctic krill. Major efforts were made to introduce new food products of higher protein quality into national diet—efforts that were met with a notable lack of success. By the early 1970s it seemed as if nutrition was nothing but evaluation of protein needs and quality. At times it appeared that, if all the plans proposed to increase the protein supply were undertaken, we would soon be awash in a sea of protein, a return, perhaps, to the pre-biotic soup. Concentration on protein led, however, to several problems, not the least of which was the almost pathologic concern for optimisation of protein intake. High quality protein is inefficient to produce and is, therefore, an expensive product requiring an inordinate use of resources. As a result, there developed over the course of years, a single-minded drive to optimise protein requirements and thus reduce the resource burden. It was also hoped that, by establishing precise levels for protein requirements, other policy areas reflecting the concern over protein could also be optimised. Thus, an infinite series of meetings were organised to discuss protein requirements, nitrogen balance methodology and other ancillary issues such as impact of infection and other stress on protein needs. The issue was of such political importance that with each modification of the protein requirement by international or national bodies, cries of political opportunism and chicanery were raised. Perhaps nowhere else was the gap between 'haves and have nots,' better exemplified than in the 1 g/kg body weight/day allowance established in the US and other western countries and the approximately 0·5 g/kg body weight/day proposed by several developing countries for the protein needs of their populations.

By the mid-seventies, however, a new set of issues had arisen and the emphasis, at least in debate, had shifted. A strong body of opinion concluded that the protein gap was a myth and does not exist. The proponents of this concept felt that the main problem of our expanding world population was primarily a food or energy gap, not protein deficiency. One of the main challenges to the 'protein gap' concept concluded that 'the adequate safe protein–energy ratio in the diets of 2–3 year old children is close to 5% and since most varieties of cereal grains appear to provide utilisable protein levels close to this amount, this lends further support to the view that primary protein deficiency is unlikely to be the main factor causing protein–energy malnutrition in communities for which cereals are the cheapest source of protein'.[1] While part of this challenge to the established dogma can be attributed to the natural iconoclastic tendency of good scientists, there was and is sufficient merit in their arguments to consider their views very seriously and, at the very least, to reconsider and justify the current position, particularly the need or drive to optimise the protein requirement.

While the truth concerning the major dietary problem facing the world population probably lies somewhere between the two concepts, there is one thing of which we are reasonably sure, i.e. the increase in the world population which has more than doubled during the last 35 years and is due to add another billion in the next 12 years.[2] The problems of maintaining a balance between world food production and increasing world population are becoming more serious, so serious that the lack of a solution to this problem carries within it the seeds of political and social disaster of enormous magnitude. The question of optimisation of protein requirements is part of the problem and of the solution and thus worthy of further review.

The issues surrounding the optimisation of protein requirements can be arbitrarily divided into two sub-questions. The first, and most fundamental, question is whether or not there is any merit in or need to optimise protein requirements in light of our current knowledge or, to put it somewhat differently, with what precision do we need to determine protein needs considering overall international food needs. The second question is concerned with the issue of whether or not there are viable ways of optimising requirements assuming there is a need to do so.

That there is a potential for widespread protein deficiency is unmistakable. While the estimation of food protein production in the early 1960s and 1970s suggested that the production of all protein commodities increased, the drought in Africa and poor crop yields (in other parts of the world) of the past 10 years have all but erased this gain. Moreover, while production of most plant proteins used for human food grew at a slower rate than did animal proteins, plant foods still supplied approximately 65–70 % of the world supply of edible protein. The largest contribution to this group was made by the cereals such as wheat, rice, maize, millet and sorghum, all traditional and widely used sources of protein of reasonable nutritional quality if sufficient amounts are available to compensate for their amino acid deficiencies. (In one sense the debate between the protein and energy proponents is one of amount, i.e. if enough of a particular cereal grain is available to meet protein needs, will this amount also meet energy requirements? This is an interesting restatement of the concepts proposed by the energy proponents.) The estimation of available protein supplies is also confounded by the inequitable distribution and consumption of food in the world. Food supplies are not only consumed disproportionately by the more affluent countries, but also by the most prosperous segments of the population in developing countries.[2] For example, a marked discrepancy exists between the availability of high quality protein in diets of developing and developed countries. Seventy percent of the total protein consumed in North America comes from animal sources compared to 16–36 % in the developing countries. It is estimated that at least 70 % of the protein consumed in developing countries comes from plants; in some areas the number may approach and even exceed 90 %. The intake of protein foods by the world population has commonly been described as falling into three categories: (1) diets consisting of cereal grains on which the greatest part of the world's population subsists; (2) diets high in animal protein which are high protein quality and used mainly in developed countries; and (3) diets based mainly on starchy roots and fruits very low in protein. It is this latter group that is the cause of most concern.

The world dietary situation as it related to the largest segment of the population has been characterised by Jansen[3] as follows: 'Food energy consumed is marginal. Protein, as a percentage of calories, is, in most cases, reasonable but protein quality is marginal as is the nutrient density of the diet in general. A very important factor, often ignored, is that most diets in less developed countries are quite low in fat'. The implication of this statement is that diets marginal in energy will also be marginal in protein even if the protein is of high quality. Thus, satisfaction of the energy need will meet protein requirements. If, however, the protein is not of high quality, as is the usual case, then the problem will be exacerbated and protein deficiency may become of principal concern. Based on these considerations, there does appear to be a need to consider protein needs in addition to energy or total food requirements. Not only are there profound biological effects of protein deficiency when energy intakes are marginal, but the efficient utilisation of either protein or energy is dependent upon the appropriate availability of each. The question then becomes 'how precise do these requirements need to be and how, in fact, do we determine them?'.

In order to assess this problem further, there is a need to examine the current status of our knowledge on human protein requirements. For the last forty years, various groups and committees throughout the world have made recommendations for the protein needs of national and international population groups. These estimates of protein needs have been called 'requirements', 'safe levels of intakes', 'allowances', and 'recommended intakes'. The definitions of these terms and the development of these recommendations have been considered in more detail elsewhere and will not be further considered in this paper, except to point out that each is based on different concerns of national health policy and is, in large measure, a reflection of national affluence. However, for the purposes of this discussion 'protein requirement' will be used to describe the overall protein needs of population groups based largely on the WHO/FAO standards.

Current estimates of protein and energy requirements are largely based on studies of healthy well-nourished population groups and usually do not make any distinction between population groups of different ethnic backgrounds, health status and environments. Moreover, in recent years, some attempt has been made to differentiate between the energy and protein required for rehabilitation and that required for maintenance and normal growth. But these considerations have not impacted very strongly, as of yet, on the recommendations of the various advisory bodies. However, a recent meeting in California of an advisory group to the UN University has begun to consider the importance of these factors in establishing protein needs.

As indicated earlier, estimates of protein and energy requirements have changed over the years. The 1973 proposal by the Joint FAO/WHO Expert Committee on Energy and Protein Requirements proposed a safe level for protein intake of 0·57 g/kg body weight/day of egg or milk protein for normal healthy adult individuals. This value is 20 % lower than the 1965 recommendation and has been challenged by a number of workers. However, recommended energy levels have remained essentially unchanged over the years. As a result, comparisons of average intake of population groups in developing countries with the new lower safe protein intake have suggested that the protein levels were less inadequate than were the caloric intakes. It was with this type of evaluation that the current controversy concerning the importance of protein or energy was begun.

## PROTEIN–CALORIE INTERRELATIONSHIPS

The topic of protein–calorie interrelationships has been reviewed extensively. Most of the early short-term metabolic nitrogen balance studies in humans, concerned with estimating protein requirements, nitrogen utilisation and dietary protein quality, were generally carried out at generous energy intakes in order to avoid weight loss and ensure that the protein would not be used for energy. Recent data suggest that a significant portion of dietary protein is used for energy irrespective of energy levels. Thus, attention has now been focused on changes in nitrogen balance as a result of change in dietary calorie levels, above and below the estimated requirement level. Controlled human studies demonstrated that, at a given level of dietary protein, addition of energy improves nitrogen balance of adults until a plateau is reached reflecting the adequacy of the protein intake. On the other hand, several recent studies carried out in adult, well-nourished men at levels of nitrogen and energy near the estimated maintenance range have yielded large variations in individuals for improvement of nitrogen balance with added energy. In most of these studies, adult males were fed high quality proteins. The results of these studies indicated that nitrogen balance can be altered by small excesses of energy above the estimated maintenance level, the improvement of nitrogen balance ranging from 0 to 0·95 mg N/kg body weight/kJ. For example, in the studies of Garza *et al.*[4] at energy intakes sufficient to meet the estimated requirements of the subjects, five of the six men were in negative nitrogen balance when they were fed 0·57 g protein/kg body weight/day. In these five subjects, nitrogen balance improved with increased energy intake until zero balance was achieved. In a subsequent long-term study, these authors demonstrated that when 0·57 g protein/kg body weight/day (the proposed FAO/WHO requirements) was supplemented with the equivalent of 0·23 g protein/kg body weight/day from a non-essential amino acid mixture, lower energy intakes were required to maintain nitrogen balance than in the earlier experiments. These results lend support to the concept that increasing the energy content of the diet from marginal sub-maintenance levels to luxus levels results in a more positive or, at least, less negative nitrogen balance. These studies also question the validity of the 1973 FAO/WHO recommendation for the safe intake of protein in man in long-term feeding studies. These studies were carried out at caloric intakes claimed to be appropriate for long-term energy balance and appeared to demonstrate that 0·57 g protein/kg body weight/day of egg protein is inadequate to maintain protein nutritional status in young men receiving generous amounts of energy. One of the authors reported that the only way most of the students in this study could be brought into zero nitrogen balance was to provide diets supplying 500 calories more than their normal requirements and thus cause a weight gain or by increasing protein intake.

Richardson *et al.*,[5] in studying the interrelationships of protein and energy in healthy young men, tested the effect of carbohydrate on nitrogen balance of men at isoenergetic intakes. They found that nitrogen balance was better when the ratio of carbohydrate:fat was 2:1 than when the energy was derived equally from the two sources. The diets supplied the FAO/WHO safe level of milk protein. An examination of the regression equations relating nitrogen balance to intake indicated that N balance increased by 2·6 and 1·3 mg/kg body weight/kcal with the high and low carbohydrate diets, respectively. Some of the subjects had energy intakes as low as 38–49 kcal/kg body weight/day. Calloway[6] reported that added energy had more impact when the level was raised from low to adequate than from adequate to luxus levels.

There are several reports indicating a protein-sparing effect of energy increases with increasing nitrogen intakes.[7,8] Inoue *et al.*[9] reported that energy increases cause a protein-sparing action when a protein of lesser quality than egg was fed at nitrogen levels higher than the FAO/WHO safe level of egg protein to young Japanese men. When energy intakes were increased by 12 kcal/kg body weight/day, the estimated requirements for egg and rice were reduced by 23 mg and 38 mg of N/kg body weight/day.

These findings underscore the importance of caloric intake as well as protein intake in meeting the protein needs of a population as well as indicating the differential effects of fat and carbohydrate sources on nitrogen balance. This latter point further emphasises the danger of considering nutrition as a series of unipolar issues.

The studies cited above have dealt mainly with the effect of energy on nitrogen balance. However, there is another side to the issue: the effect of protein intake on energy utilisation. Little attention has been given to this area. MacLean and Graham[10] demonstrated that energy utilisation was remarkably sensitive to the quantity and quality of protein. In these studies, six children, ages 4–7, were fed a constant energy level (150 kcal/kg body weight/day, recalculated daily) and cow milk protein formulas providing 4–8 % of the energy as protein. The mean weight gain was 2·8 and 6·7 g/kg body weight/day for the 4 % and 6·4 % protein–energy ratios, respectively. There was no further gain at protein–energy ratios higher than 6·4 %. These findings, if confirmed, would demonstrate the importance of the level of protein intake on the efficiency of energy utilisation.

As mentioned earlier, estimates of protein requirements have generally been based on studies of healthy well-nourished population groups and usually make no allowance for forms of stress such as infection, injury or emotional disorders. This subject has been extensively reviewed by Irwin and Hegsted.[11] Infection is known to have a profound effect on the intake, metabolism and excretion of most nutrients and hence, on nutritional requirements, and therefore will be discussed here in some detail. It is recognised that because of these negative and interactive effects, infectious states should be taken into account in considering protein requirements for populations or sub-populations at risk. The issue of the effect of infection on protein requirements is very complex, especially in populations that are poorly nourished. It is difficult to separate the nutritional effects from the environmental effects in these cases. Research has clearly indicated that during acute infectious episodes individuals are usually in negative protein and energy balance. The amount of nitrogen loss appears to vary not only with the type and severity of the infection but also with other factors such as previous nutritional status and environmental conditions.

Infection causes a significant reduction in intakes of protein and energy due to anorexia and, often, improper therapeutic practices. Equally important are other metabolic factors which affect the negative nitrogen balances associated with infections such as increased nitrogen excretion, muscle wasting, altered organ function and the diversion of ingested protein to support the host body's defence processes.[12]

While the changes in protein metabolism during infection suggest a need to increase dietary protein, not all infections appear to have the same qualitative and quantitative effects on protein metabolism. Predictions on the quantitative significance of the effects of infection on protein requirements are difficult to make. Scrimshaw[13] in discussing the effect of infection on nutrient requirements cited a limited number of studies which indicated that the nitrogen loss due to various diseased states ranged from 3·5 protein/kg body weight to 0·9 g protein/kg. The average nitrogen loss associated with diarrhoea of infectious origin was 0·9 g protein/kg body weight. Scrimshaw recommended a figure of an additional 0·3 g/kg body weight/day of high quality protein for adults to allow recovery from acute episodes of infectious disease. The United Nations University[14] stated that the estimation of quantitative needs for protein during recovery from acute infections depends on how much longer should be allowed for recovery than the duration of the acute episode and what baseline is taken for the needs of healthy individuals.

The problem of protein needs during catch-up growth is a concern of developing and developed countries, since even in the most hygienic surroundings illnesses occur. These episodes are followed by an accelerated rate of weight gain. Fomon[15] calculated the contribution made by adipose tissue and skeletal muscle to the weight deficit of a moderately malnourished child at the age of 1 year. The deficits in relation to a normal child for muscle and adiposity were 35 and 55 %, respectively. This growth deficit and its composition indicates that the dietary protein along with the intake of other nutrients must be increased to enable the child to achieve his growth potential. Several attempts have been made to estimate the protein requirements for catch-up growth. Whitehead[16] calculated the protein requirement for the catch-up growth for a 7 kg Ugandan child between 1 and 3 years of age. The normal growth rate was considered to be about 10 g/day. To grow at seven times this rate (i.e. to 'catch-up') the protein requirements would jump from 1·6 to 3·2 g milk protein/kg body weight/ day. Although many debatable assumptions are involved in this calculation, it does reflect the magnitude of the need for catch-up growth.

## PHYSICAL ACTIVITY

Studies on the effects of exercise on protein requirements have been conflicting and inconclusive. Some studies indicate lower nitrogen retention by exercising subjects as a result of increased urinary and sweat losses,[17] whereas other reports indicate no change in urinary excretion as a result of heavy training.[18] Recently, several studies have been carried out to test the adequacy of protein levels for subjects carrying out various physical activities. Torun *et al.*[19] fed normal healthy males (18–21 years old) a diet supplying 0·5 to 1 g of egg and milk protein/kg body weight/day and adequate energy for maintenance of body weight. The findings indicated that 0·5 g of egg and milk protein/kg body weight/day were not sufficient for men performing isometric exercises, whereas 1 g/kg body weight/day appeared to be adequate.

Marable *et al.*[20] studied the effect of feeding protein at 0·8 g/kg body weight/day and 2·4 g/kg body weight/day during muscle building exercises on urinary nitrogen excretion of young male students. The results indicated decreased urinary nitrogen excretion in exercising subjects. However, the decrease may have been the result of a higher intake. The highest urinary excretion was found when subjects consumed three times the recommended dietary allowance for protein, i.e. 2·4 g/kg body weight. Iyengar *et al.*[21] also studied the effect of varying energy and protein intake on nitrogen balance in adult men engaged in heavy manual labour. They concluded that, when energy level is 20 % below requirement, 1·0–1·2 g protein/kg body weight is required for nitrogen equilibrium as compared to 0·55 g/kg body weight at adequate energy levels.

## AGEING

Physiological changes that occur with age also appear to alter requirements for protein. Uauy *et al.*[22] reported that whole body protein metabolism during ageing in many may be altered. There may be a slow loss of total body protein, due to a diminution in the size of the skeletal muscle mass. These changes are accompanied by a shift in the overall pattern of whole body protein synthesis and breakdown. Unfortunately, studies carried out to determine the protein requirements of elderly people are contradictory. Researchers have estimated protein needs of the elderly to be higher, lower or equal to those of young adults.

Uauy *et al.*[23] reported large variations in the nitrogen balance of the elderly. Four of the seven men used in the study were in balance at an intake of 0·57 g/kg body weight, another at 0·85 g/kg body weight and two of the men were never in balance at the protein levels tested. In contrast, Zanni *et al.*,[24] found that the 'safe level' of egg protein (as defined by FAO/WHO) for men over age 60 was 0·59 g protein/kg ideal body weight.

Similarly, Cheng *et al.*[25] concluded that protein requirements are not affected by the ageing process. However, Uauy *et al.* recommended that for food planning purposes an appropriate allowance for the elderly should be 12–14 % protein calories since elderly persons are more likely to be affected by factors that tend to increase protein need.

These contradictory and variable results are not too surprising. As in all human enterprises, individuality plays a role. For the elderly, that role assumes an even greater importance. Previous lifestyle, injury and other factors can modify the metabolic state at the time of measurement.

Two other factors of significance for which there is limited information are the influence of climate on protein needs and adaptation to low protein intakes. Recent studies suggest that, in countries where populations are chronically undernourished, certain groups may adapt to low protein intakes and this may alter the protein requirements relative to the well-nourished individual. Nicol and Phillip[26] found that Nigerian adult males could maintain nitrogen balance on rice protein intakes of less than 0·57 g/kg body weight/day/egg protein equivalent. A more generous intake of rice protein caused a significant increase in weight gain. These findings are contrary to those reported using well-nourished Japanese males.[27] The problem in such studies is the lack of understanding of the mechanism of this adaptation and the price paid for it. One may ask if function is impaired or if ability to respond to stress is modified. Certainly the issue of 'optimisation' requires knowledge of the base from which such estimates are made.

## METHODOLOGY

One cannot talk about protein requirements without mentioning the methodology used to determine requirements. Most of the data available on protein requirements are based on results from metabolic nitrogen balance studies. However, for the infant, body weight gain has been used. A comprehensive review of the approaches used in the assessment of protein requirements has been compiled in monograph form by Irwin and Hegsted.[11] The authors pointed out significant gaps which exist in the available estimates of protein needs for the various age groups. It is important to recognise the limitations of these methods. Some of the significant limitations[14] of the nitrogen balance method are: (a) difficulty in measuring quantitatively all of the routes of nitrogen loss from the body as well as recognised cumulative errors; (2) the effects of factors other than the level of intake and nutritional value of dietary protein; and (3) body nitrogen equilibrium does not necessarily reflect the dynamics of protein metabolism which in turn define the requirement for function in each tissue system. Similarly, any consideration of protein requirements must consider protein quality. The methodology for the biological assessment of the nutritive value of dietary protein is still very controversial. This subject has been reviewed by Hegsted.[28] A precise method of evaluating protein quality is important, not only to make possible estimation of the amount of a given protein mixture required to meet physiological needs for protein, but also to monitor the nutritional adequacy of the food supply.

## EXCESS PROTEIN

One issue rarely if ever considered in the establishment of protein requirements is concerned with the effect of excess dietary protein. In general, such studies performed in the past used very large concentrations of dietary protein, ranging from 50–80 % of the diet. It is not surprising that the results of these studies suggested acute or semi-acute toxic responses to such abnormal levels of intake. Nevertheless, since these levels are so much in excess of most human diets, little attention was paid to these results in setting human protein requirements. Recently, however, studies performed with much lower levels of protein have suggested that, when consumed over a lifespan, dietary protein intakes in excess of 25 % of the diet may lead to a variety of disease states.[29] This work, done with rats, suggests that kidney disorders, liver dysfunction, and reproductive difficulties may be associated with the long-term consumption of high protein intakes. In addition, studies performed with low energy diets in which protein represents a very high proportion of the total calories (60–80 %) but not in excess of 1 g/kg body weight/day, have also suggested the possibility of pathological response, particularly cardiac arrhythmia. The use of high protein diets for weight reduction, for example, has led to a number of deaths. Unfortunately, in spite of the obvious importance of this question in the process of optimisation, not much work has been or is being done to further explore these questions. Clearly, the development of any concept of optimal protein intakes must consider this issue.


## DISCUSSION

It is clear from consideration of the many factors mentioned in this paper that the establishment of a protein requirement is a very complicated and complex affair. Not only does one need to consider the requirements for protein for normal metabolic events, but the influence of a variety of pathological states such as infection and stress must also be considered. The need for protein for growth during early childhood and for changes occurring during ageing are also factors that are well known to workers in this field. It is also clear that the requirement for protein is intimately associated with the need of the organism for energy as well as all of the other essential nutrients. The sensitivity of the protein requirements to these other factors is only recently becoming well known. It is still too early in the investigations now ongoing to even begin to indicate the extent to which these factors have to be considered in establishing protein requirements. It is also clear from the foregoing that in spite of the intense interest in protein metabolism over the last forty years, we are still very far from a point at which we can clearly and precisely define what the requirements are. We do not yet understand changes in protein metabolism with ageing nor do we understand the changes in the utilisation of protein as a result of the variety of daily stresses to which the organism must adapt. As far as the process of adaptation itself is concerned, we still do not understand the functional cost that must be paid when a population survives on protein intakes considerably lower than that of other populations. It is simply not possible to explain all of the differences on the basis of genetic variability.

On the other hand, the impact on a population of high protein intakes beyond those that are estimated to satisfy the needs of the organism is also not well understood. It may be that a price is also exacted from populations consuming these high levels. Not only is luxus consumption a waste of scarce resources but, perhaps, more importantly, it may represent an additional metabolic burden to which the organism must adapt. To properly establish the requirements for protein one must understand this process as well as the effects of deficiency states usually considered.

At this stage in the development of our knowledge, it simply is not possible to provide a precise, or, for that matter, an accurate estimate of protein requirements. The question may be asked, however, whether it is necessary to establish such precise intakes. Variations in metabolic requirements resulting from genetic variability, lifestyle differences, climatic and age differences and so on make moot the argument that we need to supply a single precise number for protein requirements. A more reasonable approach might be the establishment of a range of values in which we are reasonably certain that the population is in reasonably good nutritional status without incurring deficiency or excess nutritional stress from excess consumption. In fact, one could argue that the current policy of establishing a protein allowance of three standard deviations above the mean value in order to include 99·9 % of the population is incorrect, putting at risk those individuals whose requirements are considerably lower than this value and whose sensitivity to excess protein in their diet may be greater than other members of the population.

In large measure, it may have been the complexity of attempting to define as dynamic an issue as protein requirements which probably led to the frustration from which resulted the current controversy over the importance of protein or energy in the resolution of the world food problem. It is apparent that a great debate has erupted on this issue and is flaming throughout the scientific world. Like most debates, it has passed, we believe, from the point at which the issues are scientific into the area where the issues have become philosophic, economic, and, at times, theologic. While no one disagrees that both energy and protein are required to maintain proper nutritional status, it is not agreed which is most deficient under contemporary conditions. Those who believe in the protein theory argue that there exists a world protein crisis that must be averted by an international effort. Those who believe in the food energy crisis argue that if energy needs are met then protein needs would also be met. Unfortunately, the foundation of both arguments is based upon the assumption of a particular protein requirement. If you believe in the protein theory then you must argue that the actual protein requirement is considerably greater than that proposed by the WHO/FAO expert committees in 1973. If you believe in the energy hypothesis, you must argue that the WHO/FAO safe level is right and indeed may actually represent an excess; thus, there is less deficiency in protein than there is in energy. Unfortunately, however, there is no easy way to resolve the debate. The appropriate experiments are difficult to do.

For example, it is simply not possible to try to cure protein–calorie deficiency by feeding only protein or only calories. Thus, it becomes difficult if not impossible, to separate protein needs from those for energy. It may not be necessary to do so. What we need is a better understanding of the processes by which food intake patterns change and more efficient technology to improve relatively low quality protein sources inexpensively and in forms acceptable to target populations. It is a truism of nutrition that the more varied a diet consumed, the less is the possibility of nutrient deficiency. While this may not be possible in all parts of the world, this should, nevertheless, represent a legitimate goal for national food policies. The drive for urbanisation, for example, although resulting in a wide variety of nutritional disorders has also placed in the hands of nutritional planners a powerful tool. The need for urban populations to emulate the lifestyle of more affluent members of their societies offers to these planners a way to convince these urban dwellers to vary their diets. The problem must be then for planners to provide appropriate foods to these segments of the population at prices which they can afford, providing technology can develop products to meet these needs.

It seems clear that the evidence available today cannot support either a conclusion that the protein gap is a myth as some workers in the field would have us believe nor the view that the principal dietary need in the world today is protein. What we all can agree on, however, is that there is a food gap in the world, a gap that seems to be increasing year by year. It would be unfortunate, we believe, that after nearly 30 years of intense effort, the attention of the scientific community should be diverted from the real problems of feeding a hungry world to sterile debates about optimisation of nutrient requirements. It would be dangerous and unfortunate if national policy-makers concluded, as a result of the ongoing debate, that research in these nutritional problems is academic, irrelevant, and a waste of resources. If a conclusion is reached that the resolution of most of the problems of the world is more food of any kind of any quality then we are going to be in a dangerous situation. It is unfortunate that malnutrition in particular and nutrition in general have become such emotional and politically sensitive issues. While it is true that much could be done with existing knowledge, it is clear, we believe, from the preceding discussion that there is still much to learn. In the long run our ability to resolve this most fundamental problem of human development must be based on knowledge. We need to better understand the impact of dietary states on function, on performance, and on adaptation. We also need technology—technology to provide a variety of low cost, acceptable food products of reasonable nutritional value. In turn, new technology raises other questions, questions of bioavailability and sensitisation. John Waterlow has described nutrition as a Cinderella of sciences. If he is right then technology is the fairy godmother making it all possible.

## REFERENCES


1. PAYNE, P. R. *Am. J. Clin. Nutr.* 1975, **28**, 281.
2. SCRIMSHAW, N. S. and YOUNG, V. R. in *Evaluation of Proteins from Human*, Bodwell, C. E. (ed.), Avi Publishing Co. Inc., Westport, CT, 1977.
3. JANSEN, R. G. in *Soy Protein and Human Nutrition*, Wilcke, H. L., Hopkins, D. T. and Waggle, D. H. (eds.), Academic Press Inc., New York, New York, 1979.
4. GARZA, C., SCRIMSHAW, N. S. and YOUNG, V. R. *Br. J. Nutr.* 1977, **37**, 403.
5. RICHARDSON, D. P., WAYLER, A. H., SCRIMSHAW, N. S. and YOUNG, V. R. *Am. J. Clin. Nutr.* 1979, **32**, 2217.
6. CALLOWAY, D. H. *J. Nutr.* 1975, **105**, 914.
7. PLOUGH, I. C., IBER, F. L., SHIPMAN, M. E. and CHALMERS, T. C. *Am. J. Clin. Nutr.* 1956, **4**, 224.
8. RAO, NAGESWARA, C., NAIDU, NADAMUNI and RAO, NAGESINGA, B. S. *Am. J. Clin. Nutr.* 1975, **28**, 1116.
9. INOUE, G., FUJITA, Y. and NIIYAMA, Y. *J. Nutr.* 1973, **103**, 1673.
10. MACLEAN, W. C. and GRAHAM, G. *Am. J. Clin. Nutr.* 1979, **32**, 1381.
11. IRWIN, I. and HEGSTED, M. *J. Nutr.* 1971, **101**, 385.
12. BEISEL, W. R., SAWYER, W. D., RYLL, E. D. and CROZIER, D. *Ann. Internal Med.* 1967, **67**, 744.
13. SCRIMSHAW, N. S. *Am. J. Clin. Nutr.* 1977, **30**, 1536.
14. VITERI, F., WHITEHEAD, R. and YOUNG, V. R. (eds.), *Protein–energy requirements under conditions prevailing in developing countries: Current knowledge and research needs.* United Nations University, Tokyo, 1979.
15. FOMON, S. J., *Infant Nutrition*, 2 ed., W. B. Saunders, Philadelphia, Penn, 1974.
16. WHITEHEAD, R. G. and BIOL, F. I. *Am. J. Clin. Nutr.* 1977, **30**, 1545.
17. GONTZEA, I., SUTZESCU, P. and DUMITRACHE, S. *Nutr. Rept. Inter.* 1974, **10**, 35.
18. HEDMAN, R. *Acta. Physiol. Scand.* 1957, **40**, 305.
19. TORUN, B., SCRIMSHAW, N. S. and YOUNG, V. R. *Am. J. Clin. Nutr.* 1977, **30**, 1983.
20. MARABLE, N. D., HICKSON, J. F., KORSLUND, M. K., HERBERT, W. G., DESJARDINS, R. F. and THYE, F. W. *Nutr. Rept. Inter.* 1979, **19**, 795.
21. IYENGAR, A. and RAO, NARASINGA, B. S. *Br. J. Nutr.* 1979, **41**, 19.
22. UAUY, R., SCRIMSHAW, N. S. and YOUNG, V. R. *Proceedings of the Nutrition Society of Canada*, 1978, p. 53.
23. UAUY, R., SCRIMSHAW, N. S. and YOUNG, V. R. *Am. J. Clin. Nutr.* 1978, **31**, 779.

24. ZANNI, E., CALLOWAY, D. H. and ZEZULKA, A. Y. *J. Nutr.* 1979, **109**, 513.
25. CHENG, A. H., GOMEZ, A., BERGAN, J. G., LEE, TUNG-CHING, MONCKEBERG, F. and CHICHESTER, C. O. *Am. J. Clin. Nutr.* 1978, **31**, 12.
26. NICOL, B. M. and PHILLIP, P. G. *Br. J. Nutr.* 1976, **36**, 337.
27. KISHI, K., MIYATANI, S. and INOUE, G. *J. Nutr.* 1978, **108**, 658.
28. HEGSTED, D. M. in *Improvement of protein nutriture*, Food and Nutrition Board, Committee on Amino Acids, National Academy of Sciences, Washington, DC, 1974.
29. ROSS, M. H. in *Nutrition and Longevity in Experimental Animals in Nutrition and Aging*, Winick, M. (ed.), Wiley and Sons, New York, NY, 1976.

# 5

# Nutritional Value of Proteins and its Assessment

ARNOLD E. BENDER

*Department of Nutrition, Queen Elizabeth College,
University of London, UK*

## INTRODUCTION

Interest in proteins and protein quality originated at the beginning of the 19th century and was revived 150 years later for almost identical reasons.

Wet weather early in the 19th century led to spoilage of cereals and a scarcity of food in parts of Europe. The Academy of Paris appointed a Gelatin Commission in 1815 to inquire whether gelatinous extract of bones could properly replace meat in the diet. In 1817 the Society for the Promotion of Arts in Geneva investigated the practicability of utilising the nutriment contained in bones.[1]

In the 1950s, 'world food problems', epitomised by the term 'protein gap', stimulated research into new sources of protein—oilseed residues, fish meal, single cell protein and leaf extract. The parallel with the replacement of meat by gelatin is illustrated by the Report of the British Department of Health and Social Security of *'Foods intended to simulate meat'* which stated that the protein quality of such material should approximate to that of meat.[2]

Magendie reported on behalf of the Gelatin Commission in 1841 that dogs could not be maintained on gelatin alone or bread alone or bread and gelatin—a finding described as being astonishing to physiologists but indicating that proteins differ in nutritional value. Muscular flesh, together with fats and mineral salts, was stated as sufficient for complete and prolonged nutrition. The Commission were unable, as still today we are unable, to decide whether the results of animal experiments could be applied to man.[1]

## 'PROTEIOS'—PRIMARY

It was long known that man and his animals required a source of energy and salts together with albuminous substances—a term first mentioned by Macquer in the second edition of his dictionary in 1777 and derived from Pliny's term 'albumin' for white of egg. Macquer concluded that the gelatinous matter of animals is the true animal substance.

Mulder, the Dutch chemist, found that a wide variety of substances, including albumin, fibrin and casein, all contained the same amounts of carbon, nitrogen, hydrogen, oxygen and sulphur and so were presumably identical. At the suggestion of the Swedish chemist, Berzelius, he introduced the term 'protein' as recently as 1839 and all were regarded as being the same substance, modified only by being combined with different proportions of water.

Proteins have long been considered the most important of the nutrients, hence the name, derived from the Greek 'proteios' for primary or 'protos' meaning first. In fact the body has first call on the diet for energy and if the diet is inadequate in quantity, i.e. in the amount of energy that it provides, then the protein is oxidised to supply energy instead of being used for tissue replacement.

The importance of proteins received considerable emphasis from 1950 onwards with promotion of what was termed 'the protein gap'. It was argued that while the demands of a rapidly increasing world population for energy could be readily met, it is more difficult and more costly in land use to produce protein-rich foods. This is not correct. Cereals contain 8–12 % protein (and some varieties of hard wheats contain more protein than some meats) so if sufficient is eaten to satisfy the energy needs the protein requirements will be satisfied at the same time. The error of emphasising proteins arose for two reasons. Firstly, the *amount* of protein consumed per day was low in some developing countries. In Indonesia, for example, average consumption was only 38 g per head compared with the basic need for 45 g. However, if the intake is calculated, not in grams per day but in protein–energy percent it supplies 7·7 % protein compared with the need for 6 %. The figures show that the total amount of *food* was inadequate— the problem was one of food rather than protein as such, and protein needs could be satisfied by supplying more food of the same kind. Indeed, an inadequate diet containing 8–10 % protein can be improved by providing extra starch or fat and so sparing protein for its proper function of tissue building rather than 'wasting' it as a source of energy.

A protein gap implies an adequate intake of energy but an inadequate intake of protein; such diets are rare *among adults* if, indeed, they are found in any community since cereals are the staple food of so many countries. Even where the staple is a food very low in protein such as cassava, arrowroot and plantain, diets almost invariably include legumes and leaves and the protein proportion does reach the required levels. The problem of infants is another matter and these arguments relate only to the diets of adults.

However, it is worth emphasising here that the needs of a child for growth are very much smaller than generally thought—only a baby during the first few months of life is growing rapidly. Between the ages of 5 and 8 years a child will gain about 10 kg (20 lb) or 3·3 kg per year. Assuming that the greater part of this is muscle which is three-quarters water, the gain in protein is only 675 g per year or less than 2 g per day for growth—by far the greater part of the protein needs (even of the young child) are to maintain the existing tissues.

The second reason for promoting the protein gap concept was the political one of awakening the conscience of governments to the world food problems by a more dramatic approach—one that indeed succeeded for a time.


## PROTEIN QUALITY

An adult needs a regular supply of protein in order to replace routine losses—the obligatory nitrogen losses. If he loses an average of 6 grams of nitrogen per day, then he must replace this by 6 grams of dietary nitrogen as protein. He is, however, losing protein from human tissues and organs and replacing it with bread, legumes, milk, fish and meat, which introduces the concept of protein quality.

Protein quality discussion, with its jargon of BV (biological value), NPU (net protein utilisation), PER (protein efficiency ratio), NPR (net protein retention), GPV (grass protein value), etc. has tended to be restricted to the specialists. A major reason for this may be because protein quality was being measured before we knew what it meant. The discovery of the amino acids, starting with glycine in 1820, was not completed until the discovery of threonine in 1935.

The earliest measurements of protein quality were those of Thomas in 1909, modified by Mitchell in 1924 (the Thomas–Mitchell assay of biological value by nitrogen balance). In 1915 Osborne, Mendel and Ferry used weight gain of experimental animals per gram of protein consumed

                            *Arnold E. Bender*

(protein efficiency ratio) which was later adopted as the official Association of Official Analytical Chemists (AOAC) method of assessment of protein quality in the United States. There have been some modifications, namely simplification, of the Thomas–Mitchell method by measuring the nitrogen in the carcasses of the laboratory animals rather than the differences between intake and output;[3] and the basic errors of protein efficiency ratio were avoided by the inclusion of a control group of animals fed on a diet free from protein—net protein ratio.[4] Otherwise, these two methods remain the basis of protein quality assays.

It was not until as late as 1946 that Block and Mitchell[5] showed the relationship between amino acid composition and biological value and so explained that the quality of a protein depends on its amino acid make-up (see Figs 1 and 2).

It is interesting to note that gelatin again became a matter of interest

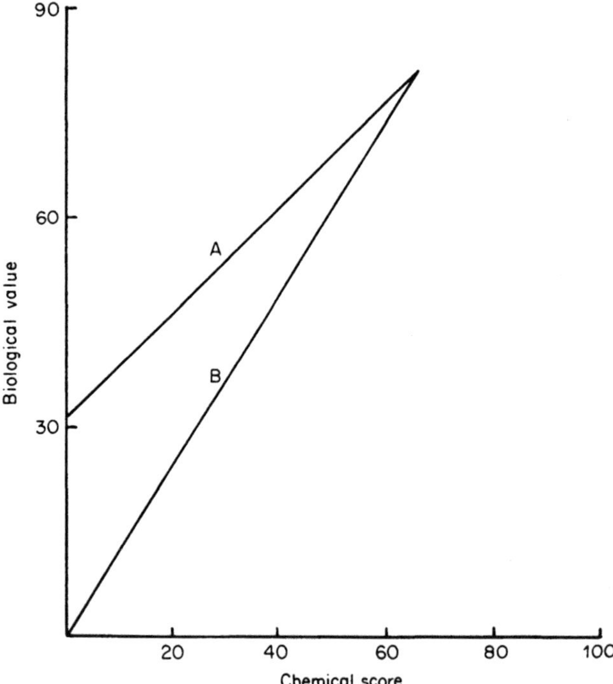


Fig. 1. Correlation between biological value and chemical composition compiled from literature data on a range of proteins.[5] Curve A: Block and Mitchell ref. 5. Curve B: theoretical relationship.

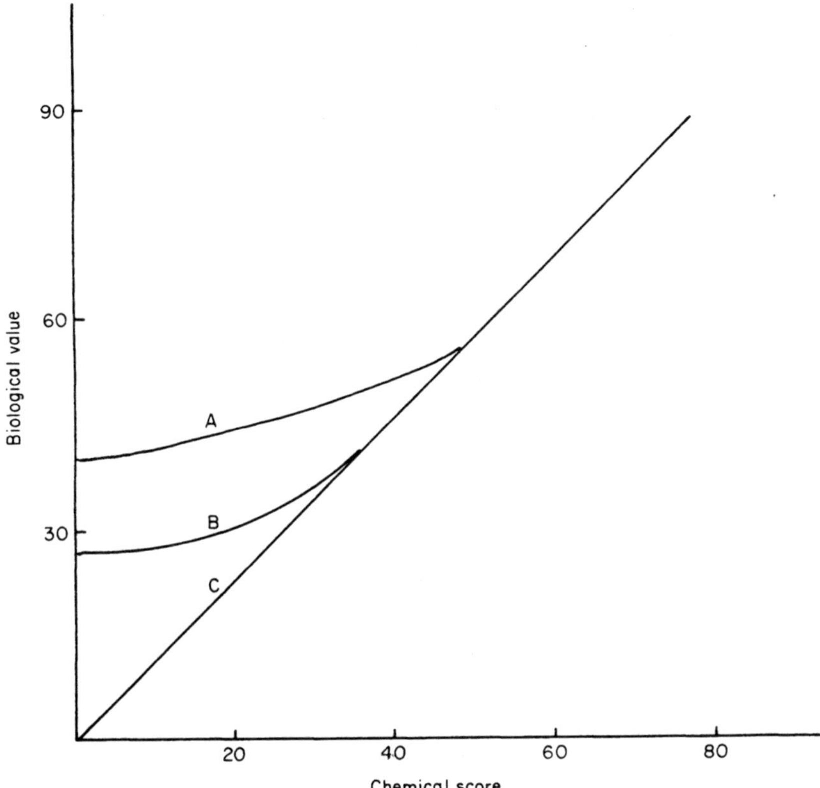



FIG. 2.   Correlation between biological value and chemical composition measured on a range of amino acid mixtures and proteins.[6–8] Curve A: protein limited by lysine. Curve B: protein limited by tryptophan, threonine, histidine, phenylalanine, leucine plus isoleucine. Curve C: protein limited by valine and sulphur amino acids.

about a century after Magendie's work. Mitchell *et al.*[9] had shown that gelatin had a biological value of 25, a value that was apparently confirmed by Mason and Palmer.[10] Later knowledge that gelatin is completely lacking in one of the essential amino acids, namely tryptophan, suggested that the biological value should, in fact, be zero. When we had established our modification of the Thomas–Mitchell technique we ascertained that the BV of gelatin was, indeed, zero.[11] The original authors had done as their forebears had more than a century earlier. The 1817 report and that of the Academy of Paris had used the 'substance formed by boiling meat, tendons, etc.' as gelatin[1] not purified bone gelatin. Similarly, Mitchell and his colleagues had assayed pork cracklings, which they described as being 'largely composed of gelatin' and was later misquoted as being gelatin.[12] The supporting evidence from Mason and Palmer was another mistake since they fed 10 % of the dietary protein as yeast extract to supply the B vitamins, which were not then available in the pure form. We were able to confirm that the addition of 10 % yeast protein to gelatin raised the BV from zero to 21.[11]

## 'BEST' METHOD OF ASSAY

Those working in the field are continually asked for the 'best' method of assay but the answer to this must always depend upon the purpose of such an assay. Proteins are required for a diversity of purposes—growth of infants, lactation of mothers, maintenance of tissues, repair during convalescence, and in the food area, for the production of meat, eggs and milk and even of wool. However, for practical reasons direct measurements can rarely be made so we require an abbreviated laboratory method (although some of the bioassays are lengthy) to serve as an index for these various purposes. It is not always recognised that a figure of BV, PER, GPV, etc. is merely such an index and cannot provide much information as to the value of the food in question when added to the existing diet of man or his infants or his domestic animals for the particular purpose intended.

It is most unfortunate that the apparent simplicity of the estimation of protein efficiency ratio misled 'Authority' into accepting this as the 'best' method and incorporating it into the AOAC standard methods. Unfortunate, because it is unreliable for the variety of reasons that Mitchell put forward many years ago[13] but particularly because the value obtained depends on the amount of food eaten. The latter varies with the apparent palatability of the diet but even more importantly, it varies with the particular batch of animals being used. The theoretical example of the errors and its experimental verification are shown in Table 1. The more the animal eats the higher the PER, and in a series of replicated assays of skim milk powder, different batches of rats fed *ad lib* ate different amounts and provided different PER values. This can be avoided by making allowances for the maintenance of the animals by including a protein-free group and is avoided in measurements of BV and NPU. However, most workers in the field have taken PER to be not only the standardised method but the one providing the ultimate answer and consequently all other methods are compared with PER. For example, at the International Congress of

TABLE 1

EFFECT OF FOOD CONSUMPTION ON MEASUREMENT OF PROTEIN QUALITY (RAT ASSAY)

*Theoretical calculation*[a]

| Protein eaten (g) | Change in body wt (g) | PER | NPR |
|---|---|---|---|
| 0 | −10 | — | — |
| 1 | −5 | — | 5/1 = 5 |
| 2 | 0 | 0/2 = 0 | 10/2 = 5 |
| 3 | 5 | 5/3 = 1·7 | 15/3 = 5 |
| 4 | 10 | 10/4 = 2·5 | 20/4 = 5 |
| 5 | 15 | 15/5 = 3·0 | 25/5 = 5 |

*Experimental verification*

(a) Dried milk

| Protein eaten (g/100 g body wt) | PER | NPR | NPU |
|---|---|---|---|
| 11·9 | 1·2 | 3·6 | 63 |
| 12·6 | 1·7 | 3·9 | 64 |
| 13·8 | 2·5 | 3·8 | 64 |
| 16·3 | 2·7 | 3·8 | 63 |
| 16·3 | 2·8 | 3·7 | 61 |
| 17·3 | 2·8 | 3·7 | 64 |

(b) Bread fortified with lysine

| | PER | NPR | NPU |
|---|---|---|---|
| 6·6 | 0·2 | 3·0 | 59 |
| 7·4 | 0·9 | 3·3 | 56 |
| 11·0 | 1·8 | 3·3 | 54 |
| 14·7 | 2·3 | 3·7 | 57 |

[a] Assume: (i) 2 g of protein required for maintenance; (ii) 1 g of protein produces 5 g body weight.

Nutrition in Rio de Janeiro in 1978 some dozen novel methods of assaying protein quality, such as the growth rate or nitrogen gain of moulds and worms, and rapid screening methods requiring only small amounts of protein, were all compared with PER.

## OTHER BIOLOGICAL METHODS

It is commonly believed that, since there are recognised drawbacks to chemical methods of determining protein quality, the ultimate in providing 'true' values is a biological method. However, all biological methods provide a figure (depending on the limiting amino acid) with *no further information*. For example, if an assay shows a BV of 0·5, this could be because one limiting amino acid is present at 50 % of the rat's requirements with all others in surplus, or that all the essential amino acids are present at this level of 50 %. They do not reveal which amino acid is limiting—unless replicated assays are carried out with supplements of the different essential amino acids. They do not provide information of the effect of adding the test protein food to a diet—which is usually the prime question. They may not even answer the question whether the foodstuff has suffered processing damage unless that damage has occurred to the limiting amino acid.

What we need to know is how much of each essential amino acid is present and *available* so that we can assess the value of such a food in supplementing a diet or to maintain or repair tissues or to provide milk, eggs or growth in domestic animals.

In view of this, more information is actually provided by an amino acid analysis—subject to the problem of the availability of the amino acids shown to be present. The development of chemical methods for available lysine has substantially improved the value of amino acid analyses, but unfortunately there are no satisfactory methods for the determination of the availability of other amino acids.

The commonly-used chemical score and protein score compare the limiting amino acid with that present in an amino acid mixture or in egg protein respectively, but this again provides only a single figure without information on other amino acids. The only method that takes into account all the essential amino acids is the essential amino acid index of Oser[14] put forward many years ago and surprisingly little used.


## RELATIVE UNIMPORTANCE OF PROTEIN QUALITY

It is, of course, true that proteins differ in their nutritional quality (Table 2). The reason, so far as evidence indicates, is the relative amount of the limiting essential amino acid. Despite possibilities of amino acid imbalance and differential rates of release of amino acids during digestion affecting quality measurements, there are virtually no quality figures which do not fit the amount of the limiting essential amino acid. Such discrepancies that have been shown and cannot be explained are very few and do not invalidate the argument.

While individual foods cover the range from BV 1·0 to zero, diets do not. According to FAO figures the diets of the best-fed western communities

TABLE 2
PROTEIN QUALITY

| Individual foods | Biological value[a] | Protein efficiency ratio[b] | Chemical score[c] |
|---|---|---|---|
| Human milk | 0·95–1·0 | 4·5 | 100 |
| Hen's egg | 0·95–1·0 | 4·5 | 100 |
| Cow's milk | 0·75 | 3·0 | 100 |
| Fish | 0·75 | 3·0 | 100 |
| Meat | 0·75 | 3·0 | 100 |
| Soy beans (heat-treated) | 0·7 | 2·5 | 63 |
| Casein | 0·7 | 2·5 | 60 |
| Rice, white | 0·6 | 1·0 | 60 |
| Wheat grain | 0·6 | 1·0 | 60 |
| Bread white | 0·5 | 0·5 | 35 |
| Peas and beans | 0·35–0·50 | 0–0·5 | 25–40 |
| Maize (corn) | 0·4 | 0 | 40 |
| Gelatin | 0 | — | 0 |

Complete diets
   Industrialised countries (based on meat, milk, wheat) BV 0·8
   Some developing countries (based largely on cereals) BV 0·7
   Some developing countries (based on cassava, etc.) BV 0·6

[a] Retained N/absorbed N-measured on rats.
[b] Weight gain per gram protein eaten (measured on rats).
[c] Limiting amino acid, determined chemically, expressed as percentage of the same amino acid in the FAO/WHO 'reference mixture' for human beings.

have BV 0·8 compared with a value of 0·7 in the developing countries (and in rare cases 0·6) (Table 3). If the addition of much-lauded foods such as meat, milk, eggs and cheese to a predominantly cereal diet—the greater part of the total diets in many developing communities—only increases the BV from 0·7 to 0·8, then the quality of supplementary protein foods cannot be of much importance in western diets. The difference it makes may be of importance in developing communities where the intake of proteins is already marginal; as shown in Table 3 the amount needs to be increased to compensate for quality. The protein quality of western diets, 0·8, is difficult to change (arithmetically it would be difficult to add or replace sufficient of the dietary protein with another of different quality so as to make any change). Further, the amount consumed is usually in excess of physiological needs.

The problem is complicated by the finding that many people in western communities at the lower end of the intake scale in the various surveys that

TABLE 3

'SAFE LEVELS' OF PROTEIN PER DAY IN DIETS OF DIFFERENT PROTEIN QUALITIES (FAO/WHO 1973 RECOMMENDATIONS)

| Age group | Body weight (kg) | Safe level of protein | | Adjusted for protein quality | | |
| --- | --- | --- | --- | --- | --- | --- |
| | | (g/kg wt) | (g/day) | Score 80 | Score 70 | Score 60 |
| Infants | | | | | | |
| 6–11 months | 9 | 1·53 | 14 | 17 | 20 | 23 |
| Children | | | | | | |
| 1–3 years | 13·4 | 1·19 | 16 | 20 | 23 | 27 |
| 4–6 | 20·2 | 1·01 | 20 | 26 | 29 | 34 |
| 7–9 | 28·1 | 0·88 | 25 | 31 | 35 | 41 |
| Males | | | | | | |
| 10–12 | 36·9 | 0·81 | 30 | 37 | 43 | 50 |
| 13–15 | 51·3 | 0·72 | 37 | 46 | 53 | 62 |
| 16–19 | 62·9 | 0·60 | 38 | 47 | 54 | 63 |
| Adults | 65 | 0·57 | 37 | 46 | 53 | 62 |
| Females | | | | | | |
| 10–12 | 38 | 0·75 | 29 | 36 | 41 | 48 |
| 13–15 | 49·9 | 0·63 | 31 | 39 | 45 | 52 |
| 16–19 | 54·4 | 0·55 | 30 | 37 | 43 | 50 |
| Adults | 55 | 0·52 | 29 | 36 | 41 | 48 |
| Pregnancy–latter half | | | +9 | +11 | +13 | +15 |
| Lactation | | | +17 | +21 | +24 | +28 |

have been reported, are, in fact, consuming less than the minimum physiological requirements with no detectable effect. The whole question of the adaptation of human beings to what are apparently inadequate diets complicates the issue, but nonetheless, protein quality and quantity are not apparently matters of any importance in western diets. Despite this, enormous efforts are devoted to producing novel sources of protein, to examining and improving their nutritional value, to supplementing protein mixtures, and certainly, to advertising to a bemused public the virtues of protein-enriched foodstuffs. There was even a peanut butter preparation on the British market that was enriched with DL-methionine—presumably in an effort to increase the nutritional value of a product which makes a smaller contribution to the average British protein intake than cabbage does.

Hence, all discussions of potential problems of replacing meat with soy preparations, all work involving the production of novel sources of protein, and all advertising claims should be scrutinised afresh.

# REFERENCES


1. McCollum, E. V. *A History of Nutrition*, 1957, Houghton Mifflin Co., Boston, MA.
2. Dept. of Health and Social Security (1980) *Novel Foods*. Her Majesty's Stationery Office, London.
3. Bender, A. E. and Miller, D. S. *Biochem. J.* 1953, **53**, viii.
4. Bender, A. E. and Doell, B. H. *Brit. J. Nutr.* 1957, **11**, 140.
5. Block, R. J. and Mitchell, H. H. *Nutr. Abstr. Rev.* 1946–7, **16**, 249.
6. Bender, A. E. *Proc. Nutr. Soc.* 1958, **17**, xxxix.
7. Bender, A. E. *Clin. Chim. Acta* 1960, **5**, 1.
8. Bender, A. E. *Meeting Protein Needs of Children*, Publ. 843, 1961, p. 407, Nat. Acad. Sci.—NRC, Washington, USA.
9. Mitchell, H. H., Beadles, J. R. and Kruger, J. H. *J. Biol. Chem.* 1927, **73**, 767.
10. Mason, I. D. and Palmer, L. S. *J. Nutr.* 1935, **9**, 489.
11. Bender, A. E., Miller, D. S. and Tunnah, E. J. *Chemistry & Industry*, Oxford University Press, 1953, p. 799.
12. American Medical Association, *Handbook of Nutrition* 2nd ed., 1951, H. K. Lewis, London.
13. Mitchell, H. H. *Indust. Engng Chem.* (*Anal*) 1944, **16**, 696.
14. Oser, B. L. *J. Amer. Diet. Assoc.* 1951, **27**, 396.

# 6

# Allergy Aspects of Food Proteins


ROLAND J. LEVINSKY

*Institute of Child Health,*
*30 Guilford Street, London, UK*


There is increasing awareness amongst the medical profession that allergy to food substances may play a major role in either causing or exacerbating a variety of diseases. There is little doubt when symptoms such as urticaria or vomiting and diarrhoea come on soon after eating a particular food substance, but when the symptoms are delayed, such as commonly occurs in atopic eczema, migraine, and even in certain types of arthritis, it is more difficult to pinpoint accurately the offending food. This paper deals with the way in which food antigens are handled by the gut in the normal healthy individual, how this differs in the food-allergic individual and how knowledge of these normal and abnormal physiological mechanisms is leading to improved methods of diagnosis and treatment.

The mucosal lining of the gut serves to provide a route for the active absorption of digested nutrients, but, in addition, it functions as a barrier between the gastro-intestinal contents and the body tissues. This barrier is incomplete for there is now plenty of evidence demonstrating that absorption of antigenically intact food proteins occurs into the circulation in all healthy children and adults.[1] The amounts absorbed are very small in comparison to that ingested, but they are sufficient to immunise since antibodies to common food protein may be found not only in secretions but also in the circulation.[2] Despite the presence of these circulating antibodies, no adverse reactions occur in the majority of individuals in response to antigen entry: protective homeostatic mechanisms prevail and hypersensitivity reactions to food substances are the exception.

# FACTORS WHICH AFFECT FOOD ANTIGEN ENTRY

## Maturity

The permeability of the gut to macromolecules differs in different species of animals, particularly during the neonatal period. This is because there are major species differences in the way in which the young animal obtains its passive immunity from its mother. Hence, in ruminants, such as cattle, there is free absorption of all macromolecular proteins for the first few days of life after which the mucosal cells come together to form tight junctions. This process, known as gut closure, is enhanced by certain factors in colostrum.[3] In rodents there is active, selective transport of immunoglobulins across the mucosal cell for the first three weeks of life but gut closure occurs prior to birth in man as the premature neonate, particularly those between 28 and 30 weeks gestation, absorb considerably more cows' milk protein antigenically intact into the circulation than full-term babies and older children. We could show no evidence that any factor in human breast milk enhanced gut closure in these premature babies.

## The Secretory IgA System

Once gut closure has occurred, the amount of antigen entering is substantially reduced since a local immune response is elicited which results in secretory IgA antibodies lining the mucosal surface, much like a coat of paint. Secretory IgA is resistant to proteolytic digestion: antigen, when trapped at the mucosal surface as an antigen/secretory IgA complex, is rendered more susceptible to degradation by pancreatic enzymes. Furthermore, such complexes may cause goblet cells to secrete mucus which additionally protects against antigen entry.[4] Peristalsis is also enhanced by the formation of these complexes. In spite of the major protection afforded by efficient intraluminal digestive processes and the mucosal barrier of secretory IgA, some macromolecular absorption still occurs in all healthy subjects.

## Antigen Transport

The transport of antigens across the enterocyte is an active one since it can be blocked by metabolic inhibitors. The protein is endocytosed at the luminal surface of the cell to form a phagosome which coalesces with lysosomes to produce a phagolysosome. Some digestion occurs within these, but the remainder of the macromolecular protein is extruded by a process of exocytosis into the lateral spaces between the mucosal cells.[5] From there absorption may occur either into the portal or lymphatic circulation. Macromolecular absorption of food and other protein is probably a vital physiological requirement for maintaining mucosal immunity at sites other than the gastrointestinal tract. Lymphocytes, primed by antigen entering *via* the gastrointestinal mucosa, home into gut-associated lymphoid tissues such as the salivary and mammary glands where the cells are capable of producing specific antibody to such enterically derived antigens. Thus, pregnant women immunised orally during pregnancy produce specific secretory IgA antibody in their colostrum soon after birth of the infant.[6] This passive acquisition of immunity to enteric antigens, in particular to bacteria, is a further reason why breast feeding is so superior to bottle feeding of infants.

**Oral Tolerance**
There is considerable animal experimental evidence showing that when the antigen is administered orally, the animal fails to mount a systemic immune response when the same antigen is subsequently given in a parenteral challenging dose.[7] This state of systemic hyporesponsiveness to an antigen initially administered orally appears to be the physiological way of preventing harmful reactions to food proteins occurring. Healthy people have serum antibodies to food proteins but only at low titres and they are not the variety which elicit hypersensitivity reactions. The mechanisms for this oral tolerance are unclear, but there is evidence that it can be transferred both by lymphocytes and by serum factors under different experimental conditions. The route of antigen entry, as well as the unique micro-environment of the gut-associated lymphoid system, are probably vital for the production of IgA antibodies. This class of antibody does not activate complement, or indeed elicit damaging reactions and therefore is probably responsible for the safe exclusion at the mucosal surface, as well as elimination from the circulation, of antigens derived from the gut. In man the physiological response to a food challenge is for IgA immune complexes to be formed in the circulation system. It is likely that these are cleared by the liver, and indeed in other species of animals such a clearance mechanism has been described by which complexed or dimeric IgA combines with secretory piece on the basement membrane of the hepatocyte.[8] The antigen is processed by the hepatocyte and the secretory IgA molecule thus released is transported into the bile where it can act again as a mucosal protective layer.

## FOOD ANTIGEN HANDLING BY ALLERGIC INDIVIDUALS

Much of atopic eczema in childhood is due to food allergy and milk and/or eggs are frequently the food substances which provoke major exacerbations. When children and adults with atopic eczema were given a cocktail of milk and eggs to drink, in contrast to the IgA immune complexes found in healthy subjects after similar ingestion, IgG, IgE and Clq binding immune complexes containing food antigens were demonstrated in the circulation.[9] In most of the patients the appearance of these complexes correlated with the development of symptoms such as skin itching and wheezing. Food allergy may be considered to be a breakdown of oral food tolerance and results in excessive entry of macromolecular food proteins into the circulation. The mucosal lining of the gut becomes more permeable to food antigens when triggering of IgE-sensitised mast cells occurs. These degranulate to release vasoactive amines which increase local vascular permeability. Evidence for this mechanism in atopic eczema is provided by the increased antigen entry after food ingestion and the development of the IgE/IgG/Clq binding immune complexes circulating at the time of clinical symptoms. The evidence thus far seems to suggest that atopic eczema may be mediated by a combination of the Type 1 (IgE) hypersensitivity reaction and the Type 3 (immune complex) reaction. Further evidence that this mechanism operates in such conditions is given by the fact that treatment with oral sodium chromoglycate, a drug which stabilises mast cell membranes at the mucosal surface and prevents degranulation, abolishes the symptoms following food challenge. Antigen entry after treatment with this drug is substantially reduced and the damaging variety of immune complexes is thereby prevented.

## SYMPTOMS ASCRIBED TO FOOD ALLERGY

A wide variety of symptoms and diseases have been ascribed to food allergy. Some have a secure basis but for many of the others the evidence is circumstantial at best and much more research needs to be done. Anaphylactoid collapse and sudden death after eating particular foods are well-described phenomena. Less dramatic is the vomiting, abdominal pain and distention which occurs within one or two hours of eating an offending food. Diarrhoea in infancy is also a common symptom and in many of these children the gastrointestinal mucosa may show loss of normal villous pattern. The skin responds to food allergy either as urticaria, angioneurotic oedema, or eczema. There is evidence that asthma may be induced by food proteins and certainly rhinitis may be alleviated by avoiding offending foodstuffs. In children, food allergy can result in failure to thrive to a major degree and, secondary to this, the child can become very anaemic due to blood loss in the faeces and oedematous due to protein loss through similar mechanisms. The association of food allergy and migraine has recently been demonstrated and there are also anecdotal reports of sensitivity to food substances causing psychiatric disturbances with the mood and behavioural changes being abolished by avoidance of the offending food. Similarly there are reports of patients with rheumatoid arthritis in whom joint symptoms are substantially improved by a diet free of certain foods. The relevance of food antigen absorption in gastrointestinal diseases such as coeliac disease, inflammatory bowel diseases and even children with malnutrition (all conditions associated with a higher incidence of precipitating anti-food antibodies) is, as yet, far from clear. Techniques are available for measuring food antigen absorption in humans and in time it may become possible to establish whether food allergy is a major part of these and other diseases.

## FACTORS PREDISPOSING TO ATOPY

There are strong genetic factors in atopy, since at least half the offspring of an atopic parent are themselves atopic. There is a strong association of allergy with immunodeficiency; for example, deficiency of IgA, the principal secretory immunoglobulin protecting at mucosal surfaces, is associated with an increased frequency of precipitins to food antigens as well as an association with other autoimmune diseases.[10] Transient IgA deficiency in the newborn period has also been shown to be associated with the later development of atopic eczema.[11] The fact that these children at six months had normal levels of IgA in their sera strengthens the view that neonatal antigen experience in an immunodeficient host is important for allergic sensitisation. Other minor immunodeficiencies associated with allergy are the increased frequency of defective yeast opsonisation (a defect of the alternative pathway of complement) and low levels of C2 (second component of complement). Alternatively, it is possible that the allergic state is based on an imbalance of T helper and T suppressor cells since the production of IgE antibody is T-cell dependent.

It is unlikely that the development of food allergies is due to excessive stimulation of the immune system by large amounts of antigen crossing the gastrointestinal mucosa. The very large amounts absorbed by premature infants are against this. It is more likely that timing of antigenic exposure, and whether the antigen is presented together with an adjuvant, are critical. Thus in rats, IgE responses may be elicited by a small dose of antigen when orally administered together with a powerful adjuvant.[12] In the bottle-fed neonate, bacterial colonisation of the gut with *E. coli* could provide the necessary adjuvant, but sensitisation would occur only in the immuno-deficient child.

There are weak associations of various syndromes of allergy with particular tissue types. The haplo-type HLA A1 and B8 are more frequent in individuals with eczema, whereas haplo-type A3 and B7 are more frequent in those with hay-fever. The inheritance of allergy, whether to foods or other antigens, must be polygenic and the fact that prevention of sensitisation during the vulnerable neonatal period can prevent the development of atopic eczema suggests that environmental factors are at least as important. Allergy is much less common in developing than in so-called developed countries. There is a lower incidence of asthma in children of African origin, born outside the United Kingdom, but offspring of parents of African origin who were born in the United Kingdom have at least as frequent an incidence as indigenous non-African children. This emphasises the environmental aspect of the development of allergy. Similarly, in Europe asthma occurs more frequently in children born in late autumn, a finding which could be explained by the high activity of *Dermatophagoides pteronyssinus* at the time when the infant is relatively immunodeficient, i.e. the period when his IgG level is in the physiological trough at three months of age.


## INFANT FEEDING


Ingestion of both food and bacterial antigens that gain entry to the circulation is important for turning on the neonates immune response. Local and systemic IgA formation occurs as well as partial tolerance. These responses may occur simultaneously to the same antigen stimulus, but an imbalance of these complex responses may well lead to allergy to food. This has led to the hypothesis that for the infant, foods regarded as especially allogenic may play a part in the general development of allergy. Eczema occurs more frequently in artificially fed babies and in a prospective study of exclusive breast feeding versus bottle feeding of infants of atopic parents it has been shown that the breast fed group had a significantly lower incidence of eczema at one year of age than the artificially fed group. Babies who received some supplementary feeding had as high an incidence of eczema as did bottle fed babies and this is probably due to the fact that when babies are supplementarily fed, the intestinal flora is identical to that of a bottle fed baby. Breast feeding induces a flora rich in lactic acid-producing bacteria, the resulting pH is acid and is not favourable for the growth of bacteria such as *E.coli*. Furthermore, human milk has the additional protection of lactoferrin which competes for the iron necessary for growth and proliferation of *E.coli*.

There is increased antigen uptake from the gut in infants recovering from gastroenteritis and it is not surprising that transient food allergy occurs frequently in these babies. This can be prevented by placing such infants on a hypoallogenic artificial feed. The development of allergy to subsequent food substances can be explained in this way. In an animal model of allergy we have shown that rats made allergic to egg albumin facilitate the entry of an unrelated protein to which the animal had not been previously exposed when the two proteins are administered simultaneously.[13] This is because the protein to which the animal is allergic causes a local anaphylactic reaction in the gut thus increasing vascular permeability and allowing greater amounts of the unrelated protein into the circulation. The animal may subsequently become sensitised to this unrelated protein in addition. In clinical practice it is extremely common for an infant allergic to one preparation of milk to subsequently become allergic to another unrelated preparation, for instance, soy bean milk.


## DIAGNOSIS OF FOOD ALLERGY


The ease or difficulty of diagnosis of food allergy depends much upon the time relationship of symptoms to ingestion of the food. Anaphylactic-type reactions which occur soon after ingestion of food are relatively easy to diagnose and by taking a careful history from the patient one can usually find the particular category of foodstuff responsible. Delayed-onset food sensitivities, i.e. those which occur hours or days after food ingestion, are much more difficult to diagnose. Goldman *et al.*,[14] laid down strict criteria for the diagnosis of milk allergy in childhood, in that he suggested that the symptoms must be shown to remit and relapse three times on withdrawal and reintroduction of food. The author feels that these criteria are unnecessarily strict, particularly as food challenge on occasion may be dangerous and in any case should only be carried out under strict medical supervision. Apart from the patient who has several very clear-cut episodes (for instance, urticaria due to tomatoes), some challenge test is necessary. It is desirable that this should be single or double blind whenever possible and the assessment of the resulting symptoms may take two days or more. Intestinal biopsy is seldom of any value as even in severely affected children this may be normal. Skin prick tests give an indication of whether the individual is atopic, but seldom give more than a pointer to which type of food substance is causing the symptoms. Similarly, IgE levels will only indicate the atopic state and specific IgE antibodies determined by the radio-allergo absorbent test (RAST) give no greater information than skin prick tests. As part of the diagnosis it may be necessary to place the patient on a hypoallergenic diet and then gradually reintroduce foods one at a time, never more than one food every three days.

## TREATMENT OF FOOD ALLERGY

The most successful means of treating food allergy is avoidance of the offending food. In all cases a nutritionally adequate diet must be ensured. Parenteral and oral hyposensitisation to foods is not effective and can be dangerous.

There is no doubt that for some individuals, where it is difficult to pinpoint accurately which food is causing the symptoms, the taking of oral sodium chromoglycate 15–30 minutes before a meal can substantially relieve symptoms. We and others have shown that such treatment substantially reduces antigen entry and also prevents the formation of those particular immune complexes which are associated with symptoms of food allergy. Oral sodium chromoglycate, however, does not work in every patient, but it is worth trying. Antihistamines have been used but with less success. Steroid therapy may be helpful in controlling protein losing enteropathy, but is not otherwise recommended. Where allergy avoidance diets are used, particularly in early childhood, it is recommended that the child should remain on the diet for at least two years before reintroduction of the food under medical supervision. Some patients with food allergy, especially those occurring following infective gastroenteritis may later be able to take the food to which they were formerly allergic. In any case most children with severe food allergies are later able to take small amounts of these foods in adult life.

## PREVENTION OF FOOD ALLERGY

Exclusive breast feeding is recommended for a minimum period of three months, particularly for the at-risk child (i.e. the child of atopic parents). Cows' milk-based supplements should be avoided if at all possible. Failing that, soy formulae may be preferable to modified cows' milk-based formulae, but in certain countries allergy to soy milk proteins is also a problem. When the child is weaned, hypoallergenic foods should be introduced first and substances such as cows' milk or whole egg should be avoided for the first year of life. These strict measures need only apply to infants of families with a strong history of similar allergies. It must be remembered also that sensitisation can occur *via* breast milk and there are now increasing reports of infants exclusively breast fed who are allergic to substances such as egg or milk proteins. In such instances it may be necessary to place the mother on a diet free of these foods while breast feeding. We have studied at least four such infants; the latest one developed a profuse colitis while on exclusive breast feeding and this colitis responded completely when cows' milk antigen ($\beta$-lactoglobulin) could be demonstrated both in the mother's serum and also in the breast milk. Fortunately such cases are very rare.

## NON-IMMUNE REACTIONS TO FOOD

Foods are capable of causing a spectrum of adverse reactions through non-immune pathways because of potent pharmacological constituents. Ingestion of sufficient quantities of pharmacologically active foodstuffs will result in symptoms in virtually all individuals, although the threshold of susceptibility may vary considerably. Many foods, such as bananas, tomatoes, cheeses and certain wines contain vasoactive amines, such as tyramine or 5-hydroxytryptamine. Chocolate, which frequently has been reported to precipitate migraine headaches, does not contain tyramine but does contain large amounts of phenylethylamine. Both agents are potent inducers of headaches. Food additives such as tartrazine or preservatives such as sodium benzoate have been implicated in causing asthma and urticarial syndromes. Hyperkinetic behaviour in children has also been attributed to these substances, but there is no controlled information to support this. Certain foods may be harmful because they contain substances that children with specific enzyme deficiencies cannot tolerate. Examples include coeliac disease which responds to a gluten-free diet; in-born errors of metabolism such as phenylketonuria, where a diet low in phenylalanine is necessary; deficiency of gastrointestinal enzymes, lactase, maltase or sucrase, which interfere with sugar digestion. In all these cases ingestion of the particular foods can cause severe symptoms which are not due to allergy.

## CONCLUSION

There is no doubt that allergy to food proteins plays a major role in several diseases and in addition possibly exacerbates many other clinical conditions. Food allergy is, however, difficult to diagnose and it is not surprising that fad diets are advocated as a treatment for many conditions by 'crank' organisations. It is only by understanding the physiology of food antigen absorption in the normal individual that we may, in time, understand what goes wrong in this major breakdown in oral food tolerance. The increasing awareness of food allergy may be a reflection of this increased research effort. The alternative view that food processing, by altering the antigenic nature of food proteins, is creating food allergies, must also be explored; the impression exists that food allergy is a much more common problem in Western industrial societies than in rural communities where unprocessed food is eaten.

## REFERENCES


1. WALKER, W. A. and ISSELBACHER, K. J., *Gastroenterology*, 1974, **67**, 531.
2. PETERSON, R. D. A. and GOOD, R. A., *Paediatrics*, 1963, **31**, 209.
3. WALKER, W. A., *Ciba Foundation Symposium*, 1979, **70**, 201.
4. WALKER, W. A., WU, M. and BLOCH, K. J., *Science*, 1977, **197**, 370.
5. WALKER, W. A., *Immunology Today*, 1981, **2**(2), 30.
6. GOLDBLUM, R. M., AHLSTEDT, S., CARLSSON, B., HANSON, L. A., JODAL, U., LIDIN-JANSON, G. and SOHL-AKERLUND, A., *Nature*, 1975, **257**, 797.
7. SWARBRICK, E. T., STOKES, C. R. and SOOTHILL, J. F., *Gut.*, 1979, **20**, 121.
8. HALL, J. G. and ANDREW, E., *Immunology Today*, 1980, **1**(5), 100.
9. BROSTOFF, J., CARINI, C., WRAITH, D. G., PAGANELLI, R. and LEVINSKY, R. J., *The Mast Cell*, J. Pepys and A. M. Edwards (eds.), Pitman Medical Publishing Co. Ltd, Tunbridge Wells, England, 1979.
10. BUCKLEY, R. H. and DEES, S., *New England J. Med.*, 1969, **281**, 465.
11. TAYLOR, B., NORMAN, A. P., ORGE, H. A., STOKES, C. R., TURNER, M. V. and SOOTHILL, J. F., *Lancet*, 1973, **2**, 111.

12. JARRETT, E. E. E., HAIG, D. M., McDOUGALL, W. and McNUTTY, E., *Immunology*, 1976, **30**, 671.
13. ROBERTS, S. A., REINHARDT, M. C., PAGANELLI, R. and LEVINSKY, R. J., *Clin. Exp. Immunol.*, 1981, **45**, 131.
14. GOLDMAN, A. S., ANDERSON, D. W., SELLARS, W. A., SAPERSTEIN, S., KNIKER, W. T. and HALPERN, S. R., *Pediatrics*, 1963, **32**, 425.

# 7

# Influence of Food Processing on Nutritive Value of Proteins

P. V. J. HEGARTY

*Department of Food Science and Nutrition,*
*University of Minnesota, St. Paul, USA*

Up to half of the total food intake in Western societies has been processed, and most of the protein-rich foods are processed in one form or another. The trends in consumption of food proteins in the US between 1965 and 1977[1] indicate a greater utilisation of processed and commercially prepared foods. At a workshop on nutrition in Karlsruhe in May 1981, H. R. Miller predicted no revolution in the patterns of food consumption up to the year 2000, although there are challenging situations appealing to food processors who are able to respond to a number of product opportunities (e.g. fast foods and 'ready to eat, ready to heat foods'). Many excellent reviews exist on the effect of various processes on the nutritive value of food proteins.[1-6] Therefore, this paper is an overview of the implications for humans of the consumption of processed proteins rather than an extensive review of the literature on the chemical changes that take place in food during processing.

The human nutritionist is more concerned with the biological significance to humans of any changes to protein quality rather than with the statistical significance of changes in protein quality induced by the processing of individual foods. Therefore it is appropriate to examine the four principles outlined by Bender[2] that should form the background to any discussion of the effects of processing treatment on the nutritive value of protein foods:

(1)  The significance of the individual food in the total diet. Individuals on restricted diets, the young being fed infant formulae (especially those recovering from protein–calorie malnutrition), the elderly, and low income groups, all may depend on a few foods. For a large

145

 *P. V. J. Hegarty*

proportion of the population in Western countries the intake of protein is up to twice the Recommended Dietary Allowance. Therefore, the effect of processing is not a significant variable in the protein quality of a mixed diet.

(2)  Inadequate methods of assessing the nutritive value of proteins can lead to false conclusions. For example, biological measurements of the nutritive value of proteins will be changed only when there is a change in the limiting amino acid.

(3)  Changes in nutritive value may be of little consequence in regions where protein consumption is more than adequate, but may be very important when the dietary intake is low.

(4)  If most diets tend to be limited by the sulphur amino acids, then only damage to these amino acids is of real consequence. Since the application of heat is one of the most common operations in the processing of food, it is worth noting that the maximum cooking loss for methionine is only 10 %.[7] In general, proteins are not damaged nutritionally unless they are severely heated or stored for long periods. Some observations made on heat-damaged proteins have been made following the use of unrealistic temperatures and/ or duration of heating.

This discussion will deal with the effect of processing on conventional foods, although the author does want to acknowledge the existence of unconventional sources of food proteins which include single cell proteins, oil seed protein residues, fish protein concentrates, leaf protein, etc. The role of these commodities as sources of food proteins has been reviewed by Worgan[8] and in other chapters in this book.


## EFFECT OF MILD HEAT


The application of heat to a food may have beneficial effects by increasing palatability and digestibility. Desirable colours, aromas and flavours may be produced by the Maillard reaction. Much of the colour and flavour is developed in baking, frying, roasting and toasting, and foods such as breakfast cereals, meat extracts and biscuits derive much of their flavour from this reaction.[9] Although there is some loss of amino acids when these flavours are produced, the loss is not usually important and is accepted as a price worth paying.[10]

Heat improves digestibility by destroying trypsin inhibitors and various toxins; trypsin inhibitor is destroyed when rice, wheat or oats are cooked[11] and the protein quality of the soy bean is improved by heat application.[12] Nowadays, several textured meat analogues contain less than 10 % of the anti-trypsin activity of raw soy bean flour.[13] Of particular significance is the reduction of trypsin inhibitor activity to less than 10 % of the original soy isolate after the processing and sterilisation of infant soy bean formulae.[14] We are concerned about the anti-trypsin activity in these human foods because pathological consequences of trypsin inhibitors, such as reduced growth rate and pancreatic hypertrophy have been demonstrated in rats. However, Liener[13] presents evidence that soy bean inhibitor is relatively ineffective against human trypsin, and therefore of little significance in humans.

Mild heat is generally harmless with regard to protein quality.[10] However, two observations from our laboratory indicate that caution must be observed with that statement. On the one hand severe heat treatment (autoclaving at 121 °C for either 30 or 90 minutes) of casein or lactalbumin produced no change in the protein efficiency ratio (PER) or the amount of available lysine in casein but reduced the PER of lactalbumin from 3·46 to 1·17 and the available lysine from 83 to 66 mg/g protein.[15] On the other hand in an experiment in which a model system consisting of 15 % soy protein isolate, 4 % glucose and microcrystalline cellulose was processed at 80 °C for a long time,[16] the amount of available lysine and the PER decreased significantly with the duration of heating up to 600 minutes. At 700 minutes there was a reversal of this trend for both available lysine and for PER. Therefore, the type of protein being heated and the duration of heating are significant variables in the evaluation of protein quality.

Mauron[4] cites four types of processing damage to proteins. The first involves mild heat treatment of foods rich in reducing sugars, such as milk-containing products; the damage is almost exclusive to lysine. This mild heat treatment creates a small change (browning) due to the Maillard reaction. The colour change usually parallels the development of a caramelised flavour. In many cases this flavour is appreciated by the consumer.[17]


## EFFECT OF SEVERE HEAT


A second type of damage occurs under severe heating conditions, either in the presence or absence of sugars, and leads to a decrease in protein digestibility and in the availability of most amino acids in addition to that of lysine. In this regard it is appropriate to examine the significance of the

Maillard reaction (non-enzymatic browning) on proteins consumed by humans. The reader is referred to the recent extensive review by Dworschak[18] for details. The Maillard reaction occurs when reducing sugars are present together with free amino groups in a food. The reaction develops slowly at room temperature, e.g. during food storage, but is highly activated by a rise in temperature. Adrian[19] identifies three nutritional and physiological repercussions of the Maillard reaction. First, amino acids, especially lysine, are destroyed. The most-sensitive proteins are albumins, globulins and globins; low molecular weight sugars (e.g. pentoses) are the most active sugars. Blocking, and subsequent changes, of lysine makes this amino acid metabolically unavailable. The amino acid balance is therefore disturbed and the protein value of the food is reduced. Drastic heat treatment, or prolonged storage, can cause lysine losses of a magnitude that makes lysine the primary limiting factor in foods where methionine is normally the limiting amino acid (milk or peanut). This can have important nutritional consequences in theory because of the high requirement for lysine during mammalian growth.[20] In practice, such intense heat treatments cause a pronounced discoloration of the product and lead to the development of a burnt flavour, thus rendering the foods unacceptable for human consumption.[17]

The second repercussion listed by Adrian[19] is the formation of soluble substances (pre-melanoidins) with antinutritional and even toxic effects on animals. PER can fall by 20–40%. However, the responsible substances correspond to the early intermediates of the Maillard reaction; their quantity and activity diminish as heating proceeds and the insoluble melanoidin stage approaches. Insoluble melanoidins are physiologically inactive. The third repercussion[9] of the Maillard reaction is flavour formation which has been discussed above.

Most of the literature associates the loss of lysine with the Maillard reaction. One recent example is a report from Sweden which examined the protein quality of milk–cereal-based foods for infants and children in relation to processing methods and composition of the products.[21] In that country, ready-to-serve milk–cereal-based gruel powders are given at 1–2 meals per day to almost all infants after the age of 4–6 months and to young children. Cereal-based baby foods often contain sucrose, which increases both the palatability and the energy density of these products. Since sucrose is a highly cariogenic substance, replacement of sucrose by the less cariogenic glucose or fructose was considered. However, the risk of decreasing available lysine through non-enzymatic browning was increased by this replacement. It was concluded that to obtain the optimal protein quality of pre-cooked milk–cereal-based baby foods it is preferred that an unsweetened product be used. If a sweetened product is required, then any reducing sugar, e.g. glucose or fructose must be added after the heat treatment, otherwise sucrose should be the sweetening agent. Lactose may be suitable in some types of dried mixes. The product should not be dried to a lower moisture content than is necessary for safe storage. Some of the lost lysine is replaced if the product is consumed with milk, which is high in lysine. The importance of the adverse consequences of protein damage in a heat treated milk–cereal–legume product was shown both chemically and clinically during a programme to combat malnutrition in Africa.[22,23] On the other hand, this statement should be counteracted by the observation in clinical studies on small infants, in which severely overheated milk powder ($\sim 50\%$ lysine deterioration) was fed for three weeks alternately with first-quality spray-dried milk powder, that no deleterious effect on growth and general well-being was observed as long as the minimal lysine requirements were met.[4]

Losses in palatability have been reported in cases where there was no loss in nutritive value.[24] Bender[2] regards this as being unimportant. Furthermore, he concludes that the losses produced by the Maillard reaction are not important to the human who consumes a mixed diet consisting of 80–90 g of protein per day. Besides the loss of lysine after high heat treatment, there are losses in the availability of methionine, cystine, tryptophan and arginine in foods.[25] The extent of the losses caused by non-enzymatic browning depends on process times and temperatures, moisture content, presence of reducing substances (for example, sugars, secondary lipid oxidation products) and pH.[25,26]

Lipids, especially those with high concentrations of unsaturated fatty acids, are prone to oxidise in the presence of air even at low temperatures. The nature of this reaction is not fully understood but its rate is increased by humidity, light and some trace elements. The steps in the pathway of lipid oxidation are outlined by Dworschak.[18] Oxidised lipids may produce browning reactions with proteins. The reactions of oxidised lipids with amino acids can be divided into two types: (a) the oxidising effect of the hydroperoxides on single amino acids (this reaction produces no organoleptic changes) and (b) the reactions of aldehydes with amino acids, which are very similar to the Maillard reaction. For example, the amount of available lysine decreases considerably as rancidity advanced in herring-meal stored at room temperature.[27] Though oxidised fats damage cystine, methionine and tryptophan in particular, the nutritional consequences are less severe than those of the Maillard reaction.[18]

Finally, on the topic of non-enzymatic browning, McWeeny[28] illustrates the difficult compromises that must be made if both microbial growth and non-enzymatic browning are storage problems in the same commodity. Conditions favourable to the control of microbial growth tend to enhance browning, and vice versa.

In summary, the extent of thermal damage to proteins is a function of time, temperature, moisture content, reducing substances and pH. One or more of these variables can cause the destruction of amino acids, formation of amino acid complexes, inter- and intramolecular cross-linking of proteins, possible formation of toxic compounds, and unavailability of some amino acids. However, the maximum cooking losses for lysine and methionine are 40% and 10% respectively, and the losses for the other essential amino acids are less than 20%.[7]

## EFFECT OF ALKALI

The third type of processing damage, as outlined by Mauron,[4] is the loss of lysine and cystine, in particular when protein is exposed to severe alkali treatment. The production of protein concentrates and isolates, spun fibres, and proteins with specific functional properties such as foaming ability, requires exposure to alkali. This type of treatment causes chemical modification of both essential and non-essential amino acids.[29] Wood-Rethwill and Warthesen[30] have reviewed the literature and indicate that alkali treatment adversely affects the nutritional quality of a food due to amino acid destruction, racemisation and crosslink formation. Furthermore, the formation of crosslinked amino acids, such as lysino-alanine (LAL) has raised some concern about the potential toxic effects of alkaline processing of food. Racemisation of amino acids and con-current formation of lysinoalanine (LAL) in food proteins can impair the nutritional quality and safety of foods by: (a) decreasing the amount of essential L-amino acids; (b) decreasing digestibility and bioavailability of proteins; and (c) forming toxic products.[31] Temperature and pH, and, to a much lesser extent, length of treatment, affect LAL formation and racemisation. Friedman *et al.*[31] have argued that even if protein-bound D-amino acids could be liberated in the digestive tract, they might not be transported into the blood stream and utilised for growth.[32] The L-amino acids are invariably taken up faster than the D-enantiomers in the intestine[33] and in the kidney.[34] Recent evidence has shown that D-methionine is poorly utilised by humans.[35,36] As an indication of the amount of racemised amino acids in commercially available foods, the level of D-aspartic acid ranged from 9 % (relative to the L-form) for textured soy protein to 17 % for a non-dairy creamer.[31] The long term consequences of the consumption by humans and animals of processed proteins containing D-amino acids is not known. Systematic surveys of commercial foods for the content of D-amino acids are required.

LAL produces cytomegalic lesions in the renal tubular cells of rats;[37] however, this problem may be specific to the rat species.[38] Other workers have confirmed that the effects of LAL in the diet depend on the animal species studied.[39] Perhaps the most important observation on the significance of LAL in the diet was made by Sternberg *et al.*[40] who showed that the occurrence of LAL was widespread in a variety of food ingredients, home-cooked foods and commercial food preparations. They concluded, therefore, that alkali treatment was not the sole factor in producing LAL. For example, milk proteins tend to develop LAL independently of the type of treatment (alkali or heat). The high values in infant formulae may be of nutritional significance. The state of the art at the moment is best summarised by Dworschak:[18] 'The possible harmfulness of LAL on humans is an open question yet'.


## USE OF MODEL SYSTEMS


A brief comment is appropriate on the use of model systems in predicting free amino acid loss as a function of process time and temperature. Wolf *et al.*[41] conclude that protein, salt, reducing sugars amd water activity must be considered in any model system used to study the influence of processing operations on the nutritional quality of food components. These authors indicate that the value of inclusion of an oil is questionable. Most likely, the oxidation of the oil prior to processing will determine its importance.

The advantage of such model systems is in reducing the substantial amount of work involved in testing empirically each food product of varying compositions under varying processing conditions. Mathematical models can be used to predict the destruction of nutrients, and to indicate how the composition of the food and the processing conditions can be altered to minimise nutrient loss. Model systems do have a role in determining the effect of processing on protein quality. An example of a good correlation between changes in available lysine with changes in PER has been presented by Wolf *et al.*[16]

                    *P. V. J. Hegarty*

Recent papers dealing with the effects of various processes on protein quality indicate that no change is induced by home-canning;[42] gamma irradiation (200 krad) and heat treatment (60 °C, 10 minutes) for the preparation of a fish protein concentrate[43] or defatted peanut flour.[44] Microwave cooking has a similar effect to home meal preparation on the nutritive value of peas. However, significant losses of some amino acids, but little loss of lysine and methionine, were found.[45] Fermentation on the other hand increases the % Relative Nutritive Value for wheat, barley, rice, maize and millet.[46] The improvement of protein quality by fermentation is of interest because it is inexpensive, does not require sophisticated equipment and consumes little energy.


## CONCLUSIONS


This paper has not attempted a comprehensive review of the literature on the effect of various processes on the nutritive value of food proteins. However, the conclusions to be drawn after such an exercise would be similar to those drawn from the evidence presented here. Namely, certain processes such as heat or alkali treatment can damage the nutritive value of individual foods. However, if these foods are minor components in a varied diet, or if the total dietary intake of protein is high, then any changes observed may be of academic rather than of practical interest. People on restricted diets, e.g. the young being fed infant formulae, the elderly, and some people from the lower income goups, may be vulnerable because of the limited variety of foods consumed.

Finally, it is appropriate to examine future trends in food processing as they affect the nutritive value of food proteins. The production of possible toxic or non-metabolisable products during processing must be monitored. Two relevant examples are lysinoalanine production and the racemisation of amino acids. With respect to whether our food in the future will be more highly processed that at present, it is appropriate to quote an answer given by Bender[47] to the question 'what will our children eat?'. 'Almost certainly the greater part of their diet will be the same as that of their parents—fish and chips, meat and potatoes, bread and jam . . .'. It would appear that no major changes are predicted in the types of food consumed, and therefore no changes in the current known alterations induced by processing on the nutritive value of food proteins.

# REFERENCES

1. HAMA, M. Y., *Family Econ. Rev.*, Spring, 1980, 4.
2. BENDER, A. E., in *Evaluation of Novel Protein Products*, Bender, A. E., Kihlberg, R., Lofqvist, B. and Munck. L. (eds.), Pergamon Press, Oxford, 1970, 319.
3. FORD, J. E., in *Proteins in Human Nutrition*, Porter, J. W. G. and Rolls, B. A. (eds.), Academic Press, London, 1973, 515.
4. MAURON, J., in *Proc. IV Cong. Food Sci. & Technol*, Instuto de Agroquinica y Technologia de Alimentos, Valencia, Spain, Vol. 1, 1974, 564.
5. TANNENBAUM, S., in *Nutrients in Processed Foods—Proteins*, White, P. L. and Fletcher, D. C. (eds.), Publishing Sciences Group, Inc., Acton, Mass., 1974, 131.
6. VAUGHAN, D. A., in *Evaluation of Proteins for Humans*, Bodwell, C. E. (ed.), AVI, Westport, Conn., 1970, 255.
7. HARRIS, R. S., in *Nutritional Evaluation of Food Processing*, Harris, R. S. and von Loesecke, H. (eds.), AVI, Westport, Conn., 1971, 2.
8. WORGAN, J. T., in *Proteins in Human Nutrition*, Porter, J. W. G. and Rolls, B. A. (eds.), Academic Press, London, 1973, 47.
9. LINKO, Y. and JOHNSON, J. A., *J. Agric. Fd. Chem.*, 1963, **11**, 150.
10. BENDER, A. E., *J. Fd. Technol.*, 1966, **1**, 261.
11. LAPORTE, T. and TREMOLIERES, J., *C.R. Soc. Biol.*, 1962, **156**, 1261.
12. LIENER, I. E., DEUEL, J. J. JR. and FEVOLD, H. L., *J. Nutr.*, 1949, **39**, 325.
13. LIENER, I. E., *Proc. Nutr. Soc.*, 1979, **38**, 109.
14. CHURELLA, H. R., YAO, B. C. and Thompson, W. A. B., *J. Agric. Fd. Chem.*, 1976, **24**, 393.
15. KEYES, S. C. and HEGARTY, P. V. J., *J. Agric. Fd. Chem.*, 1979, **27**, 1405.
16. WOLF, J. C., THOMPSON, D. R., AHN, P. C. and HEGARTY, P. V. J., *J. Fd. Sci.*, 1979, **44**, 294.
17. GROUX, M., *J. Dairy Sci.*, 1974, **57**, 153.
18. DWORSCHAK, E., *C.R.C. Critical Revs. Food Sci. Nutr.*, 1980, **13**, 1.
19. ADRIAN, J., *World Rev. Nutr. Dietet.*, 1974, **19**, 72.
20. HEGSTED, D. M., in *Proteins in Human Nutrition*, Porter, J. W. G. and Rolls, B. A. (eds.), Academic Press, London, 1973, 275.
21. ABRAHAMSSON, L., BENGTSSON, O., HAMBRAEUS, L. and HOLM, H., *J. Fd. Technol.*, 1979, **14**, 429.
22. CLEGG, K. M., *Br. J. Nutr.*, 1960, **14**, 325.
23. CLEGG, K. M. and DEAN, R. F. A., *Am. J. Clin. Nutr.*, 1960, **8**, 885.
24. REGIER, L. W. and TAPPEL, A. L., *Fd. Res.*, 1956, **21**, 630.
25. CARPENTER, K. J., MORGAN, C. B., LEA, C. H. and PARR, L. J., *Br. J. Nutr.*, 1962, **16**, 451.
26. LABUZA, T. P., *Fd. Technol.*, 1973, **27**(1), 20.
27. EL-LAKANY, S. and MARCH, B. E., *J. Sci. Fd. Agric.*, 1974, **25**, 889.
28. McWEENY, D. J., *J. Fd. Technol.*, 1980, **15**, 195.
29. WHITAKER, J. R. and FEENEY, R. E., *Adv. Exp. Med. Biol.*, 1977, **86B**, 155.
30. WOOD-RETHWILL, J. C. and WARTHESEN, J. J., *J. Fd. Sci.*, 1980, **45**, 1637.

31. FRIEDMAN, M., ZAHNLEY, J. C. and MASTERS, P. M., *J. Fd. Sci.*, 1981, **46**, 127.
32. STEGINK, L. D., in *Clinical Nutrition Update: Amino Acids*, American Medical Association, Chicago, 1977, 198.
33. FINCH, L. R. and HIRD, F. J. R., *Biochim. Biophys. Acta*, 1960, **43**, 278.
34. ROSENHAGEN, M. and SEGAL, S., *Am. J. Physiol.*, 1974, **227**, 843.
35. ZEZULKA, A. Y. and CALLOWAY, D. H., *J. Nutr.*, 1976, **106**, 1286.
36. PRINTEN, K. J., BRUMMEL, M. C., CHO, E. S. and STEGINK, L. D., *Am. J. Clin. Nutr.*, 1979, **32**, 1200.
37. WOODWARD, J. C. and SHORT, D. D., *Fd. Cosmet. Toxicol.*, 1977, **15**, 117.
38. DE GROOT, A. P., SLUMP, P., FERON, V. J. and VAN BEEK, L., *J. Nutr.*, 1976, **106**, 1527.
39. FINOT, P. A., BUJARD, E. and ARNAUD, M., in *Protein Crosslinking—B. Nutritional and Medical Conferences*, Friedman, M. (ed.), Plenum Publishing Corporation, New York, 51.
40. STERNBERG, M., KIM, C. Y. and SCHWENDE, F. J., *Science*, 1975, **190**, 992.
41. WOLF, J. C., THOMPSON, D. R., WARTHESEN, J. J. and REINECCIUS, C. H., *J. Fd. Sci.*, 1981, **46**, 1074.
42. KEYES, S. C., WOLF, I. D., ZOTTOLA, E. A. and HEGARTY, P. V. J., *J. Fd. Sci.*, 1978, **43**, 1352.
43. WARRIER, S. B. and NINJOOR, V., *J. Fd. Sci.*, 1981, **46**, 234.
44. ALID, G., YANEZ, E., AGUILERA, J. M., MONCKEBERG, F. and CHICHESTER, C. O., *J. Fd. Sci.*, 1981, **46**, 948.
45. CHUNG, S. Y., MORR, C. V. and JEN, J. J., *J. Fd. Sci.*, 1981, **46**, 272.
46. HAMAD, A. M. and FIELDS, M. L., *J. Fd. Sci.*, 1979, **44**, 456.
47. BENDER, A. E., *Chemy. Ind.*, 1980, No. 20 (October), 797.

# 8

# Milk Proteins—Chemistry and Physics

D. G. Dalgleish

*The Hannah Research Institute, Ayr, UK*

The proteins of milk can be divided into two broad classes, namely the caseins and the serum proteins. The caseins are defined as phosphoproteins which are precipitable at pH 4·7, while the serum proteins are soluble under these conditions. The two groups of proteins are chemically and physically distinct, and in fact in milk itself they exist almost completely in two different phases. The caseins exist as 'colloidal' aggregates, containing several thousand individual protein molecules, whereas the serum proteins are not aggregated in this way, and are in free solution. Of the two groups, the caseins are the more complex, since they have important metal-ion binding capability, as a result of which they undergo complex precipitation reactions. The serum proteins, on the other hand, are more typical globular proteins which, from the point of view of food science, exhibit their most important properties when they are denatured, rather than in their native states (for review, see ref. 92).

## THE PROTEINS OF MILK

Milk contains about 33 g of protein/litre of which $\sim 27$ g/litre (i.e. $\sim 80\%$) is casein. Both casein and serum protein are heterogeneous.

The casein fraction consists of four principal proteins: $\alpha_{s1}$-, $\alpha_{s2}$-, $\beta$- and $\kappa$-caseins. The $\gamma$-caseins, although they are often treated as a separate group, are produced from $\beta$-casein by limited, specific proteolysis by milk proteases (mainly blood fibrinolysin), forming, at the same time, part of the proteose-peptone fraction.[1,2] In a survey of milk from South-west Scotland[3] these five fractions were present to the extent of 38%, 10%, 36%,

13 % and 3 %, respectively, of the total casein content of the milk. They may be taken as representative of general milk composition, although there may be regional differences as a result of lactational patterns, etc.,[4,5] particularly in the relative proportions of $\beta$- and $\gamma$-caseins.

The serum protein group contains mainly $\beta$-lactoglobulin (54 % of the total serum protein) and $\alpha$-lactalbumin (21 %), the remainder being composed of serum albumin, immunoglobulins and proteose-peptones.

Qualitative comparison of the caseins and the serum proteins is possible, although both groups are heterogeneous. The major serum proteins are globular, and are capable of being crystallised; caseins are apparently non-globular and are believed to have more extended structures although detailed evidence is lacking. These structural distinctions arise, at least in part, from the presence of numerous disulphide bonds in serum proteins,[6,7] whereas, in contrast, the caseins possess either few or no half-cystine residues, and therefore lack the capacity to stabilise their structures by disulphide formation.[8-11] Those caseins which do possess cysteinyl residues appear to use them for the formation of intermolecular rather than intramolecular bonding: $\kappa$-casein can be extracted as a disulphide-linked polymer,[12] although whether the protein is present in milk in this form is not certain, and disulphide-linked dimers of $\alpha_{s2}$-casein are also known (the so-called $\alpha_{s5}$-casein).[13]

The most obvious and most important difference between the caseins and the serum proteins is that all the former are phosphorylated to some extent. Phosphorylation can occur on serine residues which are two residues closer to the N-terminal of the protein than an acid residue or another phosphorylated serine residue:[8] HN—Ser $x$–$y$—($y$ = Glu, Asp, Ser P).

All of the potentially phosphorylatable seryl residues are not in fact phosphorylated in all caseins. $\kappa$-Casein has only one phosphorylated residue, $\beta$-casein has five, $\alpha_{s1}$-casein has eight or nine[14] (the latter referred to as $\alpha_{s0}$-). $\alpha_{s2}$-Casein has from 10 to 13 phosphorylated serines ($\alpha_{s2}$-caseins share the same polypeptide chain, but differ in the extent to which they are phosphorylated: the $\alpha_{s2}$-casein group includes the caseins formerly known as $\alpha_{s2}$-, $\alpha_{s3}$-, $\alpha_{s4}$-, $\alpha_{s5}$- and $\alpha_{s6}$-caseins). Although threonine should also be capable of phosphorylation, phosphothreonyl residues are uncommon. The D-variant of $\alpha_{s1}$-casein has one such residue, and it is possible that some threonyl residues are phosphorylated in $\alpha_{s2}$-casein, but otherwise phosphorylation is restricted to seryl residues. The presence of phosphate in the caseins appears to be the source of their extensive capacity to bind divalent metal ions which, in turn, defines many of their important properties.

Both serum proteins and caseins contain components which are glycosylated. $\kappa$-Casein may possess short carbohydrate chains linked to a particular threonyl residue, but not all of the $\kappa$-casein molecules are glycosylated in this way.[15] There is a possibility[9] that $\alpha_{s2}$-caseins may also be glycosylated, but no definite structures have been identified. Likewise, certain minor serum proteins appear to be glycosylated.[16]


## STRUCTURES OF THE PROTEINS


The primary structures of all of the major milk proteins are known and, are shown in Figs 1–6. All the proteins have genetic variants, although the variations do not appear to have significant effects upon their general properties. The main points of interest, so far as the major serum proteins are concerned, are the positions of the disulphide linkages, and the fact that $\beta$-lactoglobulin contains one free sulphydryl group per molecule. This residue is of some importance during denaturation of the protein, since it appears to facilitate an —SH/S—S interchange reaction, allowing the formation of new tertiary structures. $\alpha$-Lactalbumin, whose —SH groups are all involved in disulphides, is somewhat more difficult to denature by heating. $\beta$-Lactoglobulin usually exists in a dimeric form while $\alpha$-lactalbumin is monomeric in solution.

The primary structures of the caseins confirm that, although the caseins share some properties, the group is heterogeneous as regards primary structure. Homologous structures are notably absent, with the exception of the important sequence:

$$\text{-SerP-SerP-SerP-Glu-Glu-}$$

which occurs once in both $\alpha_{s1}$- and $\beta$-caseins, and twice in $\alpha_{s2}$-casein. With this exception, it is clear that there are four almost entirely different polypeptide chains in the casein group. The individual amino acid compositions and sequences of the caseins can be used to provide some indications of the different properties to be expected of these proteins.

$\beta$-Casein is strikingly hydrophobic over much of its chain, having a calculated hydrophobicity[17] of 1334 cal/residue. Only the N-terminal 50 residues contain an appreciable proportion of charged and hydrophilic residues, and the hydrophobicity of the residues 30–209 is 1408 cal/residue, which represents a very high value for protein. Notable also is the number of prolyl residues in the protein—as much as 1/6 of the total number of

158                                           *D. G. Dalgleish*

$$^{10}$$
H-Arg-Pro-Lys-His-Pro-Ile-Lys-His-Gln-Gly-Leu-Pro-Gln-Glu-Val-Leu-Asn-

$$^{20} \qquad\qquad\qquad\qquad\qquad\qquad ^{30}$$
Glu-Asn-Leu-Leu-Arg-Phe-Phe-Val-Ala-Pro-Phe-Pro-Gln-Val-Phe-Gly-

$$^{40}$$
Lys-Glu-Lys-Val-Asn-Glu-Leu-Ser*-Lys-Asp-Ile-Gly-SerP-Glu-SerP-Thr-

$$^{50} \qquad\qquad\qquad\qquad\qquad\qquad ^{60}$$
Glu-Asp-Gln-Ala-Met-Glu-Asp-Ile-Lys-Gln-Met-Glu-Ala-Glu-SerP-Ile-

$$^{70} \qquad\qquad\qquad\qquad\qquad\qquad ^{80}$$
SerP-SerP-SerP-Glu-Glu-Ile-Val-Pro-Asn-SerP-Val-Glu-Gln-Lys-His-Ile-

$$^{90}$$
Gln-Lys-Glu-Asp-Val-Pro-Ser-Glu-Arg-Tyr-Leu-Gly-Tyr-Leu-Glu-Gln-

$$^{100} \qquad\qquad\qquad\qquad\qquad\qquad ^{110}$$
Leu-Leu-Arg-Leu-Lys-Lys-Tyr-Lys-Val-Pro-Gln-Leu-Glu-Ile-Val-Pro-Asn-

$$^{120} \qquad\qquad\qquad\qquad\qquad\qquad ^{130}$$
SerP-Ala-Glu-Glu-Arg-Leu-His-Ser-Met-Lys-Glu-Gly-Ile-His-Ala-Gln-

$$^{140}$$
Gln-Lys-Glu-Pro-Met-Ile-Gly-Val-Asn-Gln-Glu-Leu-Ala-Tyr-Phe-Tyr-

$$^{170} \qquad\qquad\qquad\qquad\qquad\qquad ^{180}$$
Pro-Glu-Leu-Gly-Thr-Gln-Tyr-Thr-Asp-Ala-Pro-Ser-Phe-Ser-Asp-Ile-Pro-

$$^{190} \qquad\qquad\qquad\qquad\qquad\qquad ^{199}$$
Asn-Pro-Ile-Gly-Ser-Glu-Asn-Ser-Glu-Lys-Thr-Thr-Met-Pro-Leu-Trp-OH

FIG. 1. Sequence of bovine $\alpha_{s1}$-casein B.[8] The asterisk denotes the site of the ninth serine phosphate residue present in $\alpha_{s0}$-casein. Genetic variant A has residues 14 through 26 deleted; variant C has Gly for Glu at residue 192; variant D has ThrP for Ala at residue 53.


$$^{10}$$
H-Arg-Glu-Leu-Glu-Glu-Leu-Asn-Val-Pro-Gly-Glu-Ile-Val-Glu-SerP-Leu-

$$^{20} \qquad\qquad\qquad\qquad\qquad\qquad ^{30}$$
SerP-SerP-SerP-Glu-Glu-Ser-Ile-Thr-Arg-Ile-Asn-Lys-Lys-Ile-Glu-Lys-

$$^{40}$$
Phe-Gln-SerP-Glu-Glu-Gln-Gln-Gln-Thr-Glu-Asp-Glu-Leu-Gln-Asp-Lys-

$$^{50} \qquad\qquad\qquad\qquad\qquad\qquad ^{60}$$
Ile-His-Pro-Phe-Ala-Gln-Thr-Gln-Ser-Leu-Val-Tyr-Pro-Phe-Pro-Gly-Pro-

$$^{70} \qquad\qquad\qquad\qquad\qquad\qquad ^{80}$$
Ile-Pro-Asn-Ser-Leu-Pro-Gln-Asn-Ile-Pro-Pro-Leu-Thr-Gln-Thr-Pro-Val-

$$^{90}$$
Val-Val-Pro-Pro-Phe-Leu-Gln-Pro-Glu-Val-Met-Gly-Val-Ser-Lys-Val-Lys-

$$^{100} \qquad\qquad\qquad\qquad\qquad\qquad ^{110}$$
Glu-Ala-Met-Ala-Pro-Lys-His-Lys-Glu-Met-Pro-Phe-Pro-Lys-Tyr-Pro-

$$^{120} \qquad\qquad\qquad\qquad\qquad\qquad ^{130}$$
Val-Gln-Pro-Phe-Thr-Glu-Ser-Gln-Ser-Leu-Thr-Leu-Thr-Asp-Val-Glu-

$$^{140}$$
Asn-Leu-His-Leu-Pro-Pro-Leu-Leu-Leu-Gln-Ser-Trp-Met-His-Gln-Pro-

$$^{150} \qquad\qquad\qquad\qquad\qquad\qquad ^{160}$$
His-Gln-Pro-Leu-Pro-Pro-Thr-Val-Met-Phe-Pro-Pro-Gln-Ser-Val-Leu-Ser-

$$^{170} \qquad\qquad\qquad\qquad\qquad\qquad ^{180}$$
Leu-Ser-Gln-Ser-Lys-Val-Leu-Pro-Val-Pro-Glu-Lys-Ala-Val-Pro-Tyr-Pro-

$$^{190}$$
Gln-Arg-Asp-Met-Pro-Ile-Gln-Ala-Phe-Leu-Leu-Tyr-Gln-Gln-Pro-Val-

$$^{200} \qquad\qquad\qquad\qquad\qquad\qquad ^{209}$$
Leu-Gly-Pro-Val-Arg-Gly-Pro-Phe-Pro-Ile-Ile-Val-OH

FIG. 2. Sequence of bovine $\beta$-casein A[2].[10] The sites of splitting to form the $\gamma$-caseins are between residues 28 and 29, 105 and 106, and 107 and 108, giving $\gamma_1$-(29–209), $\gamma_2$-(106–209) and $\gamma_3$-(108–209) caseins.

10
H-Lys-Asn-Thr-Met-Glu-His-Val-SerP-SerP-SerP-Glu-Glu-Ser-Ile-Ile-SerP-
20 30
Gln-Glu-Thr-Tyr-Lys-Gln-Glu-Lys-Asn-Met-Ala-Asn-Ile-Pro-Ser-Lys-
40
Glu-Asn-Leu-Cys-Ser-Thr-Phe-Cys-Lys-Glu-Val-Val-Arg-Asn-Ala-Asn-
50 60
Glu-Glu-Glu-Tyr-Ser-Ile-Gly-SerP-SerP-SerP-Glu-Glu-SerP-Ala-Glu-
70 80
Val-Ala-Thr-Glu-Glu-Val-Lys-Ile-Thr-Val-Asp-Asp-Lys-His-Tyr-Gln-Lys-
90
Ala-Leu-Asn-Glu-Ile-Asn-Glu-Phe-Tyr-Gln-Lys-Phe-Pro-Gln-Tyr-Leu-
100 110
Gln-Tyr-Leu-Tyr-Gln-Gly-Pro-Ile-Val-Leu-Asn-Pro-Trp-Asp-Gln-Val-
120
Lys-Arg-Asn-Ala-Val-Pro-Ile-Thr-Pro-Thr-Leu-Asn-Arg-Glu-Gln-Leu-
130 140
SerP-Thr-SerP-Glu-Glu-Asn-Ser-Lys-Lys-Thr-Val-Asp-Met-Glu-SerP-
150
Thr-Glu-Val-Phe-Thr-Lys-Lys-Thr-Lys-Leu-Thr-Glu-Glu-Glu-Lys-Asn-
160 170
Arg-Leu-Asn-Phe-Leu-Lys-Lys-Ile-Ser-Gln-Arg-Tyr-Gln-Lys-Phe-Ala-
180 190
Leu-Pro-Gln-Tyr-Leu-Lys-Thr-Val-Tyr-Gln-His-Gln-Lys-Ala-Met-Lys-
200 207
Pro-Trp-Ile-Gln-Pro-Lys-Thr-Lys-Val-Ile-Pro-Tyr-Val-Arg-Tyr-Leu-OH

FIG. 3. Sequence of bovine $\alpha_{s2}$-casein. In addition to the sites of phosphorylation shown, it is possible for phosphorylation to occur on residues 3, 31, 66, 130 and 154. Evidence of gene duplication is afforded by partial homology of peptides 50–123 and 132–209.[9]

10
PyroGlu-Glu-Gln-Asn-Gln-Glu-Gln-Pro-Ile-Arg-Cys-Glu-Cys-Asp-Glu-Arg-
20 30
Phe-Phe-Ser-Asp-Lys-Ile-Ala-Lys-Tyr-Ile-Pro-Ile-Gln-Tyr-Val-Leu-Ser-
40
Arg-Tyr-Pro-Ser-Tyr-Gly-Leu-Asn-Tyr-Tyr-Gln-Gln-Lys-Pro-Val-Ala-
50 60
Leu-Ile-Asn-Asn-Gln-Phe-Leu-Pro-Tyr-Pro-Tyr-Tyr-Ala-Lys-Pro-Ala-
70 80
Ala-Val-Arg-Ser-Pro-Ala-Gln-Ile-Leu-Gln-Trp-Gln-Val-Leu-Ser-Asp-Thr-
90
Val-Pro-Ala-Lys-Ser-Cys-Gln-Ala-Gln-Pro-Thr-Thr-Met-Ala-Arg-His-
100 110
Pro-His-Pro-His-Leu-Ser-Phe*-Met-Ala-Ile-Pro-Pro-Lys-Lys-Asn-Gln-
120 130
Asp-Lys-Thr-Glu-Ile-Pro-Thr-Ile-Asn-Thr-Ile-Ala-Ser-Gly-Glu-Pro-Thr-
140
Ser-Thr-Pro-Thr-Ile-Glu-Ala-Val-Glu-Ser-Thr-Val-Ala-Thr-Leu-Gln-Ala-
150 160
SerP-Pro-Glu-Val-Ile-Glu-Ser-Pro-Pro-Glu-Ile-Asn-Thr-Val-Gln-Val-
169
Thr-Ser-Thr-Ala-Val-OH

FIG. 4. Sequence of bovine $\kappa$-casein B.[11] The site for splitting by chymosin and pepsin is shown by the asterisk. The A variant of the protein has Thr for Ile at residue 136 and Asp for Ala at residue 148.

160    *D. G. Dalgleish*

H-Leu-Ile-Val-Thr-Gln-Thr-Met-Lys-Gly-Leu[10]-Asp-Ile-Gln-Lys-Val-Ala-Gly-
Thr[20]-Trp-Tyr-Ser-Leu-Ala-Met-Ala-Ala-Ser-Asp[30]-Ile-Ser-Leu-Leu-Asp-Ala-
Gln-Ser-Ala-Pro[40]-Leu-Arg-Val-Tyr-Val-Glu-Glu-Leu-Lys-Pro[50]-Thr-Pro-
Glu-Gly-Asp-Leu-Glu-Ile-Leu-Leu-Gln[60]-Lys-Trp-Glu-Asn-Asp-Glu-Cys-
Ala-Gln-Lys[70]-Lys-Ile-Ile-Ala-Glu-Lys-Thr-Lys-Ile[80]-Pro-Ala-Val-Phe-Lys-
Leu-Asp-Ala-Ile-Asn-Glu[90]-Asn-Lys-Val-Leu-Val-Leu-Asp-Thr-Asp-Tyr-
Lys[100]-Lys-Tyr-Leu-Leu-Phe-Cys-Met-Glu-Asn-Ser[110]-Ala-Glu-Pro-Glu-Gln-
Ser-Leu-Val-Cys[120]-Gln-Cys-Leu-Val-Arg-Thr-Pro-Glu-Val-Asp-Asp[130]-Glu-
Ala-Leu-Gln-Lys-Phe-Asp-Lys-Ala-Leu-Lys[140]-Ala-Leu-Pro-Met-His-Ile-
Arg[150]-Leu-Ser-Phe-Asn-Pro-Thr-Leu-Gln-Glu-Glu[160]-Gln-Cys-His-Ile[162]-OH

FIG. 5.    Sequence of bovine β-lactoglobulin. A.[6] Disulphide bonds are present between
residues 66 and 160, and 119 or 121 and 106.

residues. Prolyl residues do not participate in the formation of α-helix or β-sheet structures in proteins, and therefore β-casein can contain only small amounts, if any, of these regular structures. The sequence of the protein, which shows a strong tendency towards clustering of hydrophobic and hydrophilic residues, suggests that the protein may be likened to a detergent-type of molecule, with a charged, hydrophilic 'head' and a non-polar 'tail'. Analogously to detergents, β-casein monomers show a strongly

H-Glu-Gln-Leu-Thr-Lys-Cys-Glu-Val-Phe-Arg[10]-Glu-Leu-Lys-Asp-Leu-Lys-
Gly[20]-Tyr-Gly-Gly-Val-Ser-Leu-Pro-Glu-Trp-Val[30]-Cys-Thr-Thr-Phe-His-
Thr-Ser-Gly-Tyr-Asp-Thr-Glu-Ala-Ile-Val[40]-Glu-Asn-Asn-Gln-Ser-Thr-
Asp[50]-Tyr-Gly-Leu-Phe-Gln-Ile-Asn-Asn-Lys-Ile[60]-Trp-Cys-Lys-Asn-Asp-
Gln-Asp-Pro-His-Ser-Ser[70]-Asn-Ile-Cys-Asn-Ile-Ser-Cys-Asp-Lys-Phe[80]-Leu-
Asn-Asn-Asp-Leu-Thr-Asn-Asn-Ile-Met[90]-Cys-Val-Lys-Lys-Ile-Leu-Asp-
Lys[100]-Val-Gly-Ile-Asn-Tyr-Trp-Leu-Ala-His-Lys[110]-Ala-Leu-Cys-Ser-Glu-Lys-
Leu[120]-Asp-Gln-Trp-Leu-Cys[123]-Glu-Lys-Leu-OH

FIG. 6.    Sequence of bovine α-lactalbumin B.[7] Disulphide bonds are present between residues
6 and 120, 28 and 111, 61 and 77, and 73 and 91.

temperature-dependent association in aqueous solution, to form aggregates of degree of polymerisation up to about 40, depending on the conditions of experiment.[18,19,20] It is thought that these aggregates are micelle-like, having hydrophobic interiors made up of the packed C-terminal regions of the molecules and hydrophilic exteriors involving the N-terminal regions. Estimates of the charge on the protein monomers based on the amino acid composition suggest a value of about $-12$ at pH 6·8.

In contrast, the $\alpha_{s1,0}$- and $\alpha_{s2}$-caseins particularly the latter, show a more even distribution of hydrophilic residues. Also, both are more highly phosphorylated than $\beta$-casein. They have a considerably lower content of proline, so that there are increased possibilities of the formation of $\alpha$-helical or $\beta$-sheet structures; however, measurements suggest that regular structure content is still low. The hydrophobicities of the $\alpha_s$-caseins are lower than that of $\beta$-casein ($\alpha_{s1}$-casein, 1172 cal/residue and $\alpha_{s2}$-casein, 1111 cal/residue), bringing them closer to the average value for soluble proteins, and their charges are higher ($\alpha_{s1}$-casein, $-21$; $\alpha_{s0}$-casein, $-23$ and $\alpha_{s2}$-caseins between, $-16$ and $-22$, calculated at pH 6·8). Both the higher charges and lower hydrophobicities contribute to the lesser tendency of these caseins to aggregate in the absence of $Ca^{2+}$, compared with $\beta$-casein. Small oligomers of $\alpha_{s1}$-casein can be formed under appropriate conditions (up to six monomers in the absence of $Ca^{2+}$).[21]

$\kappa$-Casein is the most structurally important of the caseins: it is responsible for stabilising the micelle, arguably by acting to produce a surface layer around a core of precipitated or aggregated $\alpha_s$- and $\beta$-caseins. The acidic residues of the protein are found in the C-terminal region, with most of the basic and hydrophobic residues being in the N-terminal moiety. The stabilising action of $\kappa$-casein is lost when it is attacked by chymosin or other acid proteases, which specifically cleave between residues 105 and 106 of the polypeptide,[22] producing the highly hydrophobic peptide 1–105 (*para-$\kappa$-casein*) and the acidic, soluble peptide 106–169 (caseinomacropeptide). The *para-$\kappa$-casein* moiety remains on the casein micelle, but the macropeptide is lost into solution, and the micelles precipitate since the stabilising action of the $\kappa$-casein arises from the interaction of the macropeptide of the intact protein with the surrounding solution. In some $\kappa$-casein molecules the latter interaction is facilitated by the presence of a sugar chain which is attached to residue Thr 131.[23] However, the sugar is not essential to the stabilising action. The overall charge of $\kappa$-casein at pH 6·8 as calculated from the amino acid composition should be low (in the absence of contributions from carbohydrate). Of the individual moieties, *para-$\kappa$-casein* has a charge of $+5$ and macropeptide a charge of $-8$. The hydrophobicity of intact $\kappa$-casein is 1224 cal/residue, which makes it, overall, a hydrophobic protein. However, the distinction between the two parts of the protein is emphasised by the calculation that the hydrophobicity of *para-$\kappa$*-casein is 1310 cal/residue, whereas the macropeptide is not markedly hydrophobic at 1082 cal/residue.

$\kappa$-Casein, in addition to polymerising via sulphydryl interactions, can also undergo non-covalent association. In the presence of SH-blocking agents, a monomer–polymer equilibrium is established, the polymer containing some 30 monomers of $\kappa$-casein.[24] This polymerisation can be defined by a critical micelle concentration[25] which decreases as ionic strength is increased. Ionic strength does not affect the final aggregate size to the same extent.


## ION BINDING AND PRECIPITATION OF THE CASEINS

Probably the most important property of the $\alpha_s$- and $\beta$- caseins is their ability to bind significant numbers of divalent metal ions (principally $Ca^{2+}$), and subsequently to aggregate strongly or even to precipitate. $\alpha_{s1}$-Casein can bind up to $20\,Ca^{2+}$ per molecule of protein at high concentrations of calcium,[26] and under normal circumstances the protein can bind up to 10 moles of $Ca^{2+}$/mole protein.[27] The binding isotherms of $Ca^{2+}$ to $\alpha_{s1}$- and $\beta$-caseins (Fig. 7) demonstrate that the binding cannot be explained by a single binding constant, but rather that the affinity of the caseins for binding $Ca^{2+}$ depends on the number of $Ca^{2+}$ which are already bound to the protein.[28] Binding is also dependent on temperature, pH and ionic strength. Increasing temperature, increasing pH (in the region of pH 6–7·5) and decreasing ionic strength all increase the extent of binding of $Ca^{2+}$.[29,30] Ionic strength may alter the binding by increasing shielding of charged particles, but also possibly because of competition between $Ca^{2+}$ and monovalent cations for the binding sites. The decreasing binding of $Ca^{2+}$ as pH is reduced arises from the titration of the phosphoserine residues, rendering them incapable of binding $Ca^{2+}$, since only ionised phosphate is relevant in this context.

It seems likely that at least the first few bound $Ca^{2+}$ are attached to the phosphoseryl residues. Only at higher binding ratios is $Ca^{2+}$ likely to bind to aspartyl or glutamyl residues, the binding constant for such interactions being weaker than the binding to phosphate. Under normal conditions of milk, the extent of $Ca^{2+}$-binding is not likely to exceed the number of phosphoseryl residues in the proteins.

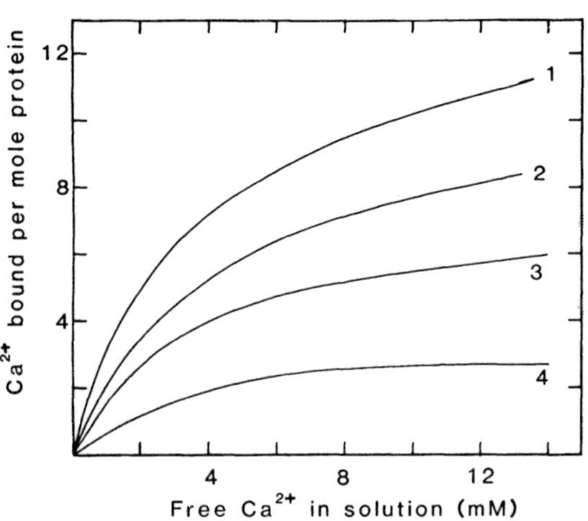

FIG. 7.   Binding of $Ca^{2+}$ to $\alpha_{s1}$- and $\beta$-caseins. The four curves shown are: 1, $\alpha_{s1}$-casein, pH 7·0, 20 °C, 50 mM NaCl; 2, $\alpha_{s1}$-casein, pH 6·5, 20 °C, 50 mM NaCl; 3, $\beta$-casein, pH 7·0, 20 °C, 50 mM NaCl; 4, $\beta$-casein, pH 7·0, 20 °C, 150 mM NaCl.

Binding of sufficient $Ca^{2+}$ to $\alpha_s$- and $\beta$-caseins causes precipitation of the protein–$Ca^{2+}$ complex. There is a threshold for precipitation, which does not increase simply with the extent of $Ca^{2+}$ binding. Only after a certain critical concentration of $Ca^{2+}$ is present does precipitation occur:[31] this critical concentration of $Ca^{2+}$ depends on the concentration of casein as well as on other conditions of the experiment.[29] Nor, once precipitation is observed, does all of the casein precipitate, especially in the case of $\alpha_{s1}$-casein, since a proportion of the casein is always left in solution, even at high concentrations of $Ca^{2+}$.[32]

The predominant effect of the binding of $Ca^{2+}$ to the caseins is to reduce considerably the charge on the protein molecules.[33] In the absence of $Ca^{2+}$, the proteins are maintained in solution by a balance of hydrophobic forces (which promote aggregation), hydrophilic interactions between protein and solvent, and repulsions between proteins as a result of the charge (even the 'micellar' aggregates of $\beta$-casein have charged surfaces). When $Ca^{2+}$ binds to the proteins, the net negative charge is reduced by two units for each $Ca^{2+}$ which is bound.[33] This reduces the electrostatic repulsion between the molecules, and thus aggregation of the proteins becomes more likely. A detailed description of the aggregation process is available for $\alpha_{s1}$-casein.[32] The first stage of the reaction is the binding of sufficient $Ca^{2+}$ to reduce the charge on the protein from $-22$ to about $-8$. When the charge has been lowered sufficiently, the precipitation reactions start by the establishment of an equilibrium between monomeric and octameric particles of $\alpha_{s1}$-casein. Only the octamers can aggregate further to form a precipitate, so that the extent of precipitation is governed by the extent of octamer formation, which in turn depends on the equilibrium constant for the reaction between monomers and octamers. This equilibrium constant is defined by the charge on the protein–$Ca^{2+}$ complex, since the repulsion between the interacting monomers is defined by their surface charges. During the establishment of equilibrium, and once equilibration has been achieved, the octameric particles themselves aggregate, by a simple Smoluchowski mechanism,[34] so that the growth of molecular weight becomes linear, once formation of octamer has attained its equilibrium value. The rate-constant for the second-stage reaction is also dependent upon the charge of the monomers, so that not only the extent of precipitation but also its rate, are defined by the extent of binding of $Ca^{2+}$ to the protein. The kinetics and the different stages of the reaction of $\alpha_{s1}$-casein with $Ca^{2+}$ are shown in Fig. 8.

The precipitation of $\beta$-casein also proceeds in a similar manner, although the intermediate precipitating particle is larger. Because of the tendency of

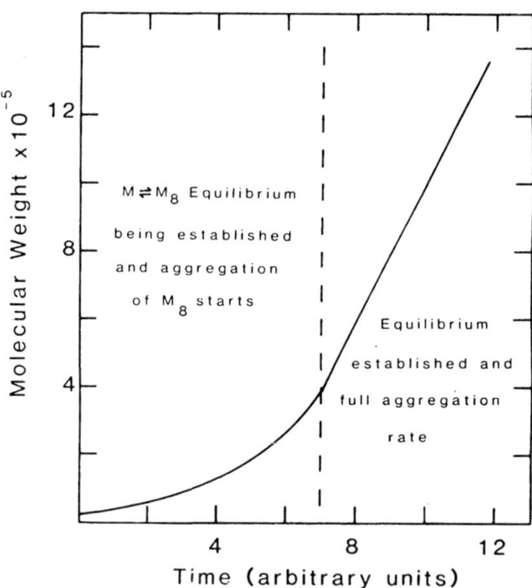


FIG. 8. Kinetics of the aggregation of $\alpha_{s1}$-casein, after addition of $Ca^{2+}$ at time zero. The curve shown is a composite of experimental results at different concentrations of casein and $Ca^{2+}$.

$\beta$-casein towards self-association, these intermediate particles are nearly formed at the time of $Ca^{2+}$ binding, and the partition between soluble and precipitable material is determined not only by the extent of calcium binding but also by the extent to which aggregation has already occurred. The rate of the precipitation is, however, determined by the charge on the $\beta$-casein–$Ca^{2+}$ complex, as is the case for $\alpha_{s1}$-casein. Since the formation of aggregates of $\beta$-casein is to some extent dependent on hydrophobic interactions, the overall precipitation by $Ca^{2+}$ is defined both by hydrophobic and ion-binding properties of $\beta$-casein. Thus, temperature, by affecting the hydrophobic interactions, is more significant in the case of $\beta$-casein precipitation than for $\alpha_{s1}$-casein.

The presence of $\kappa$-casein alters these reactions considerably, since precipitation is prevented and instead stable colloidal particles are formed.[35] Even in the absence of $Ca^{2+}$, $\kappa$-casein can associate with $\alpha_s$-casein,[36] and when $Ca^{2+}$ is present, the $\kappa$-casein stabilises the aggregating $\alpha_s$- or $\beta$-caseins, in a similar manner to its stabilising action in native casein micelles. Particles formed in this way, from simple casein mixtures and $Ca^{2+}$, are similar in size to native casein micelles, but are less stable to outside influences.[37,38] It is possible to make synthetic micelles of this type using $\alpha_s/\kappa$, $\beta/\kappa$ or $\alpha_s/\beta/\kappa$ mixtures. Their sizes are dependent on the proportion of $\kappa$-casein in the mixture, and also on the concentration of $Ca^{2+}$ which is used. The action of $\kappa$-casein appears to be most significant during the second precipitation stage, so that aggregation is no longer infinite. No significant effect on the binding of $Ca^{2+}$ by $\alpha_s$- and $\beta$-caseins is found when $\kappa$-casein is incorporated, nor is the critical concentration of $Ca^{2+}$, which is required to cause extensive aggregation, affected.


## THE CASEIN MICELLE

In freshly secreted cows' milk, the bulk of the casein is present in the form of casein micelles. These are particles which contain many individual molecules of the different caseins, complexed with calcium ions, calcium phosphate and small amounts of citrate. The function of the calcium phosphate is not fully defined, although it appears to be essential to the formation and properties of the micelles. The synthetic micelle-like aggregates described above lack colloidal calcium and phosphate, and are less stable to heat, alcohol treatment, and dissociating influences, than natural casein micelles.[37] One of the possible functions of calcium phosphate, besides playing a structural role, is to slow down exchange reactions between solution components and micellar components, thus providing for greater apparent stability. Naturally occurring micelles are polydisperse in size and molecular weight: the size ranges from 10 nm to more than 250 nm, and molecular weights range up to $10^9$ daltons. Various distribution functions have been measured,[39-41] principally by electron microscopy, and these confirm that the distribution of sizes is broad and that there are considerable differences between milks from different sources, from different animals, and even from the same animal at different stages of lactation.[42] Generally, the smallest micelles are the most numerous, and the number-frequency drops as the size increases. Apart from electron microscopy, micelle sizes have been measured by light-scattering techniques which emphasise the larger sizes of particle. Electron microscopy on the other hand is liable not to 'see' the small number of very large micelles, so that reconciling the results from the two techniques may be difficult.[40] Compared with normal proteins, the micelles are very highly hydrated, containing about 3 g of water per g of protein.[43,44] This water is relatively loosely bound, and can be squeezed out by ultra-centrifugation. This water may however, be taken up again by redispersing the micelles.

In addition to being polydisperse in size, the casein micelles are heterodisperse in composition. Most important is the proportion of $\kappa$-casein, which increases as the micelle size decreases.[45] There is in fact a close correlation between the proportion of $\kappa$-casein in the micelles and the surface-to-volume ratio of the particles.[46] Therefore, it seems reasonable to assume that at least the bulk of the $\kappa$-casein is on the surface of the micelle, where it can best exert its stabilising influence. This type of micellar structure was initially postulated by Waugh *et al.*:[47] micelles were envisaged as having a core of precipitated $\alpha_s$- and $\beta$-caseins surrounded by a coat of $\kappa$-casein, which was capable of hydrophobic interaction with the core via the *para-$\kappa$-casein* moiety, and with the solution via the hydrophilic macropeptide. The surface layer of macropeptide may present a 'hairy' appearance, with the macropeptide chains projecting into solution, providing stabilisation;[48,49] such a hypothesis explains the drop in viscosity[50] and in micelle radius[51] when casein micelles are renneted. Confirmation of the surface location of the bulk of the $\kappa$-casein can be obtained by treating the micelles with immobilised pepsin or chymosin. Under such circumstances, the enzyme is incapable of entering the micelle, and its activity is confined to the exterior. Such immobilised enzymes will hydrolyse almost all of the $\kappa$-casein of the milk, indicating that the $\kappa$-casein must be accessible and hence be located on the surface.[52]

It is yet to be confirmed whether casein micelles are composed of smaller aggregates of caseins (submicelles). Under the electron microscope the micelles do have an appearance which is consistent with them being structured in this way,[53] and rather detailed models of micelle structures have been based on such submicelles. In these models it is postulated that the submicelles are heterogeneous in composition, and consist of about 30 molecules of caseins; when micelles are formed from these, the submicelles which are relatively rich in $\kappa$-casein are found on the surface, with the relatively more hydrophobic submicelles being located in the interior.[54] While this is a satisfactory explanation of micellar appearance as observed in the electron microscope, there are some aspects which are difficult to explain. Most importantly, it is difficult to explain the stability of the submicelles themselves. If they are heterogeneous in composition (and some must be composed entirely of $\alpha_s$- and $\beta$-caseins), then it is not easy to explain how they can remain stable and unaggregated at the '30-mer' stage, rather than aggregating to give larger units. Furthermore, the dissociation of $\beta$-casein preferentially from micelles (see the next section) seems to indicate that the subunits can dissociate, or that $\beta$-casein occurs extensively in submicelles containing only that protein. This may not be completely unlikely, in view of the self-aggregating tendency of this protein, but then it must be postulated that other submicelles contain $\alpha_s$- and $\kappa$-caseins only. It is possible to produce aggregates of caseins of approximate molecular weight of some hundreds of thousands by dialysis of milk[55] but it is not possible to confirm that these aggregates are actually present in natural casein micelles. In the absence of a completely satisfactory theory of how micelles aggregate *in vivo*, it is not possible to define whether aggregates of $\alpha_s$-casein or $\beta$-casein as described above exist as fundamental building blocks for micelles, or whether the aggregation properties of all four major caseins are modified by the presence of the others. Such aggregation could lead to a granular microgel, containing regions of varying density, rather than a micelle strictly composed of subunits.[43]

Even if it is assumed that the outer coating of the casein micelles is of $\kappa$-casein, the detailed structure of the interior is not known (even in submicellar models). It contains the $\alpha_s$- and $\beta$-casein fractions, together with the colloidal calcium phosphate. Although these are held inside the $\kappa$-casein surface of the micelle, they are nonetheless accessible even to large molecules, demonstrating once again the open and highly hydrated structure of the casein micelle. For instance, the micelle may be penetrated by so large a molecule as carboxypeptidase,[56] so that the 'coat' of $\kappa$-casein is by no means complete. It is also possible to transfer material from the interior of the micelle into the surrounding solution.[57] The nature of the complex between the $\alpha_s$- and $\beta$-caseins and the colloidal calcium phosphate and their mutual effects on stability are not known. Likewise, the structure of the calcium phosphate itself is not defined. From the lack of X-ray diffraction pattern, the structure has been taken to be an amorphous form of calcium phosphate,[58] but the heat of solution suggests that hydroxyapatite is the major form.[59] Whether the colloidal calcium phosphate is directly bound to the phosphoseryl residues of the caseins is not known, although recent evidence supports the view that calcium binding and formation of micellar calcium phosphate are distinct and separate processes.[60]

## STABILITY OF CASEIN MICELLES

Considering that milk is produced by the cow for only short-term storage at about 40 °C, casein micelles are remarkably stable to processing environments. Thus they maintain stability, in the sense that they do not aggregate, for some appreciable time at temperatures above 100 °C, and the micelles resist breakdown during homogenisation (high-shear treatment). Nonetheless, some of the caseins, especially $\beta$-casein, appear to be mobile within the micellar framework. Cooling, for instance, liberates caseins into solution so that after 24 h at 4 °C as much as 50 % of the $\beta$-casein is liberated from the micelles. On rewarming, this $\beta$-casein reassociates with the micelles, although some time is required for complete reaggregation. There is no conclusive evidence that marked changes in micellar behaviour are caused by this dissociation–reassociation cycle, so that it is to be assumed that the $\beta$-casein returns to approximately the same location within the micelle, rather than, for example, binding to the surface of the micelles during reassociation.

The stability of micelles is most completely destroyed by removing or modifying the $\kappa$-casein of the surface. The traditional way of achieving this is to use proteolytic enzymes such as chymosin or pepsin; more recently, fungal or microbial proteases have been used.[61] These cleave the $\kappa$-casein at the bond between residues 105 and 106, leaving the insoluble *para-$\kappa$*-casein associated with the micelles, which thereupon aggregate. This aggregation is prevented (or dramatically slowed) by cooling the milk to 15 °C or below, which has only a small effect upon the enzymic reaction, but has a major effect on the aggregation.[62] The apparent activation energy for the aggregation of renneted micelles is very high, and the fact that the

Arrhenius plot for the reaction is curved indicates that several factors are involved in determining the temperature coefficient of the reaction. The reasons for the initial stability of the micelles, and their ability to coagulate after renneting, are not fully established. Removal of the macropeptide by renneting reduces the surface potential of the micelles by about one-third,[63,64] so that intermicellar electrostatic repulsions will be reduced. Also, the loss of the macropeptide chains may result in a loss of steric stabilisation, so that closer approach of the micelles is facilitated. The forces which hold together the aggregating, renneted micelles may be mainly hydrophobic, but ionic interactions between positively charged *para-κ*-casein and negative sections of the $\alpha_s$- and $\beta$-caseins are possible, as are bridges between micelles formed of calcium ions or calcium phosphate. None of these possible interactions can be left out of account, although the effect of cooling on the aggregation is some indication that hydrophobic interactions play a large part in the stability of the aggregate. Several mechanisms for the aggregation have been proposed, which attempt to explain the reaction in terms of the aggregability of micelles at different stages of proteolysis.[65–68]

Part of the stability of the casein micelles to processing may perhaps be explained in terms of an excess of $\kappa$-casein being present. This is demonstrated once again by the action of chymosin. We might expect that micelles would tend to greater aggregation as more $\kappa$-casein was split, and that aggregability of renneted micelles would increase approximately linearly with the extent of $\kappa$-casein destruction. This is not so: in fact the micelles will only start to aggregate when more than about 85 % of their $\kappa$-casein has been split by the enzyme[52,69] (Fig. 9). Thus, the amount of $\kappa$-casein present in milk can be seen to be apparently in excess of that simply required to maintain stability of the micelles. The clotting time of renneted milk is determined mainly by the time which is taken by the enzyme to split sufficient of the $\kappa$-casein; the contribution of aggregation to the clotting time is generally less than the time required for enzyme action.[70] At lower temperatures, the clotting time becomes increasingly dependent on the rate-constant for micellar aggregation.

Casein micelles can be dissociated by the removal of their calcium and calcium phosphate by chelating agents. When total chelation of the calcium has been achieved, the micelles are broken into small units and behave as mixtures of caseins, but when only part of the calcium is removed, the micelles lose some protein (mainly $\beta$-casein), and grow slightly in size. This suggests that micelles are a network of $\alpha$- and $\kappa$-caseins, with an interstitial 'filling' of $\beta$-casein. The slight growth in size when $\beta$-casein and $Ca^{2+}$ are

                                        *D. G. Dalgleish*

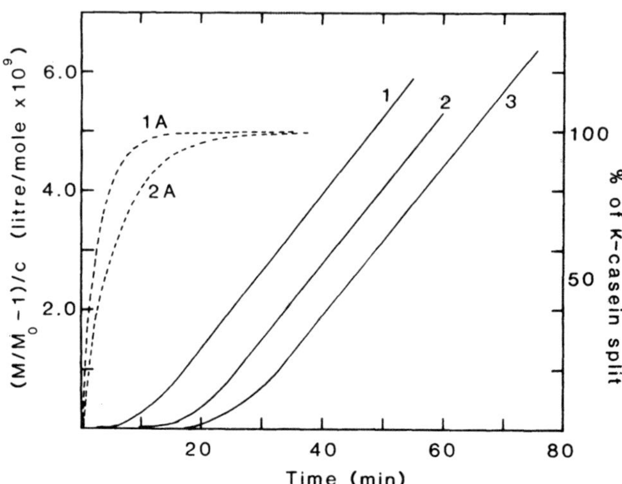

FIG. 9.  Aggregation of casein micelles after addition of rennet. Curves 1 and 2 represent the aggregation of micelles of initial molecular weight $8 \cdot 6 \times 10^8$, using two different rennet concentrations. The course of $\kappa$-casein splitting is shown by the two broken lines 1A and 2A. Curve 3 is for smaller micelles, with a lower rennet concentration. The slopes of the curves give the aggregation rate constants for the micelles, and the parallel nature of the three linear portions of the curves shows that they have the same rate constants in the limit.

removed can be at least partly explained in terms of the $Ca^{2+}$ exerting a compacting role when bound, perhaps by the formation of calcium bridges between protein molecules.

In addition to their precipitation following enzymic attack, micelles may be precipitated by the action of heat or by the addition of alcohol. It has been claimed that these two processes are related, but evidence is lacking at present. Precipitation by alcohol is markedly pH-dependent, occurring more readily below pH 6·4 than above that pH[71] (Fig. 10). Although it is casein which precipitates, the serum phase components, principally $Ca^{2+}$, are determinants of the stability: increased serum $Ca^{2+}$ destabilises the milk towards precipitation.[72] Studies in which solvents other than alcohol were used suggest that the precipitation is controlled by micellar charge (hence the importance of $Ca^{2+}$), and that the electrostatic contribution to micellar stability is dominant.[73]

Homogenisation of whole milk leads to breakdown of the micelles. Homogenisation of skimmed milk does not have this effect, so that the micelles themselves resist the high-shear conditions which obtain during this process. The micelles must therefore be disrupted by the presence of excess amounts of lipid surface which are formed by the breakdown of the

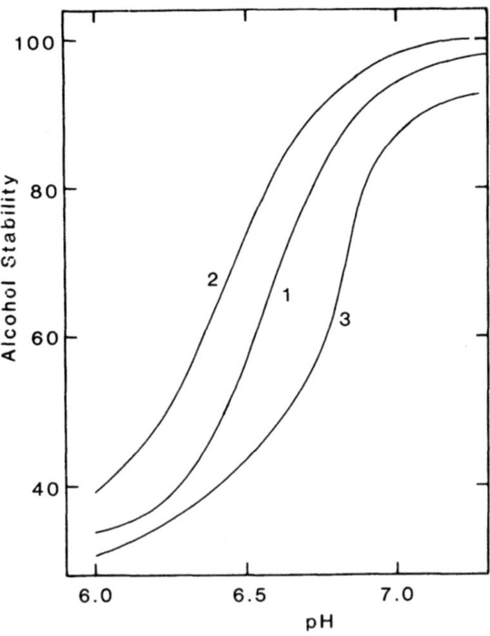

FIG. 10.   Stability of micelles to alcohol. Alcohol stability is defined as the percentage of alcohol in a water/alcohol mixture which when added to milk in a ratio of 2:1 just causes precipitation. Curve 1 represents normal milk behaviour; curve 2 shows the change when 5 mM EDTA is added to the milk, and curve 3 shows the decreased stability when 5 mM CaCl is added.

fat globules. These new areas are highly hydrophobic, and therefore present an opportunity for the re-equilibration of the hydrophobic and hydrophilic portions of the caseins.[74,75] From this stems the well-known emulsifying power of the caseins. The structure of the casein around the newly formed fat surface is undefined, but it is clear that $\kappa$-casein, as in natural micelles, plays a dominant role. The emulsified material is unstable when subjected to rennet, confirming that $\kappa$-casein is still important in maintaining the stability of the dispersion. However, the curd formed from renneted homogenised milk is not the same as the curd from whole or skimmed milk in properties: this may arise from the changes in casein structure which have been introduced by homogenisation in the presence of the fat, or alternatively that the reduced size of the fat globules causes a more even structure. In either case, it is clear that the main structural units of normal curd, that is the renneted micelles, are no longer present, and the actual aggregating units may appear to be large casein micelles, but lack

                    *D. G. Dalgleish*

their structural integrity, being no more than a layer on the surface of the fat globules. Thus although homogenised milk shares some properties of natural milk, e.g. stability and rennetability, there are significant differences in other aspects (curd formation and heat stability[76]).


## THE ACTION OF HEAT ON MILK PROTEINS

Both serum proteins and caseins are affected by heating, in several ways. Heat provides a means of denaturation, principally of the serum proteins, and also provides the energy to enable a number of chemical reactions to occur. The net result of heating milk to temperatures in excess of 100 °C is to cause eventual precipitation of the caseins and casein micelles, although the causes of such reactions are not fully determined. In addition to direct effects upon the proteins themselves, heating also increases the amount of $Ca^{2+}$ bound to the caseins, and may also initiate the precipitation or deposition of solid calcium phosphate, in addition to that already present in the micelle.

At temperatures in the region of 80 °C, $\beta$-lactoglobulin is denatured. This denaturation represents an alteration of the three-dimensional structure of the protein, with an interchange of disulphide bridges. Thus, instead of the single structure, a multiplicity of structures are formed.[77] Also, and especially when $\beta$-lactoglobulin is the only protein present, the new disulphide bridges which are formed are not only intramolecular but intermolecular, leading to aggregation and to gel formation. This process is somewhat facilitated by the fact that $\beta$-lactoglobulin has an odd number of half-cystine residues, so that one of these must be free at all times. Prior to denaturation, the free sulphydryl group is, apparently, somewhat buried in the molecular structure, but on denaturation it becomes accessible. In milk, the intermolecular formation of disulphides need not be confined to interactions between $\beta$-lactoglobulin molecules; disulphides may be formed between the denatured $\beta$-lactoglobulin and the sulphydryl groups of $\kappa$-casein and possibly also with $\alpha_{s2}$-casein.[78] It has been established that such complexes are formed, with denatured $\beta$-lactoglobulin and also $\alpha$-lactalbumin bound to the casein micelles.[79]

Thus, at temperatures in excess of 100 °C, the serum proteins may be extensively bound to the casein micelles. This in itself will lead to changes in the properties of the micelles, since they have, in effect, acquired an altered surface. This may explain certain aspects of the heat-stability of milk, especially the effect of forewarming. When milk is held for some time at

90 °C prior to extensive heating at 120 °C or 140 °C, aggregation at the higher temperature is modified. It is to be assumed that forewarming allows time for some of these denaturation reactions, and the binding of $\beta$-lactoglobulin to $\kappa$-casein, which have insufficient time to occur when the milk is heated directly to the higher temperatures.

This binding of denatured serum proteins to the casein micelles via disulphide formation need not lead to stabilisation. Thus, $\alpha$-lactalbumin and $\beta$-lactoglobulin have the effect of stabilising the micelles towards heating, but lysozyme, ovalbumin and bovine serum albumin destabilise the micelle system.[80] Lysozyme likewise predisposes micelles to rennet coagulation,[81] although it will not under those circumstances be denatured, nor will it form disulphides with the caseins. The stabilisation or destabilisation cannot be related directly to the gross properties of the soluble proteins, and should be a subject for further study.

The heat-induced precipitation of milk is pH-dependent, and this pH-dependence demonstrates the existence of two types of milk, namely those which show a minimum in the heat-stability/pH profile (Type A) and those which do not (Type B)[82] (Fig. 11). It is not possible to determine clearly what compositional differences cause these different behaviours, but mechanistically it seems that the precipitation of Type A milks within the minimum is a two-stage process rather than a single-stage precipitation of

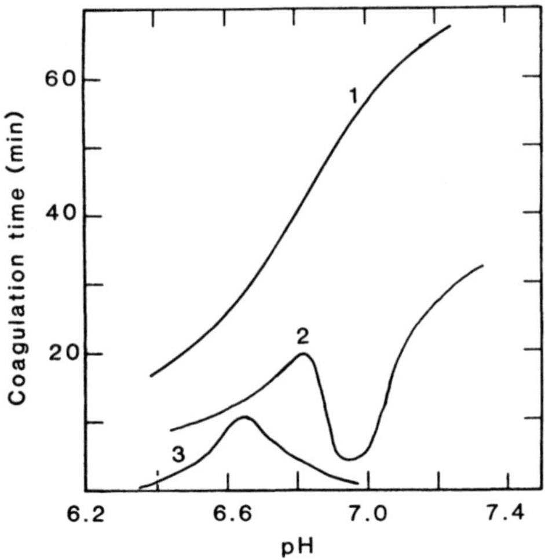

FIG. 11. Stability of milk to heat, as a function of pH. Curve 1, coagulation time of typical type B milk at 135 °C; Curve 2, coagulation of typical type A milk at 140 °C; Curve 3, coagulation of milk at 120 °C concentrated by evaporation to 22·5 % total solids.

Type B milks.[83] The minimum stability observed in Type A milks is not fixed, but can in some cases be shown to be temperature-dependent; thus a milk which shows Type A behaviour at 140 °C may show Type B behaviour at 120 °C.[83] Whether milk is Type A or Type B depends on local factors: thus the incidence of Type B milks in Ireland and Australia is around 1 %, but is 30 % of Canadian milks.[84] $\beta$-Lactoglobulin is one identifiable factor in defining heat-stability in the minimum of the clotting time–pH profile,[84] but other factors are important in defining stability. Addition of urea stabilises Type A milks outside the minimum, and a statistically significant correlation was found between heat stability and urea concentration,[85] the mechanism of which is presumably the reaction of cyanate with lysyl residues. Also, the addition of aldehydic materials considerably increases the coagulation time during heating, once again by modification of lysyl residues.[86]

The causes of heat-instability are not uniquely defined,[84] but it is possible to list several known reactions, which help to define the course of the reaction:

(i)  The alteration of surface properties by binding denatured serum proteins: hence an alteration of zeta-potential of the micelles, and possibly also in their hydration.

(ii)  The decrease in pH which accompanies heating (down to as low as 5·5).

(iii)  Modification of lysyl residues by Maillard reaction, or by reaction with cyanate.[87]

(iv)  Deposition of calcium phosphate. Milk at room temperature is saturated in calcium phosphate, and heated milk serum rapidly precipitates additional insoluble calcium phosphate. The deposition of calcium phosphate in this way may well be a contributory cause of casein precipitation.

(v)  Hydrolysis of $\kappa$-casein.[88]

It is noticeable that most of these reactions involve the possibility of modifying the surface of the micelles and thereby modifying the charge density which has already been noted as an important factor in determining protein and micellar stability.

The stability to heat is greatly influenced by the concentration of the milk: heat-stability–pH profiles of concentrated milks are not at all similar to those of unconcentrated milk[89] (Fig. 11). It appears that the extent of concentration allows different mechanisms to dominate the overall precipitation. Since the precipitation is a highly complex reaction, the overall mechanism under different conditions of temperature and concentration can be influenced by any of a number of possible reactions such as those listed above, whose relative rates can be altered almost independently.

## CONCLUSIONS

The major physical phenomenon which is important in milk proteins is that of precipitation. As has been shown, precipitation reactions involving milk proteins can be initiated by a variety of different causes, especially in the case of caseins. These different mechanisms can be described, for caseins, by considering the surface charge of the protein or micellar unit, which in most cases appears to play an important part. However, hydrophobic interactions are also important, especially in terms of micelle assembly.

Chemical effects have been less emphasised in this description, since they are manifested mainly only at high temperatures, and are relatively unexplored in comparison to the physical chemistry of the milk proteins. Modification reactions of this type not only include the effects described in the preceding section, but also reactions such as lysinoalanine formation,[91] which may alter the nutritional status of the milk protein. Maillard reaction and also destruction of lysine by urea/cyanate reaction may also be prejudicial to the nutritional value. Reactions such as these may occur, but have not yet been quantified in terms of their occurrence during milk processing, if indeed this is possible in a general rather than specific manner. Further research into the details of these reactions is required.

## REFERENCES


1. GORDON, W. G. and GROVES, M. L.. *J. Dairy Sci.*, 1975, **58**, 574.
2. ANDREWS, A. T., *Eur. J. Biochem.*, 1978, **90**, 59 and 67.
3. DAVIES, D. T. and LAW, A. J. R., *J. Dairy Res.*, 1980, **47**, 83.
4. BARRY, J. G. and DONNELLY, W. J., *J. Dairy Res.*, 1980, **47**, 71.
5. ANDERSON, M. and ANDREWS, A. T., *J. Dairy Res.*, 1977, **44**, 223.
6. BRAUNITZER, G. R., CHEN, R., SCHRANK, B. and STANGL, A., *Hoppe-Zeyler's Z. Physiol. Chem.*, 1972, **353**, 832.
7. BREW, K., CASTELLINO, F. J., VANAMAN, T. C. and HILL, R. L., *J. Biol. Chem.*, 1970, **245**, 4570.
8. MERCIER, J.-C., GROSCLAUDE, F. and RIBADEAU DUMAS, B., *Eur. J. Biochem.*, 1971, **23**, 41.

9. BRIGNON, G., RIBADEAU DUMAS, B., MERCIER, J.-C. and PELISSIER, J.-P., *FEBS Letts.*, 1977, **76**, 274.
10. RIBADEAU DUMAS, B., BRIGNON, G., GROSCLAUDE, F. and MERCIER, J.-C., *Eur. J. Biochem.*, 1972, **25**, 505.
11. MERCIER, J.-C., BRIGNON, C. and RIBADEAU DUMAS, B., *Eur. J. Biochem.*, 1973, **35**, 222.
12. SWAISGOOD, H. E., BRUNNER, J. R. and LILLEVIK, H. A., *Biochemistry*, 1964, **3**, 1616.
13. ANNAN, W. D. and MANSON, W., *J. Dairy Res.*, 1969, **36**, 259.
14. MANSON, W., CAROLAN, T. and ANNAN, W. D., *Eur. J. Biochem.*, 1977, **78**, 411.
15. JOLLES, P. and FIAT, A.-M., *J. Dairy Res.*, 1979, **46**, 187.
16. HOPPER, K. E. and McKENZIE, H. A., *Biochim. Biophys. Acta*, 1973, **295**, 352.
17. BIGELOW, C. C., *J. Theor. Biol.*, 1967, **16**, 187.
18. ANDREWS, A. L., ATKINSON, D., EVANS, M. T. A., FINER, E. G., GREEN, J. P., PHILLIPS, M. C. and ROBERTSON, R. N., *Biopolymers*, 1979, **18**, 1105.
19. EVANS, M. T. A., PHILLIPS, M. C. and JONES, M. N., *Biopolymers*, 1979, **18**, 1105.
20. PAYENS, T. A. J. and HEREMANS, K., *Biopolymers*, 1969, **8**, 335.
21. SCHMIDT, D. G., *Biochim. Biophys. Acta*, 1970, **221**, 140.
22. JOLLES, J., ALAIS, C. and JOLLES, P., *Biochim. Biophys. Acta*, 1968, **168**, 591.
23. JOLLES, J., SCHOENTGEN, F., ALAIS, C., FIAT, A.-M. and JOLLES, P., *Helv. Chim. Acta*, 1972, **55**, 2872.
24. VREEMAN, H. J., *J. Dairy Res.*, 1979, **46**, 271.
25. VREEMAN, H. J., BOTH, P., BRINKHUIS, J. A. and VAN DER SPEK, C. A., *Biochim. Biophys. Acta.*, 1977, **491**, 93.
26. WAUGH, D. F., SLATTERY, C. W. and CREAMER, L. K., *Biochemistry*, 1971, **10**, 817.
27. DICKSON, I. R. and PERKINS, D. J., *Biochem. J.*, 1971, **124**, 235.
28. SLATTERY, C. W. and WAUGH, D. F., *Biophys. Chem.*, 1973, **1**, 104.
29. DALGLEISH, D. G. and PARKER, T. G., *J. Dairy Res.*, 1980, **47**, 113.
30. DALGLEISH, D. G. and PARKER, T. G., *J. Dairy Res.*, 1981, **48**, 71.
31. NOBLE, R. W. JR. and WAUGH, D. F., *J. Amer. Chem. Soc.*, 1965, **87**, 2236.
32. DALGLEISH, D. G., HORNE, D. S. and PATERSON, E., *Biophys. Chem.*, 1981, in press.
33. HORNE, D. S. and DALGLEISH, D. G., *Int. J. Biol. Macromol.*, 1980, **2**, 154.
34. VON SMOLUCHOWSKI, M., *Z. Physik, Chem.*, 1917, **92**, 129.
35. WAUGH, D. F., in *Milk Proteins*, H. A. McKenzie (ed.), Academic Press, New York, 1971, Vol. II, Chapter 9.
36. VAN DE VOORT, F. R., MA, C.-Y. and NAKAI, S., *Arch. Biochem. Biophys.*, 1979, **195**, 596.
37. SCHMIDT, D. G., KOOPS, J. and WESTERBEEK, D., *Neth. Milk Dairy J.*, 1977, **31**, 328.
38. SCHMIDT, D. G. and KOOPS, J., *Neth. Milk Dairy J.*, 1977, **31**, 342.
39. SCHMIDT, D. G., WALSTRA, P. and BUCHHEIM, W., *Neth. Milk Dairy J.*, 1973, **27**, 128.
40. HOLT, C., KIMBER, A. M., BROOKER, B. and PRENTICE, J.-H., *J. Coll. Int. Sci.*, 1978, **65**, 555.

41. DEWAN, R. K., CHUDGAR, A., MEAD, A., BLOOMFIELD, V. A. and MORR, C. V., *Biochim. Biophys. Acta*, 1974, **342**, 313.
42. HOLT, C. and MUIR, D. D., *J. Dairy Res.*, 1978, **45**, 347.
43. HOLT, C. *Biochim. Biophys. Acta*, 1975, **400**, 293.
44. DEWAN, R. K., BLOOMFIELD, V. A., CHUDGAR, A. and MORR, C. V., *J. Dairy Sci.*, 1973, **56**, 699.
45. SCHMIDT, D. G., *Neth. Milk Dairy J.*, 1980, **34**, 42.
46. McGANN, T. C. A., DONNELLY, W. J., KEARNEY, R. D. and BUCHHEIM, W., *Biochim. Biophys. Acta*, 1980, **630**, 261.
47. WAUGH, D. F., SLATTERY, C. W., CREAMER, L. K. and DRESDNER, G. W., *Biochemistry*, 1970, **9**, 768.
48. HOLT, C., *Proc. Int. Conf. Coll. Surf. Sci.*, Wolfram, E. (ed.), Akademiai Kiado, Budapest, 1975, 641.
49. WALSTRA, P., *J. Dairy Res.*, 1979, **46**, 317.
50. SCOTT-BLAIR, G. W. and OOSTHUIZEN, J. C., *J. Dairy Res.*, 1961, **28**, 165.
51. WALSTRA, P., BLOOMFIELD, V. A., WEI, G. J. and JENNESS, R., *Biochim. Biophys. Acta*, 1981, **669**, 258.
52. DALGLEISH, D. G., *J. Dairy Res.*, 1978, **46**, 653.
53. SCHMIDT, D. G. and PAYENS, T. A. J., *Surf. & Coll. Sci.*, 1976, **9**, 165.
54. SLATTERY, C. W. and EVARD, R., *Biochim. Biophys. Acta*, 1973, **317**, 529.
55. CARROLL, R. J., FARRELL, H. M. JR. and THOMPSON, M. P., *J. Dairy Sci.*, 1971, **54**, 752.
56. RIBADEAU DUMAS, B. and GARNIER, J., *J. Dairy Res.*, 1970, **37**, 269.
57. ALI, A. E., ANDREWS, A. T. and CHEESEMAN, G. C., *J. Dairy Res.*, 1980, **47**, 371.
58. VON KNOOP, A. M., KNOOP, E., FREDE, E., PRECHT, D. and WIECHEN, A., *Milchwissenschaft*, 1979, **34**, 129.
59. LYSTER, R. L. J., *Biochem. Soc. Trans.*, 1976, **4**, 735.
60. HOLT, C., Personal communication.
61. STERNBERG, M., *Adv. App. Microbiol.*, 1976, **20**, 135.
62. BERRIDGE, N. J., *Nature*, 1942, **149**, 194.
63. GREEN, M. L. and CRUTCHFIELD, G., *J. Dairy Res.*, 1971, **38**, 151.
64. PEARCE, K. N., *J. Dairy Res.*, 1976, **43**, 27.
65. PAYENS, T. A. J. and WIERSMA, A. K., *Biophys. Chem.*, 1980, **11**, 137.
66. PAYENS, T. A. J., *Farad. Disc. Chem. Soc.*, 1978, **65**, 164.
67. DALGLEISH, D. G., *Biophys. Chem.*, 1980, **11**, 147.
68. DARLING, D. F. and VAN HOOYDONK, A. C. M., *J. Dairy Res.*, 1981, **48**, 189.
69. GREEN, M. L., HOBBS, D. G., MORANT, S. V. and HILL, V. A., *J. Dairy Res.*, 1978, **45**, 413.
70. DALGLEISH, D. G., *J. Dairy Res.*, 1980, **47**, 231.
71. HORNE, D. S. and PARKER, T. G., *Neth. Milk Dairy J.*, 1980, **34**, 126.
72. HORNE, D. S. and PARKER, T. G., *J. Dairy Res.*, 1981, **48**, 273.
73. HORNE, D. S. and PARKER, T. G., *Int. J. Biol. Macromol.*, 1981, |399.
74. MULDER, H. and WALSTRA, P. in *The Milk Fat Globule*, Commonwealth Bureau of Dairy Science & Technology, 1974.
75. OORTWIJN, H. and WALSTRA, P., *Neth. Milk Dairy J.*, 1979, **33**, 134.
76. MUIR, D. D., Personal communication.
77. HILLIER, R. M. and LYSTER, R. J., *J. Dairy Res.*, 1979, **46**, 95.
78. SMITS, P. and VAN BROUWERSHAVEN, J. H., *J. Dairy Res.*, 1980, **47**, 313.

79. SHALABI, S. I. and WHEELOCK, J. V., *J. Dairy Res.*, 1977, **44**, 351.
80. FOX, P. F. and HEARN, C. M., *J. Dairy Res.*, 1978, **45**, 159.
81. GREEN, M. L. and MARSHALL, R. J., *J. Dairy Res.*, 1977, **44**, 521.
82. TESSIER, H. and ROSE, D., *J. Dairy Sci.*, 1964, **47**, 107.
83. SWEETSUR, A. W. M. and WHITE, J. C. D., *J. Dairy Res.*, 1974, **41**, 349.
84. FOX, P. F. and MORRISSEY, P. A., *J. Dairy Res.*, 1977, **44**, 627.
85. HOLT, C., MUIR, D. D. and SWEETSUR, A. W. M., *J. Dairy Res.*, 1978, **45**, 183.
86. HOLT, C., MUIR, D. D. and SWEETSUR, A. W. M., *J. Dairy Res.*, 1978, **45**, 47.
87. DARLING, D. F., *J. Dairy Res.*, 1980, **47**, 199.
88. FOX, P. F. and HEARN, C. M., *J. Dairy Res.*, 1978, **45**, 173.
89. MUIR, D. D. and SWEETSUR, A. W. M., *J. Dairy Res.*, 1978, **45**, 37.
90. SWEETSUR, A. W. M. and MUIR, D. D., *J. Soc. Dairy Tech.*, 1980, **33**, 101.
91. ANNAN, W. D. and MANSON, W., *Fd. Chem.*, 1981, **6**, 255.
92. DE WIT, J. N., *Neth. Milk Dairy J.*, 1981, **35**, 47.

# 9

# Milk Proteins—Manufacture and Utilisation

L. L. MULLER

*Division of Food Research,
Dairy Research Laboratory, Victoria, Australia*

Development of the commercial manufacture of milk protein products has occurred almost exclusively in this century. The relative simplicity of isolating casein by acidifying milk to a pH of about 4·6 led to its small-scale production in about 1900 for use in glues.[1] The industrial applications expanded, particularly in paper-coating, as its useful adhesive and water-binding properties became more widely appreciated. World production of casein doubled between the mid-1930s and the mid-1960s, when it reached about 150 000 tonnes,[2] a level from which there has been little change in recent years.[1]

The main change has been a decline in the amount of casein used for industrial purposes and an increase in food uses to the stage where, today, 70–80 % of world production is for use as an ingredient in foods.[1,3]

The growing demand for high-quality protein as an ingredient in formulated foods, and for ingredients to provide desired functional properties, has led to the development of a wide range of milk protein products. The range now includes edible and industrial grades of acid and rennet casein, several caseinates, co-precipitates of casein and whey proteins with varying functional properties, the heat-precipitated whey protein product—lactalbumin—and, more recently, whey protein concentrates made by such fractionation processes as ultrafiltration.

It should be pointed out that a large proportion of the milk proteins used in foods is supplied by such products as dried skim milk or whey. This is illustrated in the case of the USA in Table 1.

 *L. L. Muller*

TABLE 1

ESTIMATED ANNUAL USA PRODUCTION/UTILISATION OF MILK PROTEIN PRODUCTS, 1977

| Product | Amount (*1000 tonnes*) | Protein (*1000 tonnes*) |
|---|---|---|
| Non-fat dry milk | 421 | 151·6 |
| Casein/caseinates | 66 | 62·7 |
| Co-precipitates | 5 | 4·5 |
| Partially delactosed whey | 14·5 | 2·9 |
| Partially demineralised whey | 11·8 | 1·5 |
| Whey protein concentrates | 3·6 | 1·8 |
| Whey solids in protein blends | 12·7 | 1·6 |
| Whey solids in dry whey | 210 | 27·3 |
| Total protein (1000 tonnes) | | 253·9 |

Source: reference 4

## ACID CASEIN

The development of edible grades of casein for use in foods has been achieved by paying careful attention to the hygienic aspects and by improving the approach to the essential steps of the process. These are: achievement of the correct pH of precipitation, adjustment of temperature so as to aid the expulsion of whey and the production of a curd of optimum physical characteristics for its separation from the whey, washing, de-watering, drying, and grinding. Many aspects of these essential steps apply also to the manufacture of rennet casein and co-precipitates. They are therefore discussed below in some detail.

**Precipitation and Curd Formation**
The correct pH for precipitation can be achieved through the action of lactic acid bacteria or by the addition of a mineral acid, usually hydrochloric acid.

*Lactic Casein*
   Where lactic acid bacteria are used, the product is frequently referred to as lactic casein. The original method for making lactic casein involved batch operation in a cheese vat of the stages of coagulation, cutting, 'cooking', whey drainage and washing.[5] The 'cooking' process was by indirect heating

to achieve, in about one hour, a temperature of up to 60 °C to promote syneresis of the curd.

The first step towards reducing the time and labour required in this process was to use a steam injector to draw the coagulum from the vat and simultaneously increase the temperature.[6] Later it was found[7] that the coagulum could be pumped and the cooking temperature attained by steam injection before passage through a holding pipe, which aided agglomeration.

These developments, mainly in New Zealand, made it possible for the coagulation stage to be performed in vertical tanks of up to 100 000 litres capacity followed by continuous processing at rates up to 50 000 litres/of skim milk.[1]


*Mineral Acid Casein*

In manufacturing lactic casein, a cooking temperature of 50–55 °C is generally needed to create a curd firm enough for further processing. This compares with earlier systems for acid casein manufacture in which acid was added to skim milk at a temperature of about 42 °C[8] or in which skim milk and acid were mixed continuously at 43–45 °C in a baffled chamber.[2] The differences in temperature requirement were probably associated with the rate of removal of calcium from the casein. During the several hours required for pH reduction in lactic fermentation, there is ample time to attain equilibrium between bound and free calcium. When acid was added to skim milk at 43–45 °C, it was found[9] that the coagulum began to form in less than 0·1 s, too short a time to complete the solubilisation of colloidal calcium phosphate. Inefficient mixing led to over-acidulation of some of the casein, giving small, soft particles—'fines'—which were lost in the whey, thus reducing yields. Lowering the milk temperature to 30–35 °C, spraying the acid into a zone of turbulence and then using steam injection to reach a cooking temperature of 44–46 °C overcame the difficulty. This modification was relatively simple, was capable of reducing losses of fines in whey to 1–2 % of the casein, and so was widely adopted.[2]

In practice, the efficiency of this system for acid/skim milk mixing depended on careful observance of design and operating principles in relation to milk flow rates. Variations from the optimum led to an increase in fines losses.

This problem was overcome by a further development of the principle of mixing at reduced temperature. A plant was designed[10] which ensured a constant pH by using metering pumps to give a steady flow of both milk and acid. Mixing was at a controlled temperature below 25 °C. Automatic regulation of steam injection and the use of a holding tube gave complete coagulation and a well-structured curd. This approach ensured that losses in whey were less than 1 % of the casein.[3]

## Separation of Curd, Washing

The removal of lactose, salts and free acid from casein is an essential feature in the manufacture of a high-quality product.[2] These 'impurities' are present in both the whey and the freshly precipitated curd. Their removal requires efficient separation of the curd from the whey or wash-water and the establishment of correct conditions for washing.[11] The 'impurities' in the curd diffuse from it at a rate which depends on the size and permeability of the curd particles and on the purity, amount, temperature and rate of movement of the wash-water.[12]

A number of systems have been used for the continuous separation of curd from whey or wash-water. Perhaps the most commonly used device is an inclined screen. A screen developed in Australia[13] was based on 90-mesh stainless steel gauze inclined at an angle of 30 °C. Studies in New Zealand[14] showed that the efficiency of inclined screens was a function of the aperture size, the dimensions and the angle. The screen developed for milk processing rates of about 23 000 litres/h was of 180 $\mu$m stainless steel mesh, was 0·75 m × 1·20 m and had the facility to adjust the angle within the range of 20°–45°. A screen developed in Europe[3] was based on a polyester fabric (80 $\mu$m) laid on a cascade-like profile which causes the curd to turn and roll as it travels down the slope.

High efficiency in separation of curd from whey can be achieved by the use of continuous roller presses[15] or horizontal centrifuges.[16] More whey is recovered and less wash-water is needed.

To reduce the lactose to the desired level of about 0·1 % in the dry casein, at least three washing stages are necessary, with an average holding time of about 30 min in each stage.[2] However, four or five stages are common. Continuous washing systems are now normally used with a countercurrent flow of water, as this reduces the quantity of water needed and restricts to one outlet any loss of fine casein particles. With efficiently operated inclined screens, the loss of fines in wash-water is unlikely to be more than 1–2 % of the yield.[2] However, the losses can be important, so recovery of these fines is often attempted by using hydrochlones[17] or a desludging separator[3] on the final wash-water. An alternative approach is to use a horizontal centrifuge for final separation of curd and wash-water.

Adjustment of pH, if the water has more than a given level of alkalinity,[18] and careful control of temperature are important aspects of the washing process. The most common approach when edible casein is being manufactured is to apply a temperature of 70 °C or more in the penultimate washing stage to ensure high microbiological quality. Temperatures in the preceding stages would therefore range upwards from the temperature of precipitation. The final wash temperature is normally 40–45 °C, as it is important to minimise matting of the curd during its separation by press or centrifuge.

## Drying, Tempering, Grinding

Of the several types of drying equipment, probably the most widely used in recent years is the vibratory-type developed in New Zealand.[1,2,19] With this drier, the curd passes through a mill to reduce it to evenly sized particles before they travel by means of a vibratory action over several trays of perforated stainless steel. Even with this system the moisture content of individual particles will vary, and it is necessary to blend the casein in 'tempering bins' to obtain an even moisture content before the casein can be readily ground to a fine particle size.[20]

There are substantial capital and operational costs involved in tempering and grinding, and a new drying system—attrition drying[21]—is therefore of particular interest. This drier features a fast-revolving, multi-chambered rotor and a stator which provides for grinding of the curd to very small particles. A large surface area is exposed to the hot air which conveys the curd, so that drying is very fast—1–2 s—and the product requires no grinding. This type of drier has been recently installed in several European factories. Information on consumer reaction to the product and on comparative economics will be of considerable interest.

## RENNET CASEIN

The coagulum obtained by the action of rennet on skim milk may be processed in a similar fashion to the coagulum in lactic casein manufacture. A continuous process for the separation of curd from whey, washing, pressing and drying, was adopted in the early 1960s.[22] Continuous cooking of the coagulum was introduced in the early 1970s.[23] Operating conditions were similar to those for lactic casein.

For industrial usage in the manufacture of plastics, rennet casein made by continuous cooking has better clarity as less fat is retained in the casein, though there are more pinholes in the plastic, attributable to gas incorporation.[24]

There is growing interest in the use of rennet casein as a food ingredient, partly because of its suitability for use in making cheese analogues and partly because of its good flavour stability.[25]

## CO-PRECIPITATES

As whey proteins are soluble at pH 4·6, only some 80 % of the protein of skim milk can be recovered during the manufacture of casein. The desire to improve protein recovery, as well as to increase the nutritive value and the range of functional properties in milk proteins for use in foods, led to the commercial development of co-precipitates of casein and whey proteins.[2] The early developments were mainly in the USA and the USSR.[2,26,27] The processes in the USA[28,29] involved heating the skim milk to denature the whey proteins and using acid to form the co-precipitate. The process developed in the USSR[30] was based on co-precipitating the proteins from heated skim milk by addition of $CaCl_2$.

Australian work on co-precipitates began with modification of the USSR process to produce a curd which would withstand washing.[31] The technique that had been developed for mixing acid and skim milk in the manufacture of acid casein[19] was adapted to spray a $CaCl_2$ solution into milk heated to 90–95 °C. Use of a holding tube followed the spray to provide suitable conditions for aggregation and syneresis of the curd before separation from the whey and then washing.

The level of calcium in the co-precipitate was observed to have a major effect on its functional properties, and the process was further modified to produce well-washed co-precipitates of defined calcium content.[32] The conditions used are summarised in Table 2. The terms high-, medium- and low-calcium were used to define co-precipitates with calcium contents in the

TABLE 2
CONDITIONS USED IN MANUFACTURE OF CO-PRECIPITATES[32]

| *Type of co-precipitate* | *Time milk held at 90°C (min)* | *$CaCl_2$ added (% of weight of milk)* | *Acid added* | *pH of precipitation* |
|---|---|---|---|---|
| High-calcium | 1–2 | 0·2 | No | 5·8–5·9 |
| Medium-calcium | 10–12 | 0·06 | Yes | 5·3–5·6 |
| Low-calcium | 15–20 | 0·03 | Yes | 4·6–4·8 |

ranges 2·5–3·0 %, 1·0–2·0 % and 0·5–0·8 %, respectively. Control of calcium level was achieved by varying the amount of $CaCl_2$ added, changing the length of time at which the milk was held at about 90 °C and, for low- and medium-calcium co-precipitate, varying the pH of precipitation by acid.

There have been a number of recent studies on the variables in manufacturing co-precipitates and on modifications to the process.[27] Some of these studies involved increasing the amount of whey protein in the co-precipitate by the addition of whey to skim milk or buttermilk.[33–35] Other studies related yield and composition of the co-precipitate to such factors as pH and calcium content of the heated mixture,[36] the influence of added whey[37] and pH adjustment of the skim milk.[38] In general, these studies have indicated ways in which the process can be optimised and modified to suit different approaches to the overall system of manufacture.

A considerable amount of data has been published on the properties of co-precipitates in relation to their solubility in various solvents and to their applications in foods.[2,27,32,35,36,38,39] Estimations of protein efficiency ratios confirmed the expected improvement over casein in nutritive value.[40]

## CASEINATES, MILK PROTEINATES

Many of the food applications for milk proteins require that they be in a water-soluble form. However, there are notable exceptions. For example, in the Australian milk biscuit,[41] the requirement was for a milk protein preparation which could be added at a high level to a biscuit dough without competing unduly with the flour for the small amount of available water. A high-calcium co-precipitate, carefully controlled in respect of pH and calcium level and dispersed for spray-drying in about 2 % of sodium tripolyphosphate, possessed the necessary properties.[42]

Calcium caseinate also fulfils particular functional requirements, which include low solubility. The most common technique for preparing a calcium caseinate dispersion for spray-drying involves the reaction of a good-quality acid casein with $Ca(OH)_2$ to achieve a calcium content of 1·0–1·5 % at a pH of about 6·5.[2] Care is necessary in the application of heat during the dispersion process, as there is increasing aggregation of the protein and possible gel formation as the temperature is increased.[43,44]

Soluble caseinates of importance have sodium, ammonium, potassium or magnesium as the cation. Sodium caseinate is the one most commonly used in foods. To obtain a bland-flavoured caseinate, it is normal to use a fresh acid casein curd, dissolved in the appropriate alkali and spray-dried.[2,45] The main difficulties in the manufacture of spray-dried sodium caseinate are related to the logarithmic increase in the viscosity of caseinate solutions as their concentration increases, and to the tendency for dispersion of casein to be impeded by the relatively impervious gel which forms on the surface of the particles in the presence of alkali. Dissolving techniques found satisfactory in practice[2,6,46,47] generally involve applying strong shearing forces to fresh wet curd during additions of water and alkali. Careful control of pH is essential. The minimum viscosity for sodium caseinate is in the pH range 6·6–7·0.[45,48] In practice, it is better to maintain a pH below this value during the solubilisation process and make any adjustment when the curd is finally in solution.[46,47] It is also important to avoid exposing casein to high pH and temperature during dissolving, as this can lead to loss of lysine and serine and the production of degradation products such as lysinoalanine.[49]

The viscosity of the soluble caseinates limits the concentration at which they can be spray-dried, normally to about 20 % solids at 90–95 °C. This low level of solids reduces the output from the drier. This factor, and the low bulk density of the powder, add substantially to the costs of drying, packaging and transport. A number of approaches have been used in an effort to reduce these costs:

(1) Spray-dried caseinate has been ground in a pinmill to increase its bulk density.[50]

(2) Granular caseinates have been produced by reducing the moisture content of casein curd to such a level (below 40 %) that a free-flowing mass can be maintained by agitation after alkali (carbonate or bicarbonate powder) has been added. After time for conversion, the mixture may be dried and ground in conventional casein equipment.[45,51]

(3) Finally, there is a report[21] of the first commercial use of the attrition drier for making caseinates. The drier was fed with the free-flowing curd/alkali mixture described earlier[51] to produce a highly soluble caseinate with a bulk density much higher than that of spray-dried caseinate.


## WHEY PROTEIN PRODUCTS

The manufacture of cheese, casein and similar products throughout the world results in some five million tonnes of whey solids, of which over

500 000 tonnes is protein. A major difficulty in the recovery of this protein is its low concentration in whey—4–7 g/litre. A wide range of processes have been devised to permit recovery in an economical manner.

The main traditional process results in the product, lactalbumin. This process, which has been described in detail,[52] consists of heating the whey for times and at temperatures that are related to pH and the level of calcium, and separating the precipitated proteins before washing and drying. The product is used to improve the nutritional value of such foods as speciality breads, biscuits and pasta where the lack of solubility of the protein is an advantage.[53]

Many processes have been described for whey protein recovery,[54,55] including ultrafiltration, ion-exchange chromatography and complex formation. Gel filtration has, the author believes, been used commercially in the USA to manufacture a range of speciality products. However, there has been no major development reported of the other processes, except ion-exchange. The author understands that commercial application of the 'Spherosil' (Rhone-Poulenc Industries, Paris) porous silica ion-exchange process is beginning in France. The protein is absorbed onto the ion-exchange medium and eluted with either weak hydrochloric acid, for sweet whey, or ammonia, for acid whey.

The main process being applied commercially is ultrafiltration (UF). From a recent summary of trends in whey processing[56] and developments since then of which the author is aware, UF is being used to make whey protein concentrates in over 50 dairy factories around the world.

The subject of ultrafiltration and its applications is too vast to be dealt with in this chapter. Excellent accounts are available,[57] as well as a detailed treatment of whey proteins and products in relation to end-use applications.[58]

The recent improvements in UF equipment design and the advent of robust membranes have given the dairy industry the opportunity to explore this relatively new technology. Apart from work towards developing a range of whey protein concentrates with properties which can be varied to suit particular applications in food, the industry is also exploring the applications of UF in making cheese and similar products.

Milk proteins can now be made available to the food industry in a great variety of forms. There is still much work to be done in conjunction with the manufacturers of formulated foods to ensure that these manufacturers are aware of the possibilities and that they and the dairy industry understand how best to match protein product to its food application.

## REFERENCES

1. SOUTHWARD, C. R. and WALKER, N. J., *NZ J. Dairy Sci. Technol.*, 1980, **15**, 201.
2. MULLER, L. L., *Dairy Sci. Abstr.*, 1971, **33**, 659.
3. GWOZDZ, E., *XX Int. Dairy Congr.*, 1978, Paper No. 51ST.
4. MORR, C. V., *J. Dairy Res.*, 1979, **46**, 369.
5. SPELLACY, J. R., *Casein, Dried and Condensed Whey*, Lithotype Process Co., San Francisco, 1953.
6. TARASOV, F., *Molochn. Promst.*, 1951, **12**(5), 24.
7. KING, D. W., *XVI Int. Dairy Congr.*, 1962, C, 801.
8. GORDON, D. M., *Dairy Eng.*, 1961, **78**, 87.
9. MULLER, L. L. and HAYES, J. F., *Aust. J. Dairy Technol.*, 1962, **17**, 189.
10. GWOZDZ, E. and HUTIN, G., *Rev. Lait. Fr.*, 1970, No. 277, 471.
11. ZADOW, J. G., *Aust. J. Dairy Technol.*, 1971, **26**, 9, 14.
12. MULLER, L. L., *Aust. J. Dairy Technol.*, 1959, **14**, 81.
13. MULLER, L. L., *Aust. J. Dairy Technol.*, 1960, **15**, 89.
14. HOBMAN, P. G. and ELSTON, P. D., *NZ J. Dairy Sci. Technol.*, 1976, **11**, 136.
15. HOBMAN, P. G. and ELSTON, P. D., *NZ J. Dairy Sci. Technol.*, 1976, **11**, 281.
16. HIGGS, S. L., SOUTHWARD, C. R., MARSHALL, K. R. and WEAL, B. C., *45th Ann. Rep. NZ Dairy Res. Inst.*, 1973, 90.
17. JEBSON, R. S., *NZ J. Dairy Sci. Technol.*, 1969, **4**, 165.
18. MULLER, L. L. and HAYES, J. F., *Aust. J. Dairy Technol.*, 1961, **16**, 106.
19. Practical Dairy Developments Ltd (NZ), *British Patent* 965 135, 1964.
20. KING, D. W., *NZ J. Dairy Sci. Technol.*, 1970, **5**, 100.
21. ROEPER, J., Case report, *IDF Sem. Dairy Ingred. in Foods*, 1981.
22. NEFF, E., *J. Agric. Victoria*, 1966, **64**, 157 and 269.
23. WEAL, B. C. and SOUTHWARD, C. R., *NZ J. Dairy Sci. Technol.*, 1974, **9**, 2.
24. SOUTHWARD, C. R. and WEAL, B. C., *NZ J. Dairy Sci. Technol.*, 1974, **9**, 5.
25. ROEPER, J., *NZ J. Dairy Sci. Technol.*, 1976, **11**, 62.
26. FOX, K. K., in *Byproducts from Milk*, 2nd edn., Webb, B. H. and Whittier, E. O. (eds.), AVI, Westport, Conn., 1970.
27. SOUTHWARD, C. R. and GOLDMAN, A., *NZ J. Dairy Sci. Technol.*, 1975, **10**, 101.
28. SCOTT, E. C., *US Patent* 2 623 038, 1952.
29. HOWARD, H. W., BLOCK, R. J. and SEVALL, H. E., *US Patent* 2 665 989, 1954.
30. D'YACHENKO, P. F., VLODAVETS, I. and BOGOMOLOVA, E., *Molochn. Promst.*, 1953, **14**, 33.
31. BUCHANAN, R. A., SNOW, N. S. and HAYES, J. F., *Aust. J. Dairy Technol.*, 1965, **20**, 139.
32. MULLER, L. L., HAYES, J. F. and SNOW, N. S., *Aust. J. Dairy Technol.*, 1967, **22**, 12.
33. WAKADO Co. Ltd, *NZ Patent* 150 401, 1969.
34. AIRD, R. M., *Fd. Technol. NZ.*, 1971, **6**, 29.
35. DE BOER, R., *US Patent* 3 882 256, 1975.
36. VATTULA, T., HEIKONEN, M., KREULA, M. and LINKO, P., *Milchwissenschaft*, 1979, **34**, 139.
37. KABUS, M., *Nahrung*, 1975, **19**, 963.
38. SOUTHWARD, C. R. and AIRD, R. M., *NZ J. Dairy Sci. Technol.*, 1978, **13**, 77.

39. SOUTHWARD, C. R. and GOLDMAN, A., *NZ J. Dairy Sci. Technol.*, 1978, **13**, 97.
40. LOHREY, E. E. and HUMPHRIES, M. A., *NZ J. Dairy Sci. Technol.*, 1976, **11**, 147.
41. BUCHANAN, R. A. and TOWNSEND, R. F., *Aust. J. Dairy Technol.*, 1969, **24**, 113.
42. MULLER, L. L., HAYES, J. F. and TOWNSEND, R. F., *XVIII Int. Dairy Congr.*, 1970, 1E, 429.
43. HAYES, J. F., SOUTHBY, P. M. and MULLER, L. L., *J. Dairy Res.*, 1968, **35**, 31.
44. ROEPER, J., *Proc. Jubilee Conf. Dairy Sci. Technol. NZ*, 1977, 81.
45. TOWLER, C., *Proc. Jubilee Conf. Dairy Sci. Technol. NZ*, 1977, 83.
46. BURSTON, D. O., MULLER, L. L. and HAYES, J. F., *IDF Seminar on Casein and Caseinates, Paris*, 1967, PARIS–SEM subject 4.
47. TOWLER, C., *NZ J. Dairy Sci. Technol.*, 1976, **11**, 24.
48. HAYES, J. F. and MULLER, L. L., *Aust. J. Dairy Technol.*, 1961, **16**, 265.
49. CREAMER, L. K. and MATHESON, A. R., *NZ J. Dairy Sci. Technol.*, 1977, **12**, 253.
50. BALDWIN, A. J. and SOUTHWARD, C. R., *NZ J. Dairy Sci. Technol.*, 1977, **12**, 232.
51. TOWLER, C., *NZ J. Dairy Sci. Technol.*, 1978, **13**, 71.
52. ROBINSON, B. P., SHORT, J. L. and MARSHALL, K. R., *NZ J. Dairy Sci. Technol.*, 1976, **11**, 114.
53. MORR, C. V., *NZ J. Dairy Sci. Technol.*, 1979, **14**, 185.
54. MORR, C. V., SWENSON, P. E. and RICHTER, R. L., *J. Fd Sci.*, 1973, **38**, 324.
55. MARSHALL, S. C., *NZ J. Dairy Sci. Technol.*, 1979, **14**, 103.
56. ZALL, R. R., KUIPERS, A., MULLER, L. L. and MARSHALL, K. R., *NZ J. Dairy Sci. Technol.*, 1979, **14**, 79.
57. MARSHALL, K. R., in *Proc. Workshop Treating Food Wastes for Profit*, Aust. Academy Technol. Sci., Melbourne, 1980.
58. EVANS, M. T. A. and GORDON, J. F., in *Applied Protein Chemistry*, R. A. Grant (ed.), Applied Science, London, 1980.

# 10

# Food Protein From Leaves

N. W. PIRIE

*Rothamsted Experimental Station, Harpenden, UK*

## INTRODUCTION

Nitrogen fixation was described twenty years ago[1] as '...one of the most underexploited chemical discoveries of all time...'. It is now exploited on a very large scale and is responsible for much of the recent increase in crop production. However, less attention has been given to ensuring that optimal use is made of what is produced. Fodder fractionation is one aspect of optimisation: it may now have usurped the unenviable position once held by nitrogen fixation. The basis for that suggestion is that leaf protein is potentially the most abundant of all the protein sources, and the leaf residue after protein extraction can be more economically conserved for use as ruminant fodder than the original crop.

Protein extraction from leaves has a long history.[2,3] Its use as a food was apparently suggested a century ago because Lawes[4] commented: 'It might be possible by some chemical process to produce from grass a nutritious substance which a man could use as food, but the food so extracted would be far more costly than as it existed in the grass, and no one would think of preparing such a food for oxen and sheep...'. Up to a point, Lawes was right. It may be unreasonable to feed ruminant animals on leaf protein (LP), to extract it from leaves that can be eaten after conventional cooking, or to supply it to people who make little use of green vegetables. Leaf protein should supplement vegetables rather than replace them. In spite of the doubts of Lawes and others, some interest was shown during the first half of this century and several patents were allowed. Their validity is questionable, but they are important because, having now lapsed, they protect the basic principles of LP extraction and use from further patenting.

                                    *N. W. Pirie*

Fodder fractionation produces a protein concentrate that is food for people and other non-ruminants, a partly dewatered residue which contains the unextracted part of the protein, and an aqueous solution containing carbohydrates, amides, minerals, etc. For the sake both of the local amenities and for economy, it is essential to make full use of all three products. This was emphasised in 1942;[5] failure to appreciate the point was responsible for several gloomy economic forecasts. The point has now been recognised and there is commercial fractionation in Denmark, France, Spain, the USA and possibly elsewhere. Furthermore, research on the subject is continuing in other countries.


## SOURCES OF LEAF


Protein can be extracted satisfactorily from many, but not all, species of leaf. Leaves that are hard and dry, acid, glutinous, rich in phenolic substances, or beginning to wither as a result of disease or maturity do not extract well. Work at Rothamsted is, for convenience, mainly with species used in conventional agriculture. An extensive range of tropical species has been studied in Calcutta and Aurangabad.[3] Most of the protein used in human feeding trials was made from lucerne (*Medicago sativa*). This increases the significance of the trials because LP from lucerne has a stronger and less attractive flavour than LP from many other sources. The conclusion from this work is that the annual yield of dry, 100 %, protein can be $2\,t\,ha^{-1}$ in the UK and, because of the absence of winter cessation of growth, 3 or $4\,t\,ha^{-1}$ in the tropics. However, since fertiliser must be used lavishly to get such yields it may be prudent to aim only at 1·5 and $2·5\,t\,ha^{-1}$. Leaves that are the by-product of a conventional crop are an intermittent but attractive source.

Sugar beet (*Beta vulgaris*) and potato (*Solanum*) leaves are the two most abundant by-products in the temperate zone. Yields from the former, which is largely wasted in the UK though not in Germany or Poland, depend on the weather and the date of harvest: the largest yield during casual trials at Rothamsted was $500\,kg\,LP\,ha^{-1}$. However, even if the yield in commercial practice were only half our experimental yield, the 193 000 ha on which sugar beet is grown in Britain would produce 48 000 t of extracted protein as well as a fibre residue that would be more attractive cattle fodder than the untreated tops. Much larger amounts of LP could be made in Europe: Plaz and Chartier[6] estimate that sugar beet tops in the EEC contain 13 Mt of dry matter.

The outstanding photosynthetic efficiency of sugar beet is due in part to the vertical disposition of its leaves. Towards the end of the growing season, the leaves bend over so that there is more mutual shading and a decline in efficiency. A few experiments suggest that part of the leaf can be harvested early in the year with little diminution in sugar yield—presumably because the new growth stays more nearly vertical. The idea that some leaf can be harvested from a crop without affecting the yield of the primary product deserves fuller investigation.

As a safeguard against blight, and to facilitate tuber-lifting later, potato haulm is usually destroyed mechanically or with a herbicide at the beginning of September. The main reason for the neglect in Britain of potato haulm as an animal feeding stuff, is fear of poisoning by solanin and other glycoalkaloids. These are, to a large extent, removed from both LP and the fibrous residue during processing. Leaf protein has been made regularly from potato haulm at Rothamsted since 1952; it has also been made in India, Pakistan and Poland. In nutritive value for rats, it resembles LP from other sources.[7,8]

The yield of LP depends on the potato variety and the date on which the haulm is taken.[8,9] Some early varieties yield 600 kg ha$^{-1}$; maincrop can yield 300 kg in late August, but yield diminishes to 100 kg or less by mid-September. Once the idea has been accepted that something valuable can be extracted from potato haulm, farmers would probably harvest it a little earlier than usual. But tuber weight is still increasing in September; it would therefore seldom be worthwhile to take the haulm early simply to increase the LP yield. Early potatoes occupy 30 000 ha in Britain, and maincrop 230 000. It is reasonable to conclude that 50 000 t of LP could be extracted from the haulm.

Pea vines and vegetable wastes discarded in the field can be used: leafy wastes from vegetables in markets are usually too bruised to be worth processing. As a general rule there is so much autolysis after 24 h (or less in a hot country) that leaves from any crop should be used within a few hours of being harvested, and they should be harvested by cutting rather than flailing unless they can be extracted immediately.

Much of the protein in deciduous tree leaves autolyses in autumn to products that return to the roots: the leaves finally shed are too dry to extract satisfactorily. There are chemical treatments that cause leaf fall at other times of year. This method of collection would be worth study: otherwise, collection difficulties make fully grown trees, felled in the course of conventional forestry, improbable as a source of LP. The straight, unbranched habit of coppiced trees simplifies mechanical leaf stripping.

Coppiced trees are already extensively grown for firewood and paper pulp, and several countries plan the cultivation of 'energy plantations' for industrial fuel. When species are being selected, some attention should be given to the extractability of the protein in the by-product leaves. That point applies with equal force to crops such as cassava, chicory and sorghum which are suggested as sources of fuel alcohol.

Work at Rothamsted suggests that protein tends to extract less readily from tree leaves than from the leaves of other types of plant. Elder (*Sambucus*) is the best that we have found. There are many more species that deserve study because a food-producing tree crop would be the ideal replacement for tropical rain forest. It is now obvious that ecological disaster follows attempts to cultivate annual plants in regions where there is frequent intense rain. Unless the soil is protected by the roots of perennial plants, erosion is a serious risk. The trees usually thought of as replacements of natural rain forest produce an exportable commodity such as rubber or palm oil; it would be advantageous if by-products from tree crops could be used to produce protein for local consumption.

The untended, mixed growth of weeds on land is not a potential source of LP although some individual species from the mixture may ultimately be used. The wild growth is unsuitable because it is unmanured and grows on rough sites from which collection would be difficult—otherwise they would not have weeds on them. The situation is different with water weeds. Obviously they do not suffer from drought, and they are often abundantly manured. Excessive growth of water weeds increases the waste of water by transpiration in hot countries, interferes with flow in irrigation ditches, interferes with navigation, and is a health hazard when it harbours disease vectors such as snails. Much effort is therefore expended on attempts at control. Following mechanical destruction, herbicides and 'biological control', weed residues remain *in situ*. Thus, these methods may control growth, but they do not affect eutrophication. The killed plants rot in the water: most of the elements they contain, and the excreta of an unharvested agent of 'biological control', are likely to return there. When herbicides are used, they and their breakdown products may remain in the water and so make it unsuitable, or less suitable, for irrigation. If water weeds were used, and so removed from the water they infest, that water would not only be freed from infestation, but would be depleted of the elements causing eutrophication. This would economically convert a problem into an asset.

Water hyacinth (*Eichhornia crassipes*) is the most abundant and troublesome of the weeds. It is said to cover 200 000 ha in India, and similar areas elsewhere. The total area is probably more than 1 million ha with an annual growth of 10–30 t DM ha$^{-1}$. The DM of the whole floating plant contains about 2·5 % N, the leaves contain up to 5 %. An impressive amount of protein is therefore potentially available. Unfortunately it does not extract readily unless alkali is added.[10] Extraction is being commercialised in the Philippines. Two other floating weeds, Nile cabbage (*Pistia stratiotes*) and the fern (*Salvinia auriculata*), extract as badly as water hyacinth at their natural pH; extraction with alkali has not, apparently, been tried.

It would be easy and economical to collect these floating weeds with equipment mounted on a barge, process the extract on it, and discharge the soluble material. The compact extracted fibre and LP would have to be transferred to land less often than would be necessary with bulky fresh weed. However, that method of working would obviously not control eutophication as effectively as processing on land.

Protein appears to extract more easily from rooted than from floating water weeds; there is no obvious physiological basis for this. Boyd[11] lists several rooted species that extract well; *Justicia americana* yielded 300 kg LP ha$^{-1}$ when harvested in May or June. Nothing seems to be known about the readiness with which these plants regrow after harvest. Unfortunately, reeds (*Typha* and *Phragmites*), which produce very heavy crops in suitable regions, do not extract well. So much effort is now expended on not very successful attempts to control water weeds, both floating and rooted, that sustained work on finding uses for them is likely to be profitable. The subject is now getting more attention.

It would seldom be sensible to extract LP from the leaves of crops, trees or water weeds (e.g. *Ipomoea aquatica*) that are already being eaten in the conventional way as green vegetables. Few communities eat green vegetables to the extent that is both possible and desirable; it would therefore be advantageous to popularise their use. It is unlikely that a community which is unaccustomed to eating green vegetables will take readily to LP. If it is felt that LP would be more acceptable if it were known to be made from a familiar vegetable. Further, it would be reasonable to make it from that source for children who, because of their small stomachs, cannot cope with much leaf.

A by-product leaf becomes available when the main product is harvested, and water weeds would be harvested when the infestation had become too serious. In these circumstances, those concerned with extracting LP are offered little choice: they have to make the best use possible of what is available. On the other hand, crops grown primarily for extraction can be harvested at the optimum age. Here judgement depends on the relative importance attached to the LP and to the extracted residue, because the yield of the former declines with increasing maturity, whereas the yield of the latter increases.[3] Varieties differ in the extent to which maturity affects extractability. The percentage of the protein in a leaf that is extractable by the methods normally used depends on such factors as the protein and water content of the leaf, its pH and phenolics content, and the resilience of the fibre. All these change with maturity.[12] There has as yet been little work on the selection of varieties with a set of qualities that will ensure good extraction without unreasonably frequent harvesting of very young material. It is therefore reasonable to assume that varieties can be found that will give greater yields than those that have already been tested. For example, it should be advantageous to transfer varieties to a climatic zone in which, because anthesis in that zone is delayed, they are not normally grown. Prolonged vegetative growth favours protein extraction.

## TECHNIQUE OF EXTRACTION

Dust and soil on leaf surfaces will contaminate the extracted juice. If the LP is intended as human food, the crop should therefore not be harvested with a flail, which sucks up some dirt, and it should be rinsed before extraction. This will make the juice more dilute, but will increase the percentage of the protein that is extracted. Less care is obviously needed if the LP will be used only as animal feed.

Protein is not extracted from leaves by simple pressure. The leaves must first be rubbed or disintegrated, and it is more important to maintain pressure on the pulp for several seconds, so as to allow juice to run out of the fibre, than to apply intense pressure. From adequately pulped leaves, 80 % of the juice that can be expressed with intense pressure, is expressed at $2–3\,\text{kgf cm}^{-2}$ (200 to 300 kPa). Pressure is not sufficiently prolonged with rollers, and conventional screw expellers do not disintegrate the crop sufficiently before applying pressure. Hitherto, most research on LP has been concerned with the quality of the product and the yields that could be expected from different types of leaf. The cost of the equipment and the amount of energy used in the extraction were secondary considerations. Now that the merits of LP are recognised, the economics and practicalities of extraction deserve much more attention.

Somewhat different problems arise according to whether production is envisaged on the domestic, farm or industrial scale. These scales may be roughly classified as the daily processing of 10 kg, 100 kg and 100 t of crop (wet weight) and these quantities would produce, respectively, about 250 g,

2·5 kg and 2·5 t (dry weight) of extracted protein. At present, there is no satisfactory equipment for use on the domestic scale. Juice can be made with a domestic mincer followed by hand squeezing in a cloth—but the process is tedious. The best arrangement will probably be a pestle lifted by pedal-driven prongs and allowed to drop onto leaves in a mortar with a strong, perforated bottom. More leaf would be added as the pulp came through, and juice would be expressed in a simple hand press.

Equipment for industrial production is being studied in several institutes—notably by Bruhn and his colleagues.[13] In the latter the crop is forced out through perforations in a cylindrical die by means of an internal roller, i.e. by an action similar to that of a pelleting press. Juice is then expressed in a separate unit.

For farm-scale production a modified screw expeller has been developed.[14] Juice is liberated by rubbing the crop thoroughly with a set of angled paddles which push the resultant pulp into the section of the unit in which it is pressed. That section consists of a perforated cylinder within which an auger rotates. This is mounted on a cone with its wide end at the outlet of the cylinder. A thin shell of pressed fibre emerges from the outlet while the juice comes through the holes (3 mm) in the cylinder. With a cylinder 160 mm in diameter, the unit takes about 300 W when fed at 1 to 2 kg min$^{-1}$. Power consumption and rate of working obviously depend on the texture of the crop. The unit runs at 10 to 20 rpm and so could, if need be, be powered by an animal. Equipment for production on this scale is at present in most need of skilled work on design so that a unit that is cheap, economical and robust can be supplied for research, or for routine production of LP from local material for local use.

Commercial enterprise will ensure that there is research on large scale equipment. In the author's opinion, the Bruhn technique of extrusion through a die is the best method for making pulp on a large scale: it would be difficult to manage on a small scale. In principle, a press in which there is no relative movement between compacted fibre and the pressing surface(s), i.e. a belt, 'horn angle', or double-cone V press, is preferable to a screw expeller because less power is wasted and less fibre is extruded into the juice. However, it is difficult to make flexible belts strong enough to apply the necessary pressure in a belt press, and the pressure is applied so suddenly that part of the wet pulp is apt to be squeezed back instead of going into the nip. On the other hand it is difficult to feed sloppy pulp into the 'horn angle' or double-cone V press. By contrast, expellers are simple and robust: provided juice has been fully liberated by adequate pulping, their defects do not become obtrusive at moderate pressures. The juice that runs out from

                         *N. W. Pirie*

pulp quickly with gentle pressing contains 2–5 times as great a concentration of protein as juice coming out later under heavy pressure.[15] If, for the sake of economy in drying later, maximum juice expression is required, pulp should be pressed gently in an expeller and the resulting partly dewatered and easily handled fibrous product should be pressed again in a press in which there is no relative movement between metal and fibre.

## SEPARATION OF PROTEIN FROM THE EXTRACT

Part of the protein coagulates when a leaf extract is left for a few hours or days; the time depends on species and temperature. This form of coagulation, sometimes with exclusion of air, has been advocated, but it has many defects. The curd that forms is soft and difficult to handle, much protein is lost as a result of proteolysis, and there is a serious risk that clostridia and similar harmful bacteria will grow during the coagulation. Leaf protein precipitates immediately at pH 4. This technique does not have the last two defects and is widely used in laboratory work, but the curd is so soft that it has to be separated by centrifugation. Heat coagulation is therefore generally accepted as the most satisfactory method for making a protein curd. Coagulation begins at about 50 °C and is complete at 70 °C. The circumstances in which it might be advantageous to separate fractions of differing coagulability will be discussed later. A hard, easily filtered curd is produced by injecting steam into a stream of juice so that coagulation is complete in a few seconds. Apart from the cost of steam, there is no disadvantage in heating to nearly 100 °C; this ensures more complete destruction of microorganisms and enzyme inactivation.

Leaf species vary in their enzyme activities. With slow heating there will be loss of protein in those (e.g. wheat) which contain much protease, and pheophorbide will be formed in juice from those (e.g. lucerne) which contain much chlorophyllase. Unlike the other breakdown products of chlorophyll which are made during the processing of LP, pheophorbide is absorbed by animals and makes them sensitive to sunlight. This has caused trouble with pigs fed on LP from slowly heated lucerne juice. Steam injection is not only the best method of coagulation, it is also the simplest when more than a few litres of juice are being handled. A convenient way to heat small quantities of juice suddenly is to heat a litre of water to boiling point and then, with constant stirring and continued heating, to pour in juice at such a rate that the temperature never falls below 70 °C.

Properly heated juice from species that are not glutinous filters easily through cloth; if pressed under conditions that allow the free escape of expressed fluid (often loosely called 'whey'), it is easy to get a cake containing 50% DM. With skilled pressing and a thin cake, a still-drier product can be made. In large-scale work, LP is usually collected by centrifugation which unavoidably results in a wetter cake and consequently in one that contains less true protein in the dry matter because of the quantity of the soluble components of the 'whey' entrained in it. As a rule, it is this material that is used in animal feeding. Well-pressed LP from species with an attractive flavour (e.g. maize or pea haulm) can be used as human food; LP from other species should be resuspended in water, filtered off, and pressed again. It is more easily handled if, during resuspension, it is adjusted to about pH 4. Acidification also frees the LP from the greater part of any alkaloids that may be present in weeds in the crop and enables LP to be made from leaves (e.g. potato and tomato) that are usually considered toxic. Leaf protein from different species has approximately the same buffering power and the pH is not critical.

Even when LP is to be used as animal feed, the disadvantages of leaving a large amount of the soluble 'whey' components in a product that will be dried should not be overlooked. Some of the poor results of feeding experiments with dried LP were probably the consequence of Maillard reactions between protein and 'whey' carbohydrates. A representative set of figures may make this point clear. If 'whey' contains $50\,\text{g DM litre}^{-1}$, and the LP is pressed to 40% DM, 7·5% of that DM will be contributed by soluble 'whey' components. Pressed LP has a granular texture when it contains only 20% DM. If material such as that were dried it would contain 20% of 'whey' components, half of which may be reducing sugars. By contrast, LP pressed to 50% DM would, with the same assumptions, contain only 5·3% of 'whey' components. By a similar argument, if the pressed LP is resuspended in a volume of water similar to the original volume of juice and repressed, it should contain only about 0·2% of 'whey' components. When human food is being made, adequate pressing and washing are obviously important both to preserve the nutritive value of the product and to remove material that may have an unappealing flavour. Therefore, when the procedure for processing leaf juice is being established, it is prudent to dry some samples of the fluid coming away from the LP when it is being pressed after washing. If washing has been adequate these will contain only $2\text{–}5\,\text{g solids litre}^{-1}$. Leaf protein that has been washed to that extent at about pH 4 may be a useful source of dietary iron. However, it is a poor source of the other metals and of the water-soluble vitamins.

Several amendments, some of them advantageous, have been suggested to this simple technique of extraction and processing. Protein extraction can be increased by adding water in addition to the amount that clings to the leaf surfaces after rinsing. Nevertheless, even with added water, crops that contain more than 20 % DM when harvested seldom gave a satisfactory yield of extracted protein. The only disadvantages in adding extra water are that more steam is needed to heat the juice to coagulation temperature, and the 'whey', which may be used in various ways, is diluted. Juice from most leaves is in the range pH 5·5–6·5. Protein extraction is increased by adding alkali during pulping so as to bring the juice pH to about 8·5. At that pH there is less destruction of $\beta$-carotene and chlorophyll so that LP, unless later washed with acid, instead of being a dull olive green, retains its bright, attractive, green. Juice from lucerne coagulates satisfactorily when heated at that pH but juice from some other species does not. Enzymic and non-enzymic reactions cause browning in extracts from all species: the rate of browning depends on species and on the extent to which enzymes were inactivated by heating. Phenolic compounds involved in these reactions combine to some extent with LP and make lysine less available to non-ruminants. Leaf protein should therefore be separated from 'whey' as quickly as possible. Browning can be prevented by adding sulphide or other antioxidants or enzyme inhibitors during pulping. Care is, however, needed lest, in trying to preserve lysine, agents are added that damage other amino acids such as cystine and methionine.

These suggested amendments complicate the technique to varying extents. The most radical amendment that has been suggested, i.e. to use unfractionated juice as pig feed, would greatly simplify extraction. Unfortunately, that procedure has many defects. Crops of differing age differ in water content and they may be harvested at the end of a sunny afternoon or with morning dew or even rain on them. Consequently the protein content of juice may on one day be only a fifth of what it is on another day. This complicates the management of animal diets. In warm weather, half the protein in stored juice will undergo proteolysis and the remainder will form a curd. The nutritional value of the products of proteolysis is similar to that of the original protein, but the curd will make the even distribution of juice troublesome. Some components of the juice, which would be removed in the 'whey' if LP were separated, are of doubtful nutritional value for non-ruminants, some are slightly toxic, and the concentration of cations in juice will strain an animal's kidneys if unfractionated juice is used to supply much of the daily protein intake. All these points are obvious and predictable but they were disregarded in some unsuccessful feeding experiments. It is to be hoped that such experiments will not be repeated and published, and that coagula will always be at least partly separated from 'whey'.

Air expelled from juice during coagulation makes most of the curd float and the curd can then be skimmed off. Alternatively, the heated juice can be deaerated by suction. All the curd then sinks to form a sediment which contains 10–15 % protein. That sediment is a better feeding stuff, with more constant composition, than whole juice. Furthermore, because of diminution in volume and removal of much buffering material, less preservative, e.g. formic acid or sulphite is needed for short-term conservation, should that be necessary.

## COMPOSITION OF LEAF PROTEIN

When made carefully, dry LP contains 9–11 % N: almost all of it is present as true protein although there is some nucleic acid in preparations from very young leaves. Lipids, including chlorophyll, its breakdown products and $\beta$-carotene are often underestimated. When extracted from acidified protein, or with polar solvents, they usually account for 20–25 % of the dry matter. The amount of starch depends on species and the amount of sunshine at the time of harvest. If washing and pressing have been adequate, the amount of water-soluble material will be negligible. The amount of acid-insoluble ash is usually an index of the amount of dirt retained by inadequately rinsed leaves. However, some of the grasses are rich in silica and some of it extracts along with the LP. Silica is then unavoidable, but it is harmless.

Slight differences in the amino acid composition of LP preparations can be demonstrated[16] by elaborate statistical techniques. There is no evidence that differences between preparations from different species are larger than those between preparations from the same species taken at different ages or after different types of husbandry. Until there is more evidence, it may be assumed that such differences in nutritive value as may be found between preparations are more likely to be the result of differences in the technique of preparation or storage than of real differences between species. Amino acid analyses, such as those in Table 1, suggest that LP should have better nutritive value than any of the commonly used seed proteins. It is to the latter that the customary condemnation of plant proteins as 'second class' applies. Leaf protein is as good a supplement in pig diets as fishmeal,[17] but, as would be expected from the analyses, it is not as good as egg or milk

## TABLE 1
### AMINO ACID COMPOSITION OF SOME SAMPLES OF LEAF PROTEIN[16]

| Amino acid | g/100 g recovered amino acids |
|---|---|
| Alanine | 5·9–7·05 |
| Arginine | 6·1–7·0 |
| Aspartic acid | 9·6–10·2 |
| Glutamic acid | 11·0–12·2 |
| Glycine | 5·3–6·5 |
| Histidine | 1·8–2·8 |
| Isoleucine | 4·4–5·75 |
| Leucine | 8·4–10·7 |
| Lysine | 5·2–7·3 |
| Methionine | 1·7–2·4 |
| Phenylalanine | 5·8–7·15 |
| Proline | 4·45–5·05 |
| Serine | 4·1–5·2 |
| Threonine | 4·8–5·4 |
| Tyrosine | 4·2–5·55 |

protein. Feeding trials with rats show a greater apparent methionine deficiency than would be expected from the analyses. This has not been satisfactorily explained: it could be the result of sulphoxide formation or some form of complexing that might be avoided by a change in processing technique. Unfractionated legume seed proteins are also deficient in methionine. This is unfortunate but of little consequence in communities getting most of their energy from maize or wheat, both of which are relatively rich in methionine.

About half the fatty acids in leaf lipids are doubly or trebly unsaturated. Differences between species have been suggested, but it is well known that the fatty acid pattern in plants depends on age and climate; so, as with amino acids, real species differences are doubtful. Much of the $\beta$-carotene accompanies the protein into LP. The amount of carotene in a leaf is approximately proportional to its greenness. However, there is less carotene in leaves taken at midday than in the morning or evening. The effect of this diurnal variation on the carotene content of LP, usually $1–2 \text{ mg g}^{-1}$, is not known.

Carotene in LP, as in other foods and fodders, is slowly destroyed by exposure to light or air, especially when slightly acid. If LP is regarded as something to be made on a farm, this defect of acid washing must probably be accepted. If it is regarded as an industrial product, it could be washed by centrifugation or, if washed by acidification and filtration, the washed, pressed, and crumbled coagulum could be neutralised again by exposure to ammonia vapour. These points deserve thorough investigation because, in many parts of the world, vitamin A deficiency is so common that xerophthalmina is thought to be causing 300 000 children to lose their sight annually. In such regions, the $\beta$-carotene (pro-vitamin A) is as valuable a component of LP as the protein. There is enough $\beta$-carotene in 3 or 4 g of LP to meet a child's daily requirement.

It is easy to remove chlorophyll and its breakdown products from LP by solvent extraction. If that were done, other lipids (including $\beta$-carotene) would also be removed and they are nutritionally valuable. Furthermore, it is not likely that a process involving solvent extraction could be operated in the conditions in which LP production will be most useful. A similar objection applies to proposals that leaf juice should be heated in two stages. By this technique, a dark green curd is coagulated at 50 to 60 °C; a third or quarter of the protein remains uncoagulated at that temperature but coagulates as a pale curd on further heating. If, as suggested, the green curd goes to animals, they, rather than people, would get most of the protein and all the carotene. Furthermore, although this type of separation is easy in the laboratory, it is difficult to envisage such a separation being feasible in the circumstances in which LP would be most useful. For all these reasons it seems sensible to use LP in its original green form. If a white protein is needed, one of the legume seeds would be a better starting material than a leaf crop.


## PRESERVATION


Press-cake containing 50 % or more DM and at pH 4 has the keeping qualities of cheese. If the moist cake contains 1 % acetic acid, or 20 % sodium chloride, no fungal growth or 'off' flavour is noticeable after several weeks.[18] It keeps indefinitely in a deep-freeze. For more prolonged preservation, or to save weight if LP is being transported, LP can be dried. Before drying, the moist cake should be rubbed to a fine powder and it should then be dried quickly in a current of air—it is prolonged exposure to moist heat that does most damage. Leaf protein that has been inadequately washed so that it still contains reducing sugars, and LP that has been inadequately pressed so that drying is slow, is most at risk.

## FEEDING TRIALS

Experiments with unfractionated leaf juice as a milk substitute in lamb and calf feeding have been published by institutes in Bulgaria, India, the UK, the USA and the USSR. Not unexpectedly, considering the differing concentrations of protein in the juices used, the results have varied, but the consensus seems to be that, if trouble from agents added as preservative, and from saponins in lucerne juice, can be avoided, the juice is equivalent to skim milk; however, the juice is unsuitable for non-ruminants.

After experiments with chickens, fish, mice, pigs and rats had demonstrated that carefully made LP had the nutritive value expected from its amino acid composition, human feeding trials were started. Nitrogen retention by infants was nearly as good when half the protein in their diets was LP and half milk, as when all of it was milk. Boys on a diet supplemented with LP grew better than those getting the same amount of extra protein $(10\,\mathrm{g\,day^{-1}})$ in the form of sesame flour. The signs of kwashiorkor disappeared in a few weeks when infants were given 10 g of LP daily by their mothers as a supplement to the home diet. This last trial was in Nigeria and the authors commented particularly on the improved demeanor of the children within a few weeks of getting the LP supplement. All these trials have been fully published (see references in Pirie[3]). In a more elaborate trial on 100 children in Pakistan,[19] three matched groups of children, 7–14 years old, after getting such medical treatment as was necessary, were observed for 8 months while eating either their normal unsupplemented diet, that diet supplemented with 200 ml of milk, or supplemented with LP made from mixed grasses and berseem (*Trifolium alexandrinum*). The LP supplied the same amount of nitrogen as the milk. The average increases in weight of the members of the three groups were 1·1, 2·45 and 2·6 kg, and in height 26, 48 and 53 mm. At the start, all had less than 12 g of haemoglobin in 100 ml of blood; at the end they had 11·7, 12·6 and 12·5. Clearly, in these circumstances, LP is as good a supplement as milk. It is possible that this unexpected result is the consequence of the $\beta$-carotene in LP supplying vitamin A to a marginally deficient group of children.

Only fragmentary accounts (e.g. reference 3) of a similar trial have so far been published because it is still continuing. In each of four villages near Coimbatore (India), the midday school meal for about 60 children, 2–5 years old, is supplemented with 1·3 MJ, including 10 g of protein from different sources. In another village, the meal is supplemented with tapioca to supply the amount of energy that the protein supplements contain, and a sixth village gets no supplements, but does get the medical and educational attention given to the other five. Because changing populations of children, rather than stable populations, are being studied, the experiment is difficult to score; it is, however, clear that the energy supplement makes little difference to growth or health, that milk is the best of the protein supplements, and that LP (mainly from lucerne) is as good as or better than protein in legume seeds. Other similar trials have started in various parts of the world.

In all these trials there was some coercion, or the foods containing LP were supplied as part of a 'package' which contained some desirable components. So, although those responsible for the trials in Pakistan and Coimbatore comment that within a few months the children given the LP supplement took it as regularly and willingly as those given the other supplements, these trials do not demonstrate that LP would be eaten if no inducement were coupled with it. This raises the whole question of the rationality and fixity of food habits. When adult prejudices do not interfere, there is no difficulty in giving LP to children just after weaning: their diet is changing in any event and they have not acquired any prejudices. But LP is not as digestible as milk; milk, when the supply is limited, should therefore be reserved for the youngest children. It is the older ones and nursing mothers, who are more effectively equipped with digestive proteases, who should get the LP.

Brief consideration of the differences in food prejudice in different parts of the world, and of changes in food habits in many particular regions, shows that there are no fixed standards of acceptability. A surprising range of foodstuffs can become popular if enough effort is put into popularisation. In our early experiments on presentation at Rothamsted, in spite of introducing the arbitrary constraint that there should be more than 20 % of protein in the DM of every dish we cooked, and that half the protein should be LP, we found no problems in winning acceptance for a dozen different dishes. The large amount of energy supplied in the supplement used in Coimbatore enables the presence of LP to be effectively masked by jaggery (crude sugar). The trial in Pakistan was in some ways more interesting because several different forms of presentation were tried. Some of them are listed in Table 2 along with the verdicts of an experienced tasting panel of food scientists. It is clear that, if reasonable care is taken over presentation, there is no reason to doubt the acceptability of LP in Pakistan. Different forms of presentation would have more appeal elsewhere.

In countries with European and North American food prejudices the

                    *N. W. Pirie*

TABLE 2

ORGANOLEPTIC COMPARISONS IN PAKISTAN BETWEEN DISHES
CONTAINING 8·6 g OF LEAF PROTEIN (GREEN)

| *Dish* | *Organoleptic attribute* | | | |
|---|---|---|---|---|
| | *Look* | *Feel* | *Taste* | *Smell* |
| Dahi Bhalle | 10 | 9 | 10 | 9 |
| Kachori | 9 | 8·5 | 7 | 7 |
| Laddu | 7 | 7 | 7 | 7 |
| Murmara | 9 | 9 | 8 | 8 |
| Pakoras | 9 | 9 | 8 | 8 |
| Maroonda | 7·5 | 7 | 7 | 7 |

4–6, Acceptable; 6–8, suitable for guests; 8, unfortified food.

usual objection to LP, raised by those who have not worked with it, is that it is green. Hence the misguided enthusiasm, on which I have already commented, for some form of decolorisation. Elsewhere, greenish foods are more common. Curry and similar ingredients impart a colour similar to that of LP to foods in India and Pakistan, powdered dried leaves are widely used as a relish in tropical Africa, and leaf juice (Kola-Kanda) is a traditional food in Sri Lanka which has been the subject of a recent vigorous popularisation campaign. In those countries where LP is most likely to be useful there is therefore no reason to expect problems. The absence of opposition to a novel food finds a place in 'Mother and Child Welfare Clinics' and in canteens. Use in more primitive environments, and it is in these that LP could have the greatest impact, depends on the design and supply of simple extraction equipment and on sustained personal example by the advocate.

Occasionally, doubts about the acceptability of LP depend, not on its appearance, but on the curious suggestion that it may be toxic. All the available evidence suggests that people have been eating leaves for very much longer than they have been eating cereal seeds. Considering how few people are poisoned by cereal seeds, e.g. those with coeliac disease, the number which will be incommoded by LP is likely to be very small. Many species, the crucifers for example, are well known to have slightly toxic leaves. These toxic components are soluble in water and so would, to a large extent, be separated from the LP. Therefore, leaves that are so toxic that they are never used as vegetables, can, for this reason, be potential sources of LP. Even leaves that contain thermostable skin irritants might, with some modifications to the preparative technique, become useful if the problem of harvesting them could be solved!

## BY-PRODUCTS

Fractionation by the method outlined here produces LP, extracted fibrous residue and 'whey': the DM of the original crop is distributed between these three fractions in the approximate ratios of 1 to 5 to 1. It could therefore be argued that the fibre is the main product and the LP a by-product. This is probably the situation in wealthy countries where there is no shortage of protein in the human diet, and where the number of cattle that can be kept depends on the supply of conserved winter fodder. More than half the protein can be extracted from a young, lush crop. Because various soluble components of the leaf are also extracted, the protein content of the residue from which LP has been extracted is not halved. The residue usually contains 1·5 to 2·0 % N (in the DM): it is therefore a better ruminant feed than the best hay, but not as good as most of the 'dried grass' produced commercially. Because juice has been pressed out of it, it is friable and can be dried by blowing air through it in summer, even in Britain. When fuel is used for drying, the saving can be considerable. For example, if pressing has been managed so that the fibre contains 65 % water, 1·6 t of water have to be evaporated to get 1 t of 'dried grass' containing 10 % water. A good-quality crop seldom contains less than 85 % water and may contain more than 90 % if harvested early in the day, or in wet weather, so as to keep the drying equipment in continuous use. In these circumstances, the weights of water that have to be evaporated are 5 t and 8 t respectively to get 1 t of 'dried grass'. Cattle eat the product readily and several trials (mainly in the Rowett Research Institute, Aberdeen, but also in USA, e.g. reference 20) show that it has better feeding value than a crop with the same N content initially. This is because more of the N is true protein and the fibre, being from a less mature crop, is less lignified. Obviously, it contains less protein than the original crops, but a crop such as grass or lucerne, fertilised and harvested so as to give maximum yield, contains more protein than a ruminant needs. It is, therefore, reasonable to extract the excess for use by people and other non-ruminants. During the extraction of LP, much of the soluble material is removed from the fibrous residue so that it contains less strongly flavoured or toxic material than the original crop. Residues from plants such as water hyacinth and potato, which animals are unwilling to eat in the fresh state, should therefore be acceptable. This is a point that has still to be established by experiment.

The effluent from silage is well known to be a troublesome pollutant and to kill plants in an area near the silo. There is no effluent from a silo filled with fibrous residue to which just enough 'whey' is added to prevent access of air. Silage effluent kills plants because, near the silo, it is concentrated. If 'whey' produced during LP production is spread over an area comparable to the area from which the crop was taken, it is a valuable fertiliser. It contains most of the leaf K, much of the P, N in the form of amino acids and amides, and sugars. Ultimately, when there is regular commercial production of LP, it will be used as a culture medium for microorganisms. Several papers on its merits as a substrate have already appeared. It is not likely to be feasible to use 'whey' in this way when LP is made on a domestic or village scale. It should then be put back onto the land from which the crop was taken.

## ECONOMICS

Attention has been directed in this chapter to the value of LP as a product that can be made on a domestic scale to fortify an otherwise protein-deficient diet. Many people who live in regions climatically suited to LP production use foods such as cassava and banana (which contain little protein) as their main energy sources. Domestic production is therefore the facet of work on LP that will have the most immediate bearing on human welfare. Here, strictly economic considerations do not apply. So long as the technique of preparation is not more laborious, and does not require more elaborate equipment, than other types of food production, LP could be a useful, locally made food.

By contrast, it is essential to have some idea of the economics of the process before embarking on large-scale production. Many forecasts have been made but, because of the differing significance attached to each of the many factors involved in fodder fractionation, the conclusions reflect the preconceptions of the forecaster as much as they reflect the facts of the case. Early, mainly gloomy, forecasts have already been summarised.[3] Recent forecasts have been more favourable. One of them[21] considers all the ways in which energy is used in fodder fractionation and in alternative methods for handling fodder. A forecast[22] suggesting that the return on investment from LP production could be 30–80 % depends heavily on the value of the xanthophylls. Xanthophylls are unimportant in human nutrition but give chickens' legs a colour that is thought desirable in many countries. Such a large return on capital may not be forthcoming in Ireland or the UK where little significance is attached to leg colour. Nevertheless, a third forecast,[23] concerned mainly with conditions in Aberdeenshire, assesses the increased return from fractionation at £56 ha$^{-1}$. That last forecast explains part of the reason for earlier discouragement. It states that, until the 1970s, little attention was paid to the value of the residue from which LP had been extracted, or to the fact that optimally manured grass contains more protein than a ruminant needs. Both points had been made repeatedly in earlier publications, but were apparently disregarded.

A glance at the situation of 'fish protein concentrate' shows that uncertainty about the economics of a new product is not unexpected. Various forms of dried and powdered fish have been used for 2500 years and made commercially for 50 years. Nevertheless, it was recently[24] stated that 'It is difficult, if not impossible, to predict accurately what a marketable FPC will cost at this point in its development since commercial production is yet to be realised'. Equipment used for processing leaves on a large scale was designed for some other purpose: smaller scale equipment is in its infancy. Present day attempts at costing resemble the attempts that would have been made in the 18th century to work out costs of winning minerals from deep mines with Newcomen's pumping engines. In a few years we should at least have advanced as far as Watt!

The potential abundance of LP is the main reason for confidence that it will be used in some form. Though the basic principles of production will remain the same, differing styles of production are appropriate in an isolated village and a commercial farm, or in the humid tropics and a region with a prolonged winter. Research is therefore needed in several environments, with several objectives, and with many sources of leaf. There will be little incentive to undertake such research as long as there is uncertainty about the merits of the products and the economic prospects of the process. Quality should therefore be judged first of all on material made with more care than it might be possible to deploy in actual practice: the harmful effects (if any) of simplifications should then be studied. The costs of growing and collecting the crops are known from the other aspects of agriculture; it is the costs of power, labour and equipment for processing that are uncertain. Any process can be made to seem uneconomic if a sufficiently unsuitable system is used to operate it. In principle, the amount of energy that has to be expended in liberating and expressing leaf juice is small: the problem is to find means for keeping it small in practice.

## REFERENCES


1. FLECK, A., in *Hunger: Can It Be Averted?*, British Association, 1961.
2. PIRIE, N. W., *Science*, 1966, **152**, 1701.
3. PIRIE, N. W., *Leaf Protein and Other Aspects of Fodder Fractionation*, Cambridge University Press, London, 1978.
4. LAWES, J. B., *J. R. Agric. Soc.*, 1885, **21**, 81.
5. PIRIE, N. W., *Chemy. Ind.*, 1942, **61**, 45.
6. PALZ, W. and CHARTIER, P. (eds.), *Energy from Biomass in Europe*, Applied Science, London, 1980.
7. HENRY, K. M. and FORD, J. E., *J. Sci. Fd. Agric.*, 1965, **16**, 425.
8. HANCZAKOWSKI, P. and MAKUCH, M., *Potato Res.*, 1981, **23**, 1.
9. CARRUTHERS, I. B. and PIRIE, N. W., *Biotechnol. Bioeng.*, 1975, **17**, 1775.
10. MATAI, S., in *Protein from Water Weeds*, C. K. Varshney and J. Rzoska (eds.), Junk, The Hague, 1976.
11. BOYD, C. E., in *Leaf Protein: its Agronomy, Preparation, Quality and Use*, N. W. Pirie (ed.), Blackwell, Oxford, 1971.
12. BUTLER, J. B., *J. Sci. Fd. Agric.*, in press.
13. NELSON, F. W., BRUHN, H. D., KOEGEL, R. G. and STRAUB, R. J., *Amer. Soc. Agric. Engineers*, 1978, paper 78–1526.
14. BUTLER, J. B. and PIRIE, N. W., *Expl. Agric.*, 1981, **17**, 39.
15. DAVYS, M. N. G., PIRIE, N. W. and STREET, G., *Biotechnol. Bioeng.*, 1969, **11**, 528.
16. BYERS, M., *J. Sci. Fd. Agric.*, 1971, **22**, 242.
17. DUCKWORTH, J., HEPBURN, W. R. and WOODHAM, A. A., *J. Sci. Fd. Agric.*, 1961, **12**, 16.
18. PIRIE, N. W., *Ind. J. Nutr. Dietetics*, 1980, **17**, 349.
19. SHAH, F. H., TOOSY, R. Z. and SHEIK, A. S., *Pak. J. Sci. Ind. Res.*, 1979, **22**, 76.
20. LU, C. D., JORGENSON, N. A. and BARRINGTON, G. P., *J. Dairy Sci.*, 1980, **63**, 2051.
21. McDOUGALL, V. D., *Agric. Systems*, 1980, **5**, 251.
22. ENOCHIAN, R. V., KOHLER, G. O., EDWARDS, R. H., KUZMICKY, D. D. and VOSLOH, C. J., Agric. Econ. Rep. No. 445, USDA Washington, 1980.
23. JONES, A. S., in *Grassland in the British Economy*, J. L. Jollans (ed.), Centre for Agricultural Strategy, Reading, 1981.
24. CRISAN, E. V., in *Encyclopedia of Food Science*, M. S. Peterson and A. H. Johnoson (eds.), AVI, Westport, Conn., 1978.

# 11

# Production of Textured Foodstuffs Based on Milk Proteins


K. J. Burgess

*R & D Division, Milk Marketing Board,*
*Crudington, Telford, UK*

and

G. Coton

*Dairy Crest Research & Technical Services,*
*Milk Marketing Board, Thames Ditton, UK*

## INTRODUCTION

The production of textured foodstuffs has been an integral part of the food industry for the past 3000 years since the advent of baking bread and cheesemaking.

Many of the textured foods consumed at present owe their structure to a proteinaceous matrix, e.g. meat, fish, bread, cheese; a breakdown of dietary protein intake in the UK (Table 1) reveals that around 80 % of daily protein intake is consumed in the form of textured foods.

Present consumption trends therefore indicate that new protein-based foods should be manufactured in a textured form, either to simulate meat or cereal-type protein products or as a totally new form of textured food. To date, most textured vegetable protein products have been directed at the meat market but another promising application for textured proteins is in high-protein snack foods.

A number of alternative protein raw materials may be used for texturisation. The most important are oilseed proteins, especially soy seed, and milk proteins, especially casein.

Other protein sources have also been texturised, including meat by-products,[1] bean proteins,[2] gluten[3] and a number of single cell proteins.[4] Single cell proteins, despite considerable publicity, are unlikely to find

211

TABLE 1
UK DIETARY PROTEIN INTAKE

|  | % Protein supplied | |
| --- | --- | --- |
| Milk and cream | 19·3 | |
| Cheese | 5·1 | Animal |
| Meat and poultry | 29·6 | protein |
| Fish | 4·0 | 63% |
| Eggs | 5·0 | |
| Potatoes | 3·0 | Vegetable |
| Other vegetables | 5·2 | protein |
| Cereals | 26·0 | 34·2% |
| Other foods | 2·8 | |
|  | 100·0 | |

significant use in human foods, partly because of technical and legal difficulties but more so because they are likely to be much more expensive to produce than proteins from conventional photosynthetic sources.[5] Other proteins may find commercial application in some parts of the world where special conditions exist, but in the western world generally, and particularly within the EEC, soy and milk proteins probably represent the more immediately available sources of good-quality food protein for texturisation.

Most of the impetus for research into the production of textured protein material has come from the USA and in particular from its soy bean industry in which textured vegetable protein (TVP) was regarded as a means of increasing the return on the soy bean meal remaining after extraction of soy oil. However, in spite of a great deal of research and development effort during the past 20–25 years and the large range of textured soy product (TSP) introduced onto the market, sales of TSP and TVP, in general, have been relatively small.

The relatively low rate of market penetration may be ascribed partly to conditions in the food industry and partly to the fact that most of the products have been based on soy protein. The real price of meat has not risen as fast as was predicted and the cost benefit of TVP is therefore not as great as was once expected. The use of TVP products has also been restricted to some extent by legislation.

Soy protein as a raw material has two important disadvantages; a beany flavour associated with all soy bean products and a tendency to induce flatulence. Most commercially available textured vegetable proteins are produced by the simpler texturisation methods in order to compete with the price of meat and these products are not sufficiently similar to good quality meat to enable effective substitution. Generally, TVP is used for meat extension in processed meat products and this limits the extent of the use. The fact that soy is an 'unconventional' protein source has also been a disadvantage.

Approximate figures for TVP usage in the UK are shown in Table 2. The most important aspect of this is that around half of the TVP is used in the petfood industry and only relatively small amounts are used in meat products for human consumption.

TABLE 2
USE OF TEXTURISED VEGETABLE PROTEINS IN THE UK

|  | *1 000 tonnes p.a.* |
|---|---|
| Catering/institutional | 2–3 |
| Ingredient (meat products, etc.) | 2–3 |
| Retail products | 2–3 |
| Pet food | 8–10 |
|  | 14–19 |

Within Europe, milk protein is available in large quantities and it has several potential advantages over soy protein as a raw material for texturisation. Certainly, flavour and nutritional quality are superior to those of soy[6] and it is important to note that the dairy industry itself also has a wealth of experience in protein texturisation. Cheese is a basic, very long-established example of a textured protein but others include yoghurt and dairy desserts and in the industrial sector, rennet casein plastics. However, when milk protein is considered as a raw material for texturisation, it is very important that the relative raw material costs be compared with those for alternative protein sources, e.g. soy.

A comparison of the relative costs of a number of milk and soy protein raw materials (Table 3) indicates a wide range of normalised protein prices (i.e. price/tonne protein) for both soy and milk protein sources. In particular, milk protein prices vary widely because of the various subsidies given within the EEC resulting in prices as low as £360/tonne of protein when skim milk is subsidised for animal feed, to as high as £2143/tonne for skim milk powder sold at the full 'intervention price' to the food industry. The main conclusion to be drawn from this table is that, because of price,

 *K. J. Burgess and G. Coton*

TABLE 3

APPROXIMATE PRICES OF MILK AND SOY PROTEIN RAW MATERIALS
(1980, UK)

|  | £/tonne | £/tonne protein |
|---|---|---|
| Defatted soy flour | 350 | 700 |
| Soy protein concentrate | 1 000 | 1 430 |
| Soy protein isolate | 1 400 | 1 560 |
| Skim milk |  |  |
|    (a) Powder, human consumption | 750 | 2 143 |
|    (b) Liquid, animal feed | — | 360 |
| Sodium caseinate | 1 400 | 1 560 |
| Whey protein concentrate | 1 500 | 2 500 |

milk protein is unlikely to compete with soy protein. However, with existing subsidies, caseinates are equal in price to soy isolate and skim milk for stock feed is cheaper, on a protein basis, than defatted soy flour. Within the EEC then, it is entirely justifiable to consider milk protein as a viable economic alternative to soy. The EEC currently produces approximately 2 million tonnes per annum of skim milk powder of which about 85 %, representing 750 000 tons of protein, is used in animal feeds at subsidised prices.[7]


## TEXTURISATION METHODS

The principal published techniques for protein texturisation may be divided into those which rely on mechanical structuring and those which do not (Table 4). The most important in terms of product sales is thermoplastic extrusion. Direct steam texturisation is an extension of this process in which

TABLE 4

EXISTING PROCESSES FOR PROTEIN TEXTURISATION

| *Mechanical structuring* | *Non-mechanical structuring* |
|---|---|
| Extrusion | Gel formation |
| Spinning | Freeze texturisation |
| Steam texturisation | Microwave expansion |
| Beating/shredding | Film formation |

a machine similar to a cereal gun is used. Spinning of protein fibres has also received a great deal of attention and these three processes are in fact carried out commercially in the UK.

Of the remaining methods listed, gel formation has most application for the texturisation of casein while microwave expansion has been successfully applied, using whey protein, in Ireland.[8] However, this is unlikely to reach commercial use because of the high price of the whey protein.

Most of these techniques were developed to texturise soy protein and, understandably, not all of them are suitable for the texturisation of milk protein, in particular casein. However, extrusion spinning and gel formation techniques have been used with varying degrees of success[8] together with a technique based on cheesemaking principles, developed specifically for milk protein.

### Extrusion

In thermoplastic extrusion a protein–water mixture is forced through a heated barrel by means of one or more rotating screws. Different sections of the screw first mix and compress the ingredients whilst they are heated before forcing the mix, which by this stage is thermoplastic, through a die to a region of lower temperature and pressure. This results in steam flashing off causing the product to have an expanded structure.[9]

Many extruders have only one screw, but more recently a twin-screw extruder, which has considerably more flexibility, has found application;[10] in particular, the screw speed and feed rate can be varied independently over quite a large range.[11]

The production of expanded skim-milk-based materials using this technique is a relatively simple process but unfortunately the products have several disadvantages: first, Maillard reactions between the lactose and protein of skim milk give rise to undesirable colours and flavours. Second, the products are generally very hard because of the presence of glass-like, amorphous lactose. Finally, extruded skim milk powder has no structural integrity when rehydrated, which prevents its use as a meat analogue.

Extruded soy protein products do not suffer from this latter disadvantage and their stability is believed to be the result of a re-aggregation of the soy protein *via* intermolecular peptide bonding.[12] This theory also explains the poor stability of extruded skim milk protein because the compact nature of the casein micelles means that very few protein side-groups are available for interaction with other protein molecules. This has been confirmed by work at the National Institute for Research in Dairying, Shinfield, UK, where scanning electron microscopy has shown that extruded skim milk powder is

                    *K. J. Burgess and G. Coton*

in the form of spherical, unassociated casein micelles embedded in a glass-like matrix of lactose.[13]

The extrusion of caseinate-based materials is unfortunately no more promising. Although in this case the protein is free to interact in the texturisation process, no stable structure is formed. However, caseinates may be successfully used to supplement the protein content of cereal-based snack products to around 30 % (Fig. 1). This is purely a protein fortification application since the caseinate imparts no texture to the final product.

FIG. 1.   A milk-protein-fortified cereal snack food.

**Fibre Spinning**

The production of edible protein fibres by a spinning technique was developed in the 1950s.[14] An alkaline protein solution is pumped through a disc (or spinnerette) with some thousands of very small holes (60–100 $\mu$m) into an acid/salt bath where the protein coagulates into fibres. The fibres are strengthened by stretching, and bound together to simulate meat.

Casein is a very suitable protein for spinning and indeed spun casein fibres were produced in the textile industry in the 1940s and 1950s as a wool substitute.[15] However, casein fibres produced by this technique suffer, in part, from the same disadvantage as extruded skim milk powder in that they are not stable to rehydration unless chemically treated. In edible casein fibre production, a neutral (pH 6·8) casein dope is preferably spun into a coagulation bath containing lactic acid and salt.[15]

There are a number of ways of getting over the problem of rehydration and heat stability (Table 5). Since casein is not coagulated by heat, it seems logical to incorporate a heat-coagulable protein to stabilise the fibres. Soy, whey and egg proteins are all suitable but soluble soy protein products give the strongest fibres when co-spun with casein.[16] Addition of certain polysaccharides also confers heat stability to casein fibres; sodium alginate, which can be co-precipitated with the casein by including a calcium salt in the acid bath is particularly effective.[16]

TABLE 5
TECHNIQUES FOR PRODUCTION OF STABLE CASEIN
FIBRES

1. Incorporation of heat coagulable protein
2. Incorporation of polysaccharides
3. Caseinate complex spinning
4. Dry spinning
5. Cross-linking of casein in fibres

The stability of casein fibres to rehydration can also be improved by exploiting the sensitivity of casein to coagulation by calcium. In caseinate complex spinning, for example, an artificial casein micelle mixture is prepared and spun into a hot solution of a calcium salt.[17] Calcium-induced coagulation gives fibres of greater stability than those coagulated by acid.

Dry spinning of casein fibres also imparts greater heat stability because of the use of defined levels of calcium and heat in producing the fibre.[18] Dry spinning of a casein dope into hot air also has the advantage that no coagulation bath is necessary and costs are therefore substantially reduced.

Finally, casein can be stabilised by cross-linking using a suitable agent.[18] Formaldehyde was used in the production of textile fibres; a similar agent, dialdehyde starch, also works well but, like formaldehyde, may not be permitted as a food additive.

## Gel Formation

Gel formation is an alternative method suitable for the texturisation of casein.[19] Essentially, this involves the application of heat and calcium ions to prepare a heat-stable casein gel. The major steps involved are listed in Fig. 2.

Starting with skim milk, two possible protein raw materials may be prepared: casein or co-precipitated casein and whey protein. The latter is

 *K. J. Burgess and G. Coton*

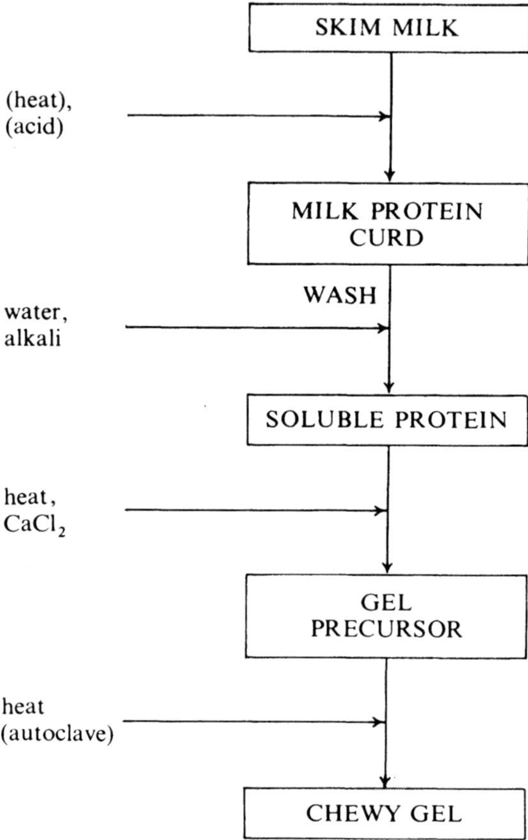



Fig. 2. Production of a milk protein gel.

formed when milk is heated to 90–95 °C and increases the protein yield by some 15–20 %.[20] After pre-heating (if used) the milk is acidified to coagulate the casein or co-precipitate. The curd is separated and resuspended in alkali at pH 7–8.

The protein solution is then heated, and concentrated calcium chloride is added to precipitate the protein as a stringy mass. After washing to remove residual lactose, the precipitate is autoclaved to produce the gel.[20]

The resultant gel contains 30–40 % solids, most of which is protein. Composition can be modified by incorporating other food materials into the gel. For example, fat can be homogenised into the protein solution before calcium addition, and polysaccharides, e.g. alginates, may also be incorporated. These can effect quite large differences in the texture of the products.

The gel can be cut into meat-like chunks (Fig. 3) but the texture is too close and homogeneous to simulate the structure of meat to a satisfactory degree.

FIG. 3.   Chunks of gelled milk protein.

## Thermoplastic Extension

Thermoplastic extension was developed in Poland specifically for the texturisation of milk protein and it incorporates many of the aspects of cheesemaking.[21] In fact, as mentioned earlier, cheese was one of the first textured protein foods and in traditional Cheddar cheese manufacture, the curd at milling is described as having the texture of chicken breast meat. The preparation of textured protein based on the cheesemaking process is outlined in Fig. 4.

The raw material is skim milk and, as with the gels mentioned earlier, the product can either be based on casein alone or, after calcium addition and heat treatment, on complexes of casein and whey protein. The next stage is the incorporation of fat which improves the 'mouthfeel' of the final product. The melted fat is homogenised into part of the skim milk which is then mixed back into the rest of the skim milk to give a fat content of 1–3 %.

After preparation of the fat-filled skim milk, the next step is the production of a rennet curd by the addition of lactic acid and rennet. A firm curd is usually formed after 20–30 min. The whey is removed and the curd washed to give what is essentially fat-filled rennet casein. This is in the form of a granular curd, which is stable to autoclaving and could therefore be used as a meat extender without further processing. However, its granular

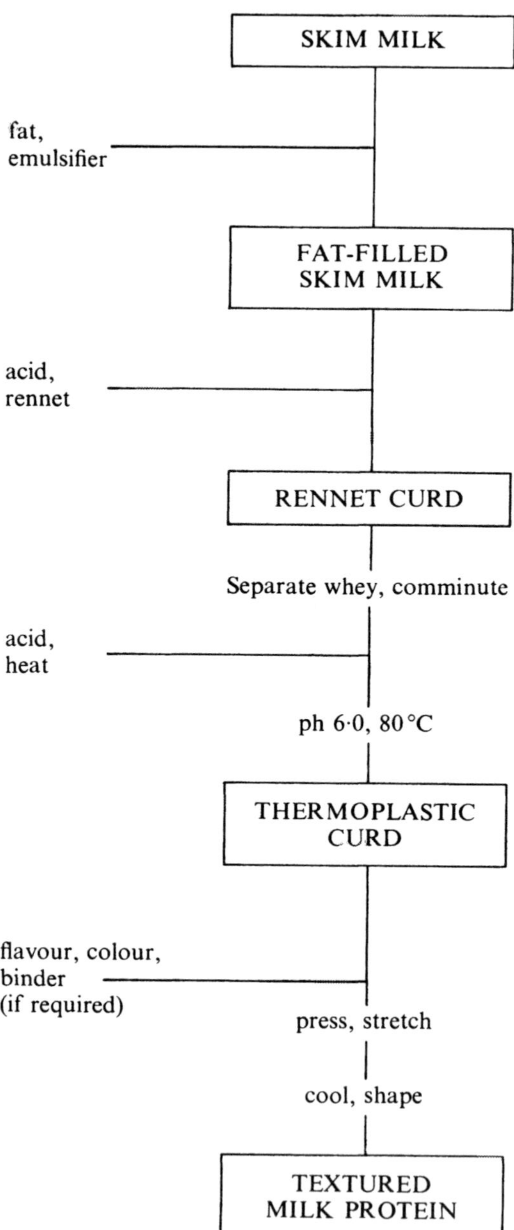



FIG. 4.  Production of textural protein from skim milk.

nature would limit its use to relatively low levels in meat extension. In any case, the curd can be further processed to give a laminated structure like that of meat.

The physical properties of the curd are very dependent on temperature and pH. If the pH is reduced to 5·8–6·0 the curd becomes thermoplastic on heating and can be worked to form a continuous structure. By compressing and stretching the curd under these conditions, a laminated structure can be produced which is stable to cooking.

In the original Polish method, working was achieved by extruding the curd through a nozzle followed by stretching. However, working with a dough-mixer gives an equally acceptable product. Finally, the textured product is conditioned and either dried or frozen prior to use.[21]

Figure 5 shows the type of material that can be produced using this technique.

The products have a stranded fibrous texture and a 'tearability' similar to that of poultry meat. They have been quite successfully used in a number of processed meat products including hamburgers, sausages and pies.[22]

However, the method as described poses some processing problems. In the first place, the whey is contaminated with non-dairy fat and this may lead to problems in whey utilisation. Also, it is very difficult to wash effectively fat-filled casein curd completely free of residual lactose. There is,

FIG. 5.   Textured milk protein prepared by a modified rennet coagulation process.

therefore, some justification for modifying this technique and this has been done by first manufacturing rennet casein curd in the conventional manner.[22] In this way, curd washing is much more efficient and the whey is not contaminated with a foreign fat. The rennet curd is then solubilised by treatment with alkali, and fat can be incorporated by conventional homogenisation.

The fat-filled rennet casein is then simply precipitated by acid injection to pH 5·5–6 and worked as before.

## CONCLUSIONS

Extrusion has proved unsuitable for milk protein texturisation in terms of end-product properties and while some of the problems associated with casein spinning have been overcome, the process is expensive and requires great technological expertise. Only dry spinning of casein fibres appears to offer any promise. Gel formation is a relatively straightforward technique but the product is limited in terms of its ability to simulate meat.

Finally, modification of rennet casein curd offers a relatively inexpensive means of creating a stable, meat-like texture using principles well known to the dairy industry. The texture achieved by this technique is not as meat-like as that of more sophisticated products based on spun fibres but is nevertheless superior to that of most extruded TVP products.

In considering methods for texturising milk protein, the major application has been assumed to be the production of meat replacement products. However, it was mentioned above that another promising application is the production of proteinaceous snack foods and that milk-protein-enriched snacks could be produced by extrusion processing. One such product is on sale in the USA under the trade name 'Moo Munchies' but within the UK such protein supplementation is generally considered as being far too expensive by the major snack producers. Milk-protein-based nutritional snacks may also be prepared by deep-fat frying[23] and as a milk biscuit[24] but these have not been commercially attractive.

Assuming, then, that the main market for textured protein material is in meat replacement, it is important to assess how much textured protein is likely to be used.

Table 6 lists total meat sales in the UK.[25] The total amount of meat protein sold on the retail market exceeds the total EEC production of milk protein in skim milk powder. Even a slight penetration of the European meat market by textured milk protein would therefore create a substantial increase in the utilisation of skim milk within the EEC.

TABLE 6

MEAT CONSUMPTION IN THE UK DURING 1979†

| *Meat* | *Thousand tonnes* |
|---|---:|
| *Boned* | |
| Bacon and ham | 513 |
| Pork | 716 |
| Beef and veal | 1 260 |
| Mutton and lamb | 400 |
| Poultry | 750 |
| Offal | 257 |
| | 3 896 |
| *Processed* | |
| Frozen meat and convenience foods | 111 |
| Pork sausages | 146 |
| Beef sausages | 134 |
| Pies, sausage rolls | 58 |
| Corned meat | 46 |
| Other meat products | 202 |
| | 697 |

† Calculated from Meat and Livestock Commission figures.[25]

It has been the intention of this paper to give an overall view of the use of milk protein as a raw material for texturisation. In order to facilitate the use of substantial quantities of protein in the food industry, marketing considerations indicate that protein must be presented in a textured form. As far as raw materials are concerned, milk and soy protein prices in the EEC are very similar. The added advantages of nutritional excellence and good flavour make milk protein a very real, if not superior, alternative to soy protein.

One of the drawbacks of milk protein texturisation in the past has been the instability of spun fibres and extruded material to rehydration. Newer technologies involving dry spinning and rennet casein modification techniques have now indicated how this problem may be overcome and allow the production of textured milk protein materials which retain their structural integrity on cooking.

Finally, as a raw material for a protein-based food, skim milk contains only 35 % protein on a dry basis, the major constituent being lactose. Isolation of the protein prior to texturisation, therefore, leaves a whey stream which requires profitable utilisation. Dairy research workers have spent considerable effort in developing whey (or lactose) treatment methods with some success.[26] Of the possibilities, we believe hydrolysis to galactose and glucose is most promising and this is now at a small commercial scale in the UK.[27] Processing of skim milk into textured protein foodstuffs and sweet whey syrups therefore means that profit is generated from both of the major constituents of skim milk and such a process scheme offers considerable potential for improving the rate of use of skim milk solids in the human food sector.

## REFERENCES


1. HAWKINS, A. E., British Patent 1 506 846, 1978.
2. SCHMANDKE, H. and SCHMIDT, G., *Nahrung*, 1977, **21**, 209.
3. KELLOGG, J. H., US Patent 869 371, 1906.
4. SCHMANDKE, H., *Nahrung*, 1978, **22**, 207.
5. NEWMARK, P., *Nature*, 1980, **287**, 6.
6. DOWNEY, G. and BURGESS, K. J., *Ir. J. Fd. Sci. Technol.*, 1979, **3**, 33.
7. *EEC Dairy Facts and Figures*, Economics Division, Milk Marketing Board of England and Wales, 1980.
8. BURGESS, K. J., DOWNEY, G. and TUOHY, S. *Farm and Food Research*, 1978, **9**(3), 54.
9. CLARK, J. P., *J. Texture Studies*, 1978, **9**, 109.
10. EVANS, J. M., *Scientific & Technical Survey No. 115*, Leatherhead Food Research Association, 1979.
11. O'KEEFE, J., *Food in Canada*, 1979, 16.
12. BURGESS, L. D. and STANLEY, D. W., *Canadian Institute of Food Science and Technology Journal*, 1976, **9**, 228.
13. ANDREWS, A., Personal communication, 1979.
14. BOYER, R. A., US Patent 2 682 466, 1952.
15. FERETTI, A., British Patent 483 809, 1938.
16. BURGESS, K. J., Unpublished results, 1980.
17. CO-OPERATIVE CONDENSFABRIEK FRIESLAND, British Patent 1 398 179, 1975.
18. KURARAY CO. LTD, British Patent 2 005 187, 1979.
19. ANSON, M. L. and PADER, M., US Patent 2 802 737, 1957.
20. KRUGER, W. and QUADE, H. D., German Democratic Republic Patent 119 953, 1976.
21. SMIETENA, F., POZNASKI, S., HOSAYA, M. and KOZLOWSKA, H., *Milchwissenschaft*, 1978, **33**, 601.
22. EARLY, R., Unpublished results, 1980.
23. PINKSTON, P. J. and CLAYDON, T. J., *Fd. Product Development*, 1971, **5**(1), 30.
24. BUCHANAN, R. A. and TOWNSEND, R. F., *Aust. J. Dairy Technol.*, 1969, **24**, 113.
25. ENGLISH, A. F., *J. Soc. Dairy Technol.*, 1981, **34**(2), 70.
26. COTON, S. G., *J. Soc. Dairy Technol.*, 1980, **33**(3), 89.
27. COTON, S. G., *IDF Seminar on Dairy Ingredients in the Food Industry*, Luxembourg, 1981, Paper No. 3.

# 12

# Isolation and Utilisation of Food Proteins*

R. A. LAWRIE

*School of Agriculture,
University of Nottingham, Loughborough, UK*

## INTRODUCTION

It is implicit in the title of this chapter that sources of protein do not *per se* necessarily constitute human foods and that the proteins of many sources must be extracted and manipulated before they become organoleptically acceptable.

The preparation of bread from cereal grains may be regarded as one of the earliest examples of man's attempt to enhance the acceptability of food sources which, whilst readily digested in the raw form by animals, cannot be effectively attacked in the human intestinal tract. Other examples, more pertinent in the present context, include the fermentation techniques which have been traditionally employed in Asia to make the soy bean more palatable for human food—albeit that those who applied them were unaware they were also making a nutritionally valuable protein more available thereby.

The deliberate application of sophisticated technology to enhance palatability by producing novel forms of food has been attempted for more than a century.[1,2] One of the first patents was awarded to Kellogg in 1907 for the fabrication of organoleptically desirable products from wheat gluten and casein. During the past 30 years there has been a very great increase in the variety of fabricating procedures and in the sources from which proteins have been extracted and upgraded. This has reflected the

* This paper is an enlarged version of a talk given at the Institute of Food Science and Technology annual symposium 'Food Technology in Europe' and published in *Institute of Food Science and Technology (UK) Proceedings*, Vol. 14, No. 3, September 1981.

world shortage, maldistribution and cost of animal protein, due to which the proteins of the plant kingdom have been intensively investigated as sources of direct human nourishment for both developing and developed countries. Products designed for the latter type of market have tended to simulate meat.[3-7]

Although soy protein has an excess of lysine and a relative deficiency of methionine, it contains all the essential amino acids and its nutritive value compares well with that of FAO reference protein.[8] This feature, together with its relative ease of cultivation in quantity, has caused the soy bean to become a vigorously exploited source of protein. But there have been many other sources including field beans, groundnut, sesame, cottonseed, and safflower.

More exotic sources of protein, such as algae,[9] leaves[10] and bacteria[11] are also under consideration. Indeed the level of wastage and under-utilisation of expensively produced animal protein has caused a rein-vestigation, in the present context, of such sources as whey,[12] abbatoir offal[13] and bone, both in respect of adhering meat[14] and of the protein components of the marrow.[15,16]

Apart from their nutritional value, proteins are important in modifying the textural properties of the foods to which they are added and this has been a further reason for seeking also to retain their functional properties during isolation and/or recovery.


## RECOVERY PROCEDURES

**Solid Sources**

Since the original objective in treating the protein-bearing oil seeds was the recovery of vegetable oil, and that in dry-rendering abattoir waste was the preparation of lard, tallow and suet, the heat treatments by which this objective was achieved were not concerned either with the nutritive value or the functionality of the residual protein.

Azeotropic distillation affords a means of simultaneously recovering fat and protein without major loss of their useful properties. Recently the procedure has been used to produce meat protein in a form suitable for human food.[17,18] Fat and water are removed, azeotropically, from meat waste using ethylene dichloride (which has a low, constant boiling temperature) or other suitable organic solvents. The residue is further extracted with isopropanol to remove traces of fat. The dry concentrate thus produced has a protein content of about 80 % and its nutritional value compares well with that of whole egg,[19] although its functional properties are not ideal. Indeed, Gault[20] reported substantial loss of solubility and of emulsifying capacity in the proteins recovered from offal tissues by azeotropic distillation, although the water-holding capacity was retained or enhanced. On the other hand, using a mixture of isopropanol and hexane to extract protein from waste beef at low temperature, Toledo[21] obtained a product with an emulsifying capacity comparable to that of fresh beef. The possibility of residual traces of solvent in the product, however, is a contraindication to the use of such processes.

Another approach to protein recovery which aims to retain the functionality and nutritive value of protein, is the use of various proteolytic, collagenolytic and elastolytic enzymes. Enzyme action can liberate proteins from intractable matrices and aid their solubility without involving extensive breakdown. The process must be carefully controlled, however, to avoid the latter eventuality, by subsequent inactivation of the enzymes employed. Nevertheless, although their functionality would be largely lost if proteins were completely hydrolysed, such hydrolysates are nutritionally superior to those obtained by acids, which destroy tryptophan.[22]

Traditionally, the acceptability and digestibility of soy bean as human food was achieved in the Far East by various fermentation processes based on the enzymes of the mould *Aspergillus oryzae*,[23,24] whereby such products as *miso* and *shoyu* (*soy sauce*) were derived. The Japanese have also been prominent in employing enzymes to develop new products. Thus, using pronase (from *Str. griseus*), a liquefied fish protein concentrate (85 % crude protein) may be produced, the enzymic phase being followed by azeotropic distillation; 90 % of the total protein can be recovered in this way.[25] Its biological value is higher than that of casein and it contains more protein and less lipid than the older product, fish protein concentrate (Table 1).[26]

Fungal and bacterial proteinases have also been used to solubilise the muscle protein of meat, poultry and bone meals[27,28] through their action

TABLE 1

COMPOSITION OF LIQUEFIED FISH PROTEIN (LFP) AND FISH PROTEIN CONCENTRATE (FPC)[26]

|      | Moisture (%) | Crude Protein (%) | Lipid (%)   | Ash (%)    |
|------|--------------|-------------------|-------------|------------|
| LFP  | 6·1–8·6      | 74·3–90·7         | 0·0–0·04    | 4·6–6·1    |
| FPC  | 6·3–10·7     | 78·0–82·3         | 0·13–0·10   | 12·9–13·8  |

                          *R. A. Lawrie*

on the collagenous and elastic tissues in which it is embedded. The recovery of muscular tissue from bones, on the other hand, can be achieved by various mechanical procedures, and, provided the bones are treated hygienically throughout, such mechanically recovered meat is not objectionable.[29]

Apart from the use of added enzymes to produce organoleptically enhanced human food, the action of indigenous enzymes of animal tissues has been exploited to produce protein-rich feeds for livestock from material which would otherwise deteriorate microbiologically and chemically. Such ensilage processes, employing inorganic or organic acids to prevent bacterial spoilage, were first applied to fish in Finland by Vertanen 60 years ago. More recently there has been renewed interest in the procedure. It represents a relatively cheap way of avoiding waste of animal protein which cannot be immediately consumed in areas of high temperature and

TABLE 2

EFFECT OF TEMPERATURE ON RECOVERY OF PROTEIN AND COLLAGEN FROM BOVINE LUNG AND RUMEN BY EXTRACTION AT pH 10 FOR 2 H FOLLOWED BY PRECIPITATION AT pH 4·5[35]

| *Tissue* | *Extraction temperature (°C)* | | | |
|---|---|---|---|---|
| | *0* | *20* | *40* | *60* |
| (a) *Protein* (as % *initial in tissue*) | | | | |
| Lung | 32 | 36 | 42 | 45 |
| Rumen | 27 | 32 | 45 | 56 |
| (b) *Collagen* (*mg/g recovered protein*) | | | | |
| Lung | 3·0 | 2·3 | 3·1 | 3·8 |
| Rumen | 0·4 | 1·2 | 1·0 | 1·9 |

humidity, where refrigeration is unavailable, but where conditions foster speedy enzyme action. The production of fish silage, using 3 % formic acid, involves grinding the waste fish and lowering its pH to 4.[30] The enzymes mainly responsible for liquefaction are those of the gut, skin and parts of the fish other than the flesh.[31] At 30 °C most of the enzymic changes have reached maximum after 10 days. The feasibility of using ensilage processes to minimise wastage of animal protein is currently being investigated.

Since proteins are amphoteric it is not surprising that both acidic and alkaline aqueous media have long been used to extract them from such sources as milk, eggs, maize, soy and wool as a preliminary to their subsequent manipulation.[32,33] As applied to abattoir waste, for example, 60–90% of the proteins may be extracted at pH 10–11, at room temperature.[34] Moreover, if extraction is performed at 60°C, the yield is twice that at 0°C, although this is partly a reflection of increased solubility of collagen (Table 2);[35] the latter increases four-fold if the pH is raised from 10 to 11. However, as the temperature of alkaline extraction is raised, and as the time of extraction is increased, greater quantities of lysinoalanine are formed. Thus, whereas this compound was not detected in the proteins extracted from bovine lung over 8 h at pH 10 and 20°C, appreciable quantities developed after only 2 h at 60°C.[35]

Although lysinoalanine does not appear to develop with the milder alkaline extraction procedures (and, in any case, any detrimental effect it may have on the health of human consumer is unproven,[36] and although it has been suggested that collagen, despite its defective content of essential amino acids, may nevertheless make some contribution to human nourishment,[37] alternative procedures for extracting protein which avoid the use of alkali have been considered. Thus, anionic detergents, such as sodium dodecyl sulphate (SDS), extract substantially greater quantities of protein from lung, rumen and intestinal offal at neutral pH than does alkali at pH

TABLE 3

EFFICIENCY OF NITROGEN RECOVERY FROM EXTRACTS OF OFFAL[38]

| *Species* | *Tissue* | *N in extract as % total nitrogen in original tissues* | |
| --- | --- | --- | --- |
| | | *(a) Alkaline extraction (pH 10·5)* | *(b) Extraction with sodium dodecyl sulphate (0·01 M)* |
| Bovine | Lung | 55 | 69 |
| | Rumen | 60 | 68 |
| | Small intestine | 72 | 82 |
| | Large intestine | 58 | 70 |
| Ovine | Lung | 58 | 68 |
| | Stomach | 54 | 71 |
| | Small intestine | 81 | 85 |
| | Large intestine | — | — |
| Porcine | Lung | 71 | 81 |
| | Stomach | 74 | 78 |
| | Small intestine | 67 | 80 |
| | Large intestine | — | — |

10 (Table 3).[38] The SDS–protein complex is precipitated with 0·005–0·05 M $FeCl_3$ and the residual SDS removed by washing with 40 % methanol or 60 % acetone in the presence of 10 % KCl.[39,40] An ion exchange resin can also be used to remove SDS from its association with recovered protein[41] but removal is less complete.

## Fluid Sources

Where proteins are already in solution or suspension they can be recovered by such well-established procedures as spray-drying or freeze-drying, but these have the disadvantage of being non-specific. Since sources are liable to contain various soluble or suspended matter other than protein, a procedure which recovers protein preferentially is required for functional and nutritional reasons. Flocculation, ultrafiltration and ion exchange chromatography have been applied to this end. Such operations serve the dual purpose of recovering protein and diminishing environmental pollution.

As applied to abattoir effluent, a 60 % reduction in biological oxygen demand may be achieved by the addition of sulphuric acid to pH 4·2 and a further 75 % reduction by the addition of lignosulphonic acid.[42] Hopwood and Rosen[43] removed proteinaceous matter and fat from aqueous effluents by precipitation of protein with sodium lignosulphonate under acidic conditions. The precipitate flocculated readily, enmeshing finely-divided fat. Separation of the flocs was achieved by dissolved air flotation. A similar process was applied by workers in New Zealand.[44] They found that the efficiency of the process was much enhanced if the pH was first lowered to 4, when haemoglobin was split into haem and globin, and some proteins were precipitated, especially in the presence of polyvalent coagulants. On raising the pH to 6·5, the globulin became insoluble; serum proteins could then be removed by adjusting the pH to 3·5 in the presence of 0·05 % sodium hexametaphosphate.

Polyelectrolytes such as sodium alginate, carboxymethyl-cellulose and polyacrylic acid have been used to precipitate proteins from milk whey and dilute solutions of soy, casein, edestin and yeast protein,[45,46] the recovered proteins having excellent functional properties.[47] More recently poly-uronates, including crude pectates from waste citrus peel, have been used to recover 90 % of the proteins from blood serum.[48]

With blood itself, in which the initial protein content is high, a heat coagulation process may be applied,[49] using direct steam injection at 90–95 °C. This causes considerable losses, however, and in New Zealand fluidised bed drying (100–200 litres per hour) has proved an effective alternative. Membrane processes, such as ultrafiltration, have been applied to achieve preliminary concentration and desalting of dilute protein-containing fluids.[50] Typical results for dilute effluent are shown in Table 4.[51] As the concentration ratio rises above 2·5:1, however, there is an increase in viscosity which limits the process to a ratio of about 3:1.[52] Using Iopor membranes with clean, homogenised, citrated or defibrinated blood, the protein content could be increased from 17% to 29% at a flow rate of 12 litres $m^{-2} h^{-1}$.[51] Blood plasma concentration could be increased

TABLE 4

CHARACTERISTICS OF FEED AND PERMEATE STREAMS FOR ULTRAFILTRATION OF BLOOD AND PAUNCH EFFLUENTS[51]

|  | *Dilute blood* | *Filtrate from pressed beef paunches* |
|---|---|---|
| Initial feed concentration, % total solids | 0–22 | 1–2 |
| Initial feed, chemical oxygen demand, mg/litre | 3 000 | 15 000–20 000 |
| Permeate, chemical oxygen demand, mg/litre | 150 | 4 500–5 000 |
| Permeate colour | colourless | very light brown |

from 7% to 20% protein. A typical procedure for recovering both serum protein and globin in a functional form is shown in Fig. 1.[53] In considering blood as a source of utilisable protein for human food it is clearly most important to ensure hygienic collection. Currently, this involves the use of a hollow knife which severs the blood vessels and collects the blood directly from them without its exposure to the environment.[54]

Ultrafiltration is also a useful procedure for the extraction of protein from milk whey.[55] Microbiological hazards are reduced by avoiding the temperature range of those organisms capable of rapid growth and by minimising time of throughput. Plant for whey recovery is operated at $\sim 10\,°C$ or $\sim 50\,°C$—flux rate at the latter being markedly higher—and can handle about 45 000 litres per day. Concentrations of $\sim 25$–30-fold can be achieved.

A combination of flocculation and ion exchange chromatography is effective. As applied to abattoir effluent (containing 0·5 mg/ml protein), mean residual protein values, after flocculation, ion exchange and a combination of the two, were 35%, 28% and 6% respectively.[56]

R. A. Lawrie

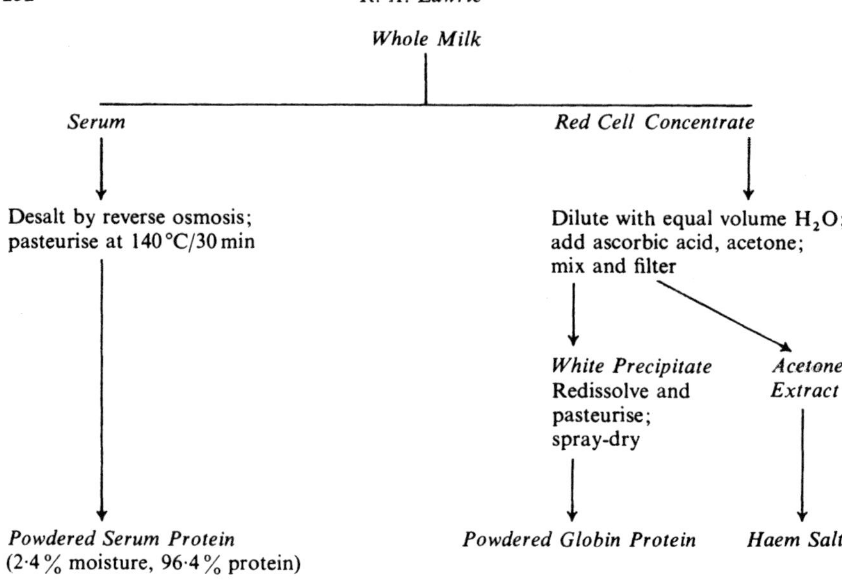


FIG. 1. Protein isolation from whole blood[53]

Ion exchange chromatography itself affords another approach to the recovery of protein from fluid media, whereby clear, virtually protein-free, effluents result. Adsorbed protein may be removed from the columns with aqueous sodium chloride or through pH control. Column packing materials may be chosen for various characteristics, and range in protein uptake from 300 to 900 mg/g. The technique may be employed in a continuously operating system. Typical data for a plant operating continuously over a three-month period on abattoir effluent are shown in Table 5.[57] Although discontinuous operation can yield a higher protein concentration, adsorption at higher pH and desorption at low pH, in a

## TABLE 5

ABATTOIR EFFLUENT TREATED BY CONTINUOUS ION EXCHANGE SYSTEM[57]

| Sample | Total N (g/litre) | Non-protein N (g/litre) | Protein (g/litre) |
|---|---|---|---|
| Raw effluent | 0·42 | 0·11 | 1·91 |
| Treated effluent | 0·11 | 0·09 | 0·13 |
| Concentrates | 1·28 | 0·08 | 7·50 |

process involving several stages for each operation, has been successfully applied to the isolation of protein from milk whey and from the effluents of soy, gelatine, vegetable and fish plants.[57] For example, with a flow rate of 450 000 litres per day, 6 % solutions of protein can be produced from milk whey.

Since the concentration of protein in blood is initially high, it is feasible to consider separating the individual proteins or groups of proteins by ion exchange chromatography with a view to utilising properties specific to each. Thus, using DEAE-Sephadex on citrated blood, three fractions of plasma proteins have been prepared.[58] One of these includes the fibrinogen and one the serum albumin. Their properties will be considered below.

## TEXTURISATION

The ultimate purpose in seeking to isolate protein from sources wherein its initial state is inaccessible or unattractive to human consumers, is to utilise it thereafter in fabricating new aesthetically acceptable products or to add desirable organoleptic properties to existing ones. To this end three broad processes have been employed: fibre-spinning, thermoplastic extrusion and heat-gelling.

### Fibre-Spinning

The term 'meat-analogue' is applied generally to those products which are textured to resemble various meats and which may replace meat completely.[59] A major advance in this field was made when techniques, hitherto employed in the textile industry, were applied to soy protein,[4] although egg and serum albumins had been spun (to make 'artificial silk') by Millar as early as 1898. In the Boyer process, proteins extracted from defatted plant matrices by food-grade alkali at pH 10–11, and concentrated to about 14–15 %, are held at 40–50 °C until the viscosity of the so-called 'spinning dope' reaches predetermined values. The dope is then fed under pressure through a porous membrane (or spinneret) with several thousand apertures ($\sim 0.2$ mm diameter) in a non-corrosive plate, into a bath containing brine and acid (pH 3–4.5). The protein precipitates as long, thin filaments which are gathered together longitudinally over godet wheels as a 'tow' of fibres. The latter may then proceed to a second bath in which the fibres are stretched and bound together by added fat, egg albumin or polysaccharides (such as carrageenan, pectate or alginate) and associated with flavour and colourants. A typical fibre would contain 50–70 % moisture, and (on a dry weight basis) 60% protein, 20% fat, 17% carbohydrate and 3% ash.[60] The texture and appearance of the fibres can be altered by controlling the viscosity of the dope, the rate of flow of dope, the rate of removal of precipitated fibres from the bath and the tension applied to them, as well as by the temperature, pH, salt concentration and by the nature of the ions in the coagulating bath. The product can be made to resemble beef, pork, chicken, turkey and other meats. The Boyer-type process has been successfully applied to the proteins extracted from soy, casein, cotton-seed, safflower, sesame and wheat gluten.[61]

In attempting to apply it to abattoir waste, it proved necessary to establish the fairly critical conditions in the alkaline protein solution which would permit fibre-spinning.[13,34,62] With blood plasma, concentrated by freeze-drying to a protein content of about 110 mg/ml, a viscosity of 250 poise was found to be sufficiently stable for the purpose. Such stability was achieved by first adding NaOH so that the ratio of NaOH/protein was 1:10. This initiated a marked increase to 50 P in 15 minutes, following which acetic acid was added to reduce the pH from 12 to 11. The protein dope thus formed was pumped at 20–30 lb/in$^2$ into a bath containing 20% NaCl in 1 M acetic acid. Protein fibres produced in this way contain 17·3% protein and 73% moisture.[34] They have a high ash content but this can be reduced to that characteristic of lean meat by washing.[63]

Such fibres can also be spun from the proteins extracted by alkali from lung and stomach but the behaviour of the extracted proteins is substantially different. Thus, the exponential rise in viscosity (terminating in gelation) which is characteristic of alkali-treated plasma, and which necessitates stabilisation with acid, does not develop. The fibres spun from lung and stomach proteins also differ from those spun from blood plasma protein in being markedly less elastic and more brittle. By mixing the proteins isolated from lung and stomach with those extracted from blood plasma, however, fibres of better texture than those spun from the solid offal tissues alone can be produced.[62,64] The incorporation of a proportion of lung proteins yields fibres with a meat-like colour which will not subsequently leach out[64]—an advantage over most vegetable sources for the production of simulated meats.

Fibres containing 65–95% protein and 5–35% polysaccharide, may be produced from a protein spinning dope containing polyuronates.[65] For coagulation, a pH range of 1–3 is required when the polysaccharide is carrageenan (which has strongly acidic sulphate groups); a bath of 5–10% $CaCl_2$ is required for alginate. The strength of the fibres produced—and their relative content of protein and polysaccharide—varies with pH.[66]

## Thermoplastic Extrusion and Heat Gelation

Extruder devices have been used over the past 50 years to fabricate products which are predominantly based on cereals.[67] The discovery by Atkinson,[68] however, that extruders could be used to plasticise concentrates and isolates of vegetable protein and of flours, by the controlled application of pressure and temperature whereby starch is gelatinised and proteins are denatured and cross-linked to form textured structures, has led to an entirely new range of products. These are predominantly protein-based.

Modern extruders are high-temperature, short-time cookers.[7] In a typical process, the protein flour, together with added fat, flavours and colours (and with or without steam preconditioning at 65–100 °C and atmospheric pressure) is fed into the first section of the extruder, where it is worked into a dough at about 85–110 °C. Thereafter the screw in the extruder barrel conveys the dough into a cone-shaped section where it is raised to temperatures between 120 and 175 °C during a brief period (12–20 s), it leaves the extruder and enters the atmosphere *via* a die (by which it is shaped). The sudden drop of pressure, and flash evaporation of moisture, cause the development of an 'open-cell' structure.

The 'cells' are oriented in the direction of flow of the plastic mass. By this process pieces of analogue are produced which, after rehydration, have a chewy texture and appearance like that of meat. The amount of water initially added to the mix affects the degree of expansion on extrusion.[7,68] The temperature at extrusion should be between 120 and 150 °C to permit flash evaporation and expansion.

Before rehydration, a typical extruded product made from soy flour would have about 50 % protein, 8 % moisture, 1 % fat, 3 % fibre and 5 % ash. Following rehydration the moisture content would be about 70 %.

Further improvement in the texture of meat analogues, even when derived from such varied sources as soy, cottonseed, rapeseed, peanuts and wheat, can be achieved by a double process,[69,70] using two extrusion cookers in series. The first moistens the input material, denaturing the protein constituents, and feeds it into a second extruder wherein it is reheated, stretched, cooled and formed into uniform, parallel layers. Relatively low pressures are employed. The portions (layers of parallel fibres) are free from the undesirable 'springiness' which may be associated with high-pressure, high-temperature extrusion.

A major advantage of thermoplastic extrusion is the possibility it affords of using flours in which protein is present at relatively low concentration in comparison with protein isolates. On the other hand, when applied to proteins recovered from abattoir waste, the presence of a cereal carrier is necessary to develop desired ranges of texture, e.g. an input mixture of 35 % protein isolate from bovine stomach and 65 % soy grits.[71] With protein isolated from such sources the flash evaporation phase of the extrusion procedure largely removes undesirable odours to give bland products. Indeed, the undesirable flavours, which may be associated with vegetable proteins, are also lost during thermoplastic extrusion.

Another approach to the production of chewy meat analogues is based on thermal gelation of an aqueous dispersion of vegetable protein, wheat flour dough, emulsified fat, colours and flavours. On steam-heating at one atmosphere for 10 minutes the protein is transformed from a viscous solution to a gel which binds together the fat, water and other constituents. The product can be manipulated to resemble various meats and meat products.

## UTILISATION

The improved utilisation of food proteins—whether as isolates or in association with polysaccharides and other edible components, and whether or not textured—has significance for the organoleptic satisfaction of consumers as well as for their health.

### Organoleptic and Functional Aspects

In terms of organoleptic satisfaction, by far the most exacting requirements in utilising recovered proteins arise where these are intended to simulate the appearance and textural properties of flesh foods. According to Bourne[72] meat analogues should be deformable, non-rigid and non-brittle and should have acceptable shear strength, requiring chewing during mastication. The success which has been achieved in this field during the past 30 years is reflected by the availability of products which resemble such specific commodities as beef, pork, bacon, turkey, chicken and lobster. Apart from the predominant protein content, these include edible binders, colours, flavourings and seasonings. Meat analogues have been marketed as sliced and canned products, in pies, as spreads and luncheon meats, and as various savoury snack items. They tend to have the advantages of lower cost, longer shelf life and inexpensive packaging.

Nevertheless, meat analogues tend to lose their juiciness and flavour after the first few bites and, thereby, most are distinguishable from the meat they are intended to simulate.[7] Moreover, the proteins used may be associated with relatively strong unusual or undesirable flavours—either naturally present or developing during processing—some of which arise from oxidation of trace amounts of lipid and can prove most persistent. For example, soy proteins bind and retain bean-like flavours, due to the presence of hexanal and ethyl vinyl ketone.[73]

Where exacting textural properties are not so critical, as in the use of recovered proteins to extend the protein content of, and add functional attributes to, existing foods, undesirable flavour characteristics can be removed by processes such as proteolysis—albeit at increased cost. There is a wide range of purposes in the food industry which recovered proteins can serve other than as simulators of existing products. Some of these are listed in Table 6.[74]

TABLE 6

FUNCTIONAL PROPERTIES OF SOY BEAN PROTEINS IN FOOD SYSTEMS[74]

| Properties | Protein form used† | Purpose in food system |
|---|---|---|
| Emulsification | F, C, I | Formation and stabilisation |
| Fat absorption | F, C, I | Promotion or prevention |
| Water absorption | F, C | Controlled uptake and retention |
| Solubility | I | Homogeneity |
| Texture | F, C, I | Viscosity, gelation, formation of structure |
| Dough formation | F | Cohesion, elasticity |
| Colour formation | I | Bleaching, browning |
| Aeration | F, C, I | Whippability, foam stabilisation |
| Antioxidation | F, C, I | Rancidity protection |

† F = flours, C = concentrates, I = isolates

The proteins of blood plasma can be used in partial substitution for the meat protein in sausages but the quantity and state of the proteins is important for the retention of quality. Thus, sausages (containing only 7 % total protein) can be produced when 20 % of the meat protein is replaced by unspun plasma protein but not at a 40 % replacement level since, on frying the latter (at $170 \pm 5\,°C$ over 5 minutes) holes form inside the mass because binding capacity is low.[75] On the other hand, satisfactory sausages can be made at the higher level of replacement when the added protein is in the form of co-spun plasma–alginate fibres or as unspun plasma protein and unspun alginate (Table 7). When unspun plasma protein and unspun

## TABLE 7

THE EFFECT OF 40% REPLACEMENT OF MEAT PROTEINS BY PLASMA PROTEINS IN VARIOUS FORM ON SAUSAGE PARAMETERS[75]

| Substituent | Objective evaluation | | Subjective evaluation | |
|---|---|---|---|---|
| | Tensile strength (arbitrary units) | Compressive strength (kg) | Texture‡ | General acceptability†† |
| Nil (control) | 0·79 | 2·05 | 3·4 | 5·9 |
| Plasma–alginate fibres | 0·55 | 1·52 | 3·2 | 5·3 |
| Unspun plasma + alginate fibres | 0·70 | 2·00 | 2·6 | 5·3 |
| Unspun plasma + unspun alginate | 0·32 | ND† | 5·0 | 3·9 |

†ND: Compressive strength could not be measured due to sample softness.
‡ Scale: 1 (soft)–5 (ideal)–9 (hard).
†† Scale: 1 (dislike extremely)–9 (like extremely).

alginate are added together, at a 40 % replacement level, however, an unsatisfactory product results.

The use of proteins extracted from underutilised sources as meat analogues and protein extenders in meat products, of course, is only one of many potential outlets. For many years the baking industry has been interested in the development of effective substitutes for egg albumin which would be less expensive. Although whole blood plasma has been considered in this context[76] and used to improve binding quality in meat analogues made from soy protein,[77] the possibility that its constituent proteins might have distinctive properties has not been investigated hitherto. Howell,[58] however, has recently made a detailed study of gelling, foaming and emulsifying properties of whole plasma, serum and three fractions of proteins separated from plasma by DEAE-cellulose chromatography, both *in vitro*, and when added to cake-type model systems, at different times and temperatures of heating. Thus the plasma fraction which comprised mainly albumin was found to be responsible for most of the gelling behaviour of whole plasma and serum. One of the other fractions, however, enhanced gelling at a significantly higher temperature—and at only half the concentration of the others or of the whole plasma. Moreover, the 'fishy' odour which tends to develop during storage of whole blood plasma appeared to be absent from the three plasma fractions. It also became clear that there is considerable interaction between the proteins present in other ingredients and those of various plasma fractions. These interactions differ in nature and degree, as reflected by various organoleptic parameters, and suggest they could be selected for specific culinary purposes.

**Health Aspects**
The nutritive value of the protein from vegetable sources is variable. There tends to be a relative deficiency of one or other of the essential amino acids (usually lysine and methionine) although soy protein has a particularly good complement, comparable with that of meat, eggs and casein (Table 8). Nevertheless the Food Standards Committee[78] recommended that, when soy or field bean proteins are used as *substitutes* for meat, these should contain 2·6 g methionine per 100 g protein (and not less than 10 mg iron and 5 $\mu$g vitamin $B_{12}$ per 100 g dry matter) and that the *replacement* of meat by such novel proteins of this specified nutritive value should not exceed 30 % of the statutory minimum meat content. While the nutritive value of protein is important, the ability of meat analogues to supply other essential nutrients in the quantities provided by the meat for which it is substituting cannot be neglected. On the other hand, such considerations can be

## TABLE 8

CONTENT OF ESSENTIAL AMINO ACIDS OF VARIOUS PROTEINS (AFTER FOOD STANDARDS COMMITTEE[78]) (AS g/100 g PROTEIN)

| Amino acid | Soy | Field bean | FAO/WHO reference | Beef | | | | | | | |
| --- | --- | --- | --- | --- | --- | --- | --- | --- | --- | --- |
| | | | | | Isolates | | | Fibres | | |
| | | | | Lean | Lung | Rumen | Plasma | Lung | Rumen | Plasma |
| Phenylalanine | 4·9 | 5·2 | 6·0† | 4·5 | 6·0 | 4·7 | 6·2 | 5·1 | 4·0 | 5·4 |
| Threonine | 4·0 | 3·8 | 4·0 | 4·3 | 4·9 | 5·2 | 6·7 | 5·1 | 4·7 | 7·3 |
| Lysine | 7·4 | 8·0 | 5·5 | 8·8 | 10·2 | 9·5 | 11·0 | 8·4 | 8·5 | 11·2 |
| Leucine | 7·1 | 8·3 | 7·0 | 7·7 | 11·6 | 9·6 | 9·8 | 10·8 | 8·5 | 11·2 |
| Isoleucine | 4·5 | 4·9 | 4·0 | 5·0 | 4·4 | 5·6 | 3·3 | 5·3 | 5·2 | 3·6 |
| Histidine | 3·2 | 3·2 | — | 3·7 | 3·1 | 3·0 | 4·7 | 2·1 | 1·8 | 2·8 |
| Methionine | 1·4 | 0·8 | 3·5‡ | 2·6 | 1·7 | 3·3 | 1·2 | 2·3 | 2·7 | 1·3 |
| Tryptophan | 1·3 | 1·0 | 1·0 | 1·3 | 2·4 | 1·8 | 2·4 | 2·9 | 1·6 | 1·8 |
| Valine | 5·0 | 5·5 | 5·0 | 5·1 | 5·9 | 5·5 | 8·0 | 4·7 | 3·5 | 6·6 |

† Phenylalanine + tyrosine
‡ Methionine + cysteic acid

ignored when recovered proteins are added to staple diets of poor nutritive value, or supplied as 'artificial milks', since the short-term nutritional benefits far outweigh them.

The proteins isolated from meat offal have a profile of essential amino acids which compares favourably with beef itself (Table 8). On the other hand, fibres spun from blood plasma proteins tend to have relatively low titres for isoleucine and methionine. Judicious mixing of proteins from various offal sources permits the spinning of fibres which have an excellent amino acid content (as well as suitable texture and colour).[62,64]

As judged by rat-feeding trials, the spinning of fibres enhances the net protein utilisation in comparison with the original protein isolates, despite a slight lowering of the methionine content.[79] This suggests that, in preparing the fibres, some inhibitors are removed or destroyed. Such might include trypsin inhibitors.

The possibility that the proteins recovered from the Plant Kingdom might be naturally associated with toxic principles is even greater and this is clearly a matter of concern when such proteins are to be used directly for human food instead of first proceeding through an animal intermediary.[80] These include a number of proteinase inhibitors, some of which are present, for example, in the soy bean. Processing, including roasting, can remove or destroy them with varying success.

Amongst other antinutritional factors, goitrogens, oestrogens, haemaglutinins, haemolysins and various allergens have been found in association with plant proteins. When the oligosaccharide, stachyose, is present, digestive upset is caused.

If the proteins are recovered from unicellular sources, it is conceivable that various known—or as yet unrecognised—toxins could be associated

TABLE 9



| Source | Psychrotrophs (7°C) | Mesophiles (30°C) |
|---|---|---|
| Fresh rumen | 10 000 | 400 000 |
| Fresh lung | 600 | 27 000 |
| Fresh blood plasma | 60 | 170 |
| Rumen fibres | 20 | 40 |
| Lung fibres | 10 | 60 |
| Plasma fibres | 30 | 350 |

with them and survive processing, although there has been little evidence for this to date.

The processing used in recovering and texturising proteins can be expected to bring about a substantial reduction in the number of microorganisms originally present in the source material. Thus the extraction by alkali of proteins from lung and rumen, and their subsequent precipitation by acid and salt in spinning fibres, are both accompanied by a reduction in psychrotrophs and mesophiles (Table 9).

Not surprisingly, the subsequent storage life of these materials is also greatly enhanced. Proteins extracted from rumen and lungs, for example, were stable for up to three weeks at 0 °C, in comparison with only three days for the sources themselves. Moreover, spinning the proteins into fibres increases the storage life even more and fibres are stable for at least three years at 0 °C.

## CONCLUSION

It is not only the need to avoid waste which justifies continuing attempts to improve the utilisation of protein for human food. A range of new products, and of improvements in existing products, can be envisaged which will extend the range of our organoleptic enjoyment of food as well as ensuring our nourishment. As technological means are developed for separating and concentrating the individual proteins from those complexes in which they occur naturally, it seems most likely that their properties will permit substantial exploitation in the interests of the consumer.

## REFERENCES


1. HARTMANN, W. E., 1976, cited by Kinsella.[7]
2. GERRARD, M. P. E., French Patent 178 367, 1886.
3. WRENSHALL, C. L., US Patent 2 560 621, 1951.
4. BOYER, R. A., US Patent 2 682 466, 1954.
5. ANSON, M. L., in *Processed Plant Protein Foodstuffs*, A. M. Altschul (ed.), Academic Press, New York, 1958, 277.
6. GUTCHO, M., *Textured Foods and Allied Products*, Noyes Data Corp., New Jersey, 1973.
7. KINSELLA, J. E., *C.R.C. Crit. Rev. Fd. Sci. Nutr.*, 1978, **10**, 147.
8. RACKIS, J., ANDERSON, R., SESAME, H., SMITH, A. and VAN ELTON, C., *J. Agr. Fd. Chem.*, 1961, **9**, 409.
9. JORGENSEN, J. and CONVIT, J., in *Algal Culture from Laboratory to Pilot Plant*, J. S. Barlow (ed.), Carnegie Inst., Washington, DC, 1953, Publ. No. 600.

10. PIRIE, N. W., *Proc. Nutr. Soc.*, 1956, **15**, 154.
11. IMRIE, F. K. E. and VLITOS, A. J., *Proc. 2nd Int. Symp. SCP*, Mass. Inst. Technol., Cambridge, Mass., 1973.
12. HOLSINGER, V. A., POZATI, L. P. and DE VILBUSS, E. D., *J. Dairy Sci.*, 1974, **57**, 849.
13. YOUNG, R. H. and LAWRIE, R. A., *J. Fd Technol.*, 1974, **9**, 69.
14. FIELD, R. A., *Wld. Rev. Anim. Prod.*, 1976, **12**, 73.
15. JOBLING, A. and HUGHES, R. B., *J. Sci. Fd. Agric.*, 1977, **28**, 1042.
16. JOBLING, A., *J. Sci. Fd, Agric.*, 1978, **29**, 1090.
17. LEVIN, E., US Patent 2 503 313, 1950.
18. LEVIN, E., *Fd. Technol. (Champaign)*, 1970, **24**, 19.
19. OLSEN, F. C., *Proc. Meat. Ind. Res. Conf.*, Amer. Meat Inst. Found., Chicago, 1970, 23.
20. GAULT, N. F. S., Ph.D. Dissertation, Univ. Nottingham, 1977.
21. TOLEDO, R. T., *J. Fd. Sci.*, 1973, **38**, 141.
22. ADLER-NISSEN, J., *J. Agric. Fd. Chem.*, 1976, **24**, 1090.
23. WATANABE, T., EBINE, H. and OKADA, M., in *New Protein Foods*, Vol. IA., A. M. Atschul (ed.), Academic Press, New York, 1974, 415.
24. WATANABE, T., *Proc. 5th Intl. Congr. Food Sci. Tech.*, *Kyoto*, Vol. 2, Elsevier, Amsterdam, 1979, 66.
25. HIGASHI, H., MARUYAMA, S., ONISHI, T., ISEKI, S. and WATANABE, T., *Bull. Tokai Reg. Fish Res. Lab.*, 1965, **43**, 77.
26. ISEKI, S., WATANABE, T. and KINUMAKI, T., *Bull. Tokai Reg Fish Res. Lab.*, 1969, **59**, 81.
27. CRISWELL, L. G., LITCHFIELD, J. H., VELY, V. G. and SACHSEL, G., *Fd. Technol. (Champaign)*, 1964, **18**, 1493.
28. TAKAHASHI, S., *Bull. Ehime Prefect. Inst. Chem. Technol.*, 1971, **9**, 20.
29. NEWMAN, P. B., *Meat Sci.*, 1981, **5**, 171.
30. TATTERSON, I. N. and WINDSOR, M. L., *J. Sci. Fd. Agric.*, 1974, **25**, 369.
31. BACKHOFF, H., *J. Fd. Technol.*, 1976, **11**, 353.
32. MILLAR, A., British Patent 60, 1898.
33. WORMELL, R., *New Fibres from Proteins*, Butterworth, London, 1954.
34. YOUNG, R. H. and LAWRIE, R. A., *J. Fd. Technol.*, 1974, **9**, 171.
35. SWINGLER, G. R. and LAWRIE, R. A., *Meat Sci.*, 1979, **3**, 63.
36. STERNBERG, M., KIM, C. Y. and SCHWANDE, F. J., *Science*, 1975, **190**, 992.
37. KOFRANYI, E. and JEKAT, F. K., *Hoppe-Seyl. Z. Phsiol. Chem.*, 1964, **338**, 154.
38. GAULT, N. F. S. and LAWRIE, R. A., *Meat Sci.*, 1980, **4**, 167.
39. ELLISON, N. S., GAULT, N. F. S. and LAWRIE, R. A., *Meat Sci.*, 1980, **4**, 80.
40. LUNDGREN, H. P., *Textile Res. J.*, 1945, **15**, 335.
41. KAPP, O. H. and VINOGRADOV, S. N., *Anal. Biochem.*, 1978, **91**, 230.
42. TONSETH, E. I. and BERRIDGE, H. B., *Eff. Wat. Treat. J.*, 1968, **8**, 124.
43. HOPWOOD, A. P. and ROSEN, D., *Process Biochem.*, 1972, **7**, 15.
44. DENMEAD, C. F., COOPER, R. N., RIGG, W. J., KINDRED, I. J., MEMS, M. A., REDPATH, D. M., FILTON, D. H. S., WADE, M. H. and WHITE, M. D., *Ann. Res. Rept.*, Meat Ind. Res. Inst. NZ Inc., 1973–74, 51.
45. HANSEN, P. M. T., HIDALGO, J. and GOULD, I. A., *J. Dairy Sci.*, 1971, **54**, 830.
46. SMITH, A. K., NASH, A. M., ELDRIDGE, A. C. and WOLF, W. J., *J. Agric. Fd. Chem.*, 1962, **10**, 302.
47. MORR, C. V., SWENSON, P. E. and RICHTER, R. L., *J. Fd. Sci.*, 1973, **38**, 324.

48. IMESON, A. P., WATSON, P. R., MITCHELL, J. R. and LEDWARD, D. A., *J. Fd. Technol.*, 1978, **13**, 329.
49. PILKINGTON, D. B., *17th Meat Ind. Res. Conf. Hamilton, NZ*, Meat Ind. Res. Inst. NZ Inc., Hamilton, 1975, 65.
50. PORTER, M. C. and MICHAELS, A. S., *Chemtech.*, 1971, **1**, 440.
51. FERNANDO, T., TOWNSEND, R. N., ASTON, T. D. and AVERILL, J. A., *Ann. Res. Rept.*, *Meat Ind. Res. Inst. NZ Inc.*, 1975–76, 66.
52. ECKELSSON, G. and VON BOCKELMANN, I., *Process Biochem.*, 1975, **10**, 11.
53. DILL, C. W., *Proc. 29th Ann. Recip. Meat Conf. Provo, Utah.* Nat. Livestock & Meat Bd., Chicago, 1976, 162.
54. HEINZ, G., *Fleischwirts*, 1969, **49**, 613.
55. COTON, S. G., in *Food from Waste*, G. G. Birch, K. J. Parker and J. T. Worgan (eds.), Applied Science, London, 1976, 221.
56. GRANT, R. A., in *Food from Waste*, G. G. Birch, K. J. Parker and J. T. Worgan (eds.), Applied Science, London, 1976, 200.
57. JONES, D. T., in *Food from Waste*, G. G. Birch, K. J. Parker and J. T. Worgan (eds.), Applied Science, London, 1976, 342.
58. HOWELL, N., Ph.D. Dissertation, University of Nottingham, 1981.
59. DUDA, Z., *Vegetable Protein Meat Extenders and Analogues*, AGA/MISC/74/7, FAO, Rome, 1974.
60. SMITH, A. K. and CIRCLE, S. J., *Soyabeans: Chemistry & Technology*, Vol. 1, AVI, Westport, Conn., 1972, 368.
61. HORAN, F. E., in *New Protein Foods*, Vol. IA, A. M. Altschul (ed.), Academic Press, New York, 1974, 367.
62. YOUNG, R. H. and LAWRIE, R. A., *J. Fd. Technol.*, 1975, **10**, 453.
63. SWINGLER, G. R. and LAWRIE, R. A., *Meat Sci.*, 1977, **1**, 161.
64. SWINGLER, G. R. and LAWRIE, R. A., *Meat Sci.*, 1978, **2**, 105.
65. GIDDEY, C., US Patent 2 947 644, 1960.
66. IMESON, A. P., LEDWARD, D. A. and MITCHELL, J. R., *Meat Sci.*, 1979, **3**, 287.
67. HARPER, J. M., *C.R.C. Crit. Rev. Fd. Sci. Nutr.*, 1978, **11**, 155.
68. ATKINSON, W. T., US Patent 3 488 770, 1970.
69. SMITH, O. B., in *New Protein Foods*, Vol. 2, A. M. Altschul (ed.), Academic Press, New York, 1976, 86.
70. WENGER, D., 1977, cited by Kinsella.[7]
71. MITTAL, P., Ph.D. Dissertation, Univ. Nottingham, 1981.
72. BOURNE, M. C., in *Fabricated Foods*, G. E. Inglett (ed.), AVI, Westport, Conn., 1975.
73. ARAI, S., *Agric. Biol. Chem.*, 1970, **34**, 1420.
74. BROOKWALTER, G. N., in *Nutritional Improvement of Food and Feed Proteins*, M. Friedman (ed.), Plenum Press, New York, 1977, 749.
75. RUSIG, O., *Meat Sci.*, 1979, **3**, 295.
76. BROOKS, J. and RATCLIFF, P. W., *J. Sci. Fd. Agric.*, 1959, **9**, 486.
77. GORDON, A., *Fd. Trade Rev.*, 1971, **4**, 29.
78. FOODS STANDARDS COMMITTEE, *Report on Novel Protein Foods*, HMSO, London, 1974.
79. SWINGLER, G. R., NEALE, R. J. and LAWRIE, R. A., *Meat Sci.*, 1978, **2**, 31.
80. LIENER, I. E., *Toxic Constituents of Plant Foodstuffs*, Academic Press, New York, 1969.
81. SWINGLER, G. R., NAYLOR, P. E. and LAWRIE, R. A., *Meat Sci.*, 1979, **3**, 83.

# 13

# Muscle Proteins and Muscle Structure

ALLEN J. BAILEY

*Agricultural Research Council,
Meat Research Institute, Bristol, UK*

## INTRODUCTION

Muscles are designed to produce force and movement in mammals, but in the context of this conference we are interested in muscle in the rigor|state as a food. However, to understand the factors determining the eating quality of the meat it is necessary to acquire as complete a knowledge of the structure and action of the living muscle as possible, and at the same time, details of the biochemical and structural changes involved in the rigor process. This basic knowledge has already proved invaluable to our present understanding and subsequent control of pre-slaughter stress, electrical stimulation, cold shortening, conditioning by enzymes and the water-holding capacity and tenderness of the meat (for reviews see refs. 1–3).

Muscles are usually classified according to their degree of structural order; striated muscles have a high degree of order and this group, which constitutes 40 % of the mass of the animal, includes the skeletal and cardiac muscles. Non-striated muscles exist in the walls of vascular systems and the gut and have a much less regular structure, which is not as well understood as striated muscle. For the striated muscle to produce movement the force generated by the contractile elements of the muscle must be transmitted to the skeleton without loss of energy and this is achieved by the tendons which are comprised of the inextensible connective tissue protein, collagen. The striated muscles constitute the muscles that ultimately provide meat, hence this paper will deal with the structural proteins of the striated muscle cell and its associated extracellular connective tissue.

The intact muscle is composed of numerous fibres that generally lie parallel to the long axis of the muscle although in some muscles the fibres are at an acute angle. These fibres, which may number as many as a million, are between 20–100 μm in diameter and vary in length from a few millimeters up to tens of centimeters. The fibres themselves contain about one thousand myofibrils of about 1–2 μm in diameter which run the whole length of the muscle fibre.

In the phase contrast light microscope the myofibrils reveal alternating light and dark bands. The dark bands (A-bands) contain a higher concentration of fibres than the light bands (I-bands), and the segment between two consecutive Z-lines, which bisect the light band, is called a sarcomere. The sarcomere can be regarded as the basic mechanical unit of the myofibril since it is the length of the sarcomere that changes during contraction and this occurs simultaneously in all the sarcomeres along the length of the myofibril.

Contraction occurs through an electrical stimulus to the cell membrane, and this action potential is spread throughout the muscle cell via the transverse tubules. These are invaginations of the cell membrane at regular intervals along the muscle. In mammals they occur at two points per sarcomere, i.e. at the A–I junctions, forming a network between the myofibrils which penetrates to the centre of the fibre. These transverse tubules conduct the action potential to the system of membrane-lined channels called the sarcoplasmic reticulum which surrounds every myofibril. The effect of the action potential is to depolarise the membrane and release $Ca^{2+}$, which is then bound by the myofibrils causing them to contract. After the passage of the action potential the membrane recovers and takes up $Ca^{2+}$ from the sarcoplasm causing the myofibrils to release $Ca^{2+}$ and relaxation occurs. With skeletal muscle an increase in force can be achieved by bringing more fibrils into play, whilst in the case of cardiac muscle all the myofibrils contract and hence to increase force an influx of more $Ca^{2+}$ is necessary.

At higher magnification in the electron microscope the fine details of the sarcomeres within the myofibrils can be seen more clearly and reveal the presence of two sets of filaments, the thick and thin filaments (Fig. 1).[4]

The thick filaments are about 15 nm in diameter and 1·6 μm in length, and the thin filaments are about 8 nm in diameter and 1·0 μm in length. The structure is divided up into zones for convenience, the A-band consisting of overlapped thick and thin filaments, the H-zone in the centre of the A-band consisting of thick filaments only, the I-band which consists of thin filaments only and the Z-line. At the centre of the A-band, the thick filaments are connected by a structure called the M-line. Examination of transverse sections reveals the filaments to be arranged in a regular

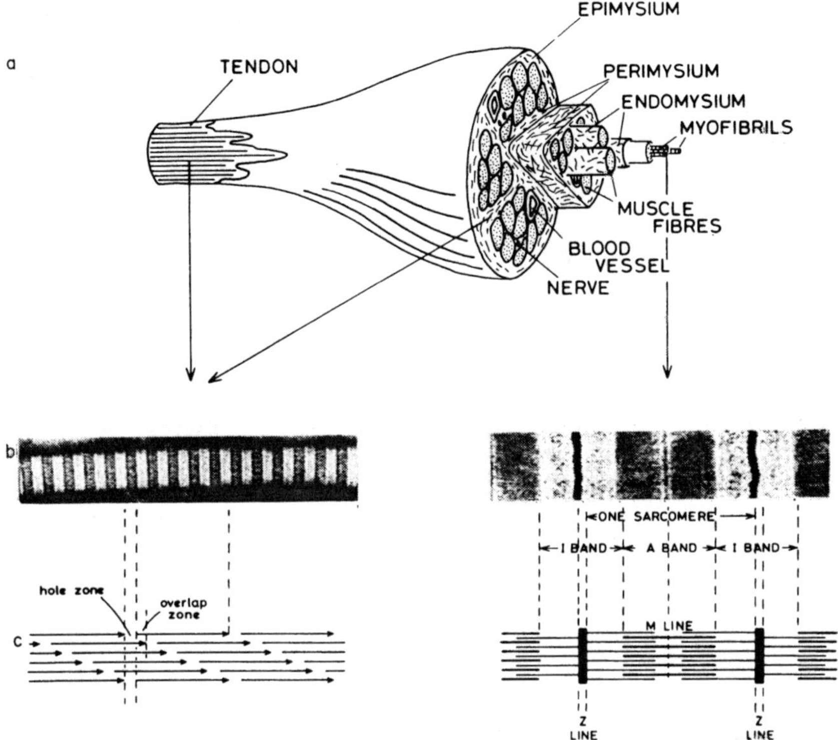



FIG. 1. Diagrammatic representation of the structure of muscle. (a) Illustrates the presence of muscle bundles, fibres and myofibrils, and the classification of the collagen into epimysium, perimysium, endomysium and tendon; (b) an electron micrograph of a single collagen fibre and of a myofibril; (c) diagrammatic representation of the organisation of the molecules in the collagen fibre and myofibril.

hexagonal lattice. Where the thick and thin filaments overlap in the A-band each thick filament is surrounded by six thin filaments. In the I-band the geometry of the thin filaments is not quite so regular since the lattice changes from hexagonal in the overlap region of the A-band to a square arrangement at the Z-line.

It is now generally agreed that contraction or extension of muscles takes place by the sliding of the thick filaments past the thin filaments without a change in the length of either filament.[5] To produce a tension by such a sliding mechanism a longitudinal force must be developed between the thick and thin filaments. Indeed such structural links, or cross-bridges, between the thick and thin filaments, can be seen in high resolution electron micrographs.

The precise mechanism of this contractile process, involving cross-bridges, is still a major problem to muscle biochemistry, although the basic concept is accepted. The evolution of this proposal has involved a detailed study of the structure of the muscle proteins over the past two decades.

## THE PROTEINS OF MUSCLE

The complexity of the structure of the myofibrils suggests that, in addition to the major constituents actin and myosin, a number of other proteins are present. All the major proteins (Table 1) have now been identified but it is probable that other minor proteins remain to be discovered.

### Thick Filament Proteins

Myosin is the major protein of the thick filaments and as such comprises about half of the protein of the myofibril. Its outstanding properties are that it possesses the only ATPase activity of the myofibril and that it binds strongly to the thin filament. Myosin therefore provides the link between structure and energetics. The shape of an extracted myosin molecule can be visualised in the electron microscope as a long thin molecule (2 nm × 160 nm) with two globular heads (19 nm long) attached at one end of the long tail. The molecules combine the features of a fibrous and globular

TABLE 1

MUSCLE MYOFIBRILS: PROTEIN COMPOSITION AND THEIR LOCATION

| *Protein* | *Location* | *Fraction of myofibril (%)* |
|---|---|---|
| Myosin | Thick filament | 55 |
| C-protein | Thick filament | 2 |
| H-protein | Thick filament | 0·3 |
| Actin | Thin filament | 23 |
| Tropomyosin | Thin filament | 6 |
| Troponin | Thin filament | 6 |
| α-actinin | Z-line | 1 |
| M-protein | M-line | |
| MM creatine kinase | M-line | |
| Desmin | Connection between myofibrils | ? |
| Titin (connectin) | Connection between Z-lines | ? |

protein. The rod-like tail is known to be a 2-chain coiled-coil $\alpha$-helix with a flexible region in the tail and the globular heads attached to the end of the molecule. The ATPase activity and thin filament binding site reside in the heads of the molecule.

About 300 myosin molecules are packed into the thick filament. Huxley[5] pointed out that in the centre of the thick filament there was a zone free of cross-bridges whereas the rest of the filament was covered with them. He interpreted this in terms of a bipolar arrangement of the molecules, such that the bare zone contains only the overlapped tails of the myosin molecules derived from both sides of the filament. This reversed polarity arrangement neatly accounts for the fact that the thin filaments over-lapping the two opposite ends of the thick filament must be propelled in opposite directions during movement.

Precipitation of myosin molecules from solution results in the formation of thick filaments rather similar to those extracted from myofibrils in possessing projections along their length apart from the bare zone in the middle, except that their length is highly variable. Myosin thus has the ability of self-assembly without the directing influence of a template. To date, the precise packing arrangement of the myosin molecules is not known but recent electron micrographs of native thick filaments have provided considerably improved details. High resolution pictures by Knight and Trinick of the Meat Institute (personal communication) reveal pairs of heads projecting out a considerable distance from the backbone and in some instances one can see the connection between the heads and the backbone. X-ray diffraction studies indicate that the cross-bridges are arranged approximately in a helical manner with a repeat every 429 Å, and the axial interval between groups of cross-bridges is one-third of this repeat (143 Å) (Fig. 2(a)). For some time the number of myosin molecules at each axial level has been controversial but recently it has been shown that at very low ionic strength the thick filament shaft dissociates into three subfila-ments.[6] It would appear therefore that the tails of the myosin molecules are packed into a subfilament approximately 80 Å in diameter and that three subfilaments are packed side by side to form the shaft, i.e. three myosin molecules every 143 Å. If each molecule is tilted slightly with respect to the filament axis, the end of the tail lies near to the axis of the filament while the junction between the light meromyosin and subfragment-2 moieties lies on the surface (for review see ref. 7).

Another structural component of the thick filament has recently been identified by Trinick.[8] When the filaments were frayed into subfilaments, structures were observed at the ends of the filaments which join the

     *Allen J. Bailey*

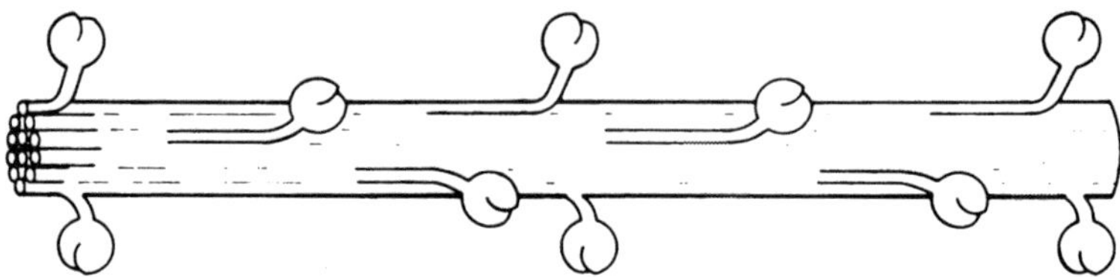



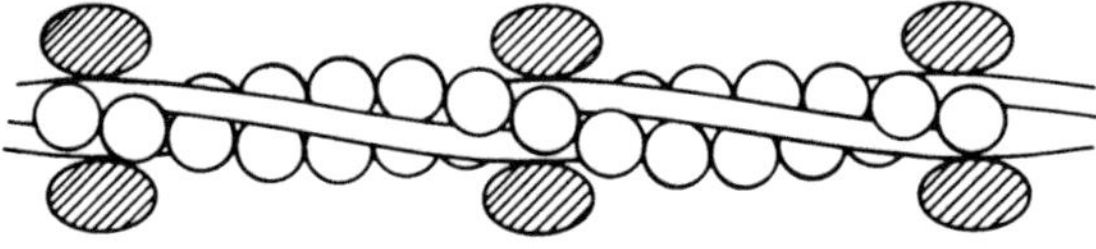

FIG. 2.   Schematic representation of the proposed organisation of the myofibrillar filaments. (a) The myosin molecules in the thick filament; (b) the actin molecules in the thin filaments, containing the rod-like tropomyosin in the grooves of the helix and the troponin complex every seventh actin molecule.

subfilaments together. These structures, termed end-filaments, are 850 Å long and 50 Å wide with a periodicity of 40 Å, and with a globular structure, about 150 Å × 100 Å, at one end. Over the past few years several suggestions have been made that there might be a third set of filaments joining the ends of the thick filaments to the Z-line. For example, by stretching muscle fibres, Locker *et al.*[9] observed thin filaments extending to the Z-line from the thick filament, and Toyoda and Maruyama[10] have proposed the presence of an additional protein, called connectin, which may be in the thick filaments. More recently, Wang *et al.*[11] have isolated a protein, termed titin, which has now been shown to be the same protein as connectin. Imunofluorescence studies indicated that this protein is located at the end of the A-band. If the structures observed in normal fibres by Trinick[8] really do exist they may act as anchorage sites joining the thick filaments to this third set of filaments.

Although myosin is the main protein of the thick filament, other proteins have now been identified. The M-line contains M-protein, MM creatine kinase,[12] the latter probably forming M-bridges connecting the thick filaments. Two additional proteins, C-protein[13] and H-protein (Offer and Starr, unpublished data) have been isolated from the thick filament. The location of these proteins has now been determined using immunological methods. Seven stripes, 430 Å apart, are labelled in each half of the A-band when the section is stained with antibody to C-protein. The protein appears to be an elongated ellipsoid and its high proline content suggests it is not $\alpha$-helical. Antibodies to the H-protein label a single stripe in each half of the A-band nearer to the M-line. The function of these proteins is unknown; it has been suggested that the C-protein may influence the movement of the cross-bridges, although it may just be present to stabilise the backbone of the thick filaments during contraction.

## Thin Filament Proteins

The main protein of the thin filament is a globular protein which polymerises at physiological salt concentration to form long filaments consisting of two long-pitched helical strands of subunits. The globular monomer, G-actin, has a molecular weight of 41 700 and contains one bound $Ca^{2+}$ ion and one non-covalently bound ATP. During polymerisation the bound ATP is converted to ADP, and polymerisation is reversible by low ionic strength solutions containing ATP.

## Regulatory Proteins

Purified actin and myosin are unaffected by $Ca^{2+}$ but impure preparations show some activity, suggesting the presence of other proteins. This led to the discovery of tropomyosin and troponin, which together function as a regulatory system. In each of the two grooves of the actin filament, strings of tropomyosin molecules, each about 410 Å long, are joined end to end along the length of the thin filament. Like the myosin tail, tropomyosin consists of two helical polypeptide chains wound in a coiled-coil (Fig. 2(b)). Each tropomyosin molecule is long enough to interact with seven actin molecules.

Attached to each tropomyosin molecule is a troponin molecule; troponin is a slightly elongated molecule of molecular weight 80 000 which binds to tropomyosin but not to actin. By the use of antibodies, the troponin has been shown to be present at discrete intervals along the thin filament; the stripes seen in the I-band of sectioned muscle are due to troponin. The molar ratio of actin, tropomyosin and troponin is 7:1:1. Troponin actually exists as a complex of subunits, their composition being revealed by SDS electrophoresis as three separate polypeptide chains: TN–T (30 500) which binds to tropomyosin; TN–I (20 900) which inhibits actomyosin ATPase; and TN–C (17 800) which binds $Ca^{2+}$. TN–C has been shown to possess four $Ca^{2+}$ binding sites,[14] similar to calmodulin, a sarcoplasmic protein.[15]

The troponin molecule is the $Ca^{2+}$-sensitising protein of the myofibril but its influence on the actin is transmitted via tropomyosin.

When troponin binds $Ca^{2+}$ it causes tropomyosin to move in the grooves of the actin helix, which probably leads to exposure of the site for the myosin heads to interact with the actin.[16] The effect may be demonstrated by X-ray diffraction but the precise structural changes involved have not yet been elucidated.

The troponin complex is different in different muscles, e.g. in fast muscle, TN–C possesses four $Ca^{2+}$ binding sites whilst in slow muscle it possesses only three binding sites. Similarly, TN–T and TN–I have been shown to have slightly different amino acid sequences in fast and slow muscles and are therefore different gene products. Antibodies to these proteins isolated from fast and slow muscles, are seen to stain specific cells in the muscle. Precisely what induces the cell to synthesise the different troponins is not clear but is certainly related to nerve innervation. Cross-innervation of rabbit soleus results in a change in these myofibrillar components. The regulation of muscle contraction, i.e. the actin–myosin interaction is clearly a complex process involving the troponin–tropomyosin complex, ATPase activity and $Ca^{2+}$ ions.[17] More recently Perry[18] reported a second form of regulation by reversible phosphorylation of the myosin light chain.

## Mechanism of Contraction

The attachment of the globular heads of the myosin molecule to specific sites on the actin must in some way produce tension and generate a sliding force. The precise mechanism by which this occurs is still unknown. The current working hypothesis was suggested by Huxley in 1969.[5] He proposed that the heads attach to the thin filaments at approximately 90° and then tilt back towards the tail to an angle of 45°, thus causing movement of the thin actin filaments (Fig. 3). Although it is the major working hypothesis, there is little or no evidence for the theory.

Clearly a considerable amount of work is required to elucidate the precise mechanism of contraction, how contraction is regulated, and the role of the new minor proteins in the myofibrillar structure. Detailed investigation of the chemical and functional properties of the various components of the myofibrils, particularly from muscles with diverse physiological properties, should provide a deeper understanding of these topics.

## Connective Tissue Proteins

Contraction of the myofibrils is, in the majority of muscles, transmitted to the skeleton through the tendons. The ability of the tendon to transmit this force with little loss in energy is dependent on the virtually inelastic

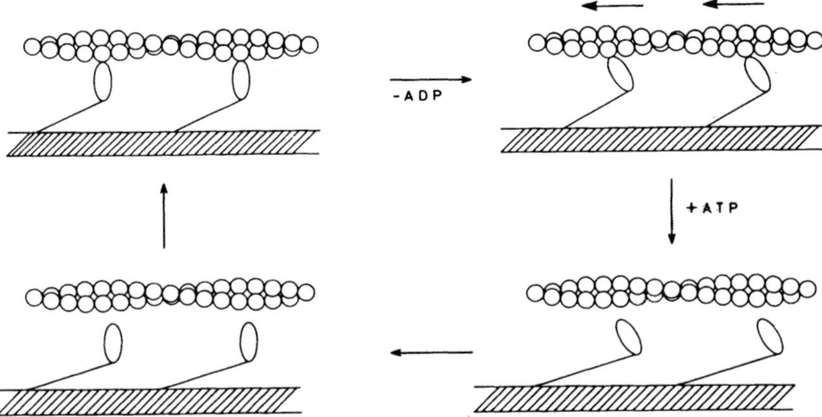


FIG. 3.    Simplified model of the stages in the interaction of the myosin heads with the actin
filament leading to mechanical movement by sliding of the filaments.

properties of collagen fibres. Collagen is also the major supporting
framework of the muscle and, in addition to transmitting force, prevents
over-extension and damage to the fragile contractile elements of muscle.

The basic structure of collagen has been known, primarily from X-ray
diffraction and electron microscopy, for some twenty years.[19,20] The
amino acid sequence is unique in that every third residue is glycine and in
the repeating sequence gly-X-Y, X and Y are frequently proline and
hydroxyproline respectively. The presence of these pyrollidine residues
causes the polypeptide chains to adopt a polyproline type of helix rather
than the α-helix common in other proteins. The slow pitch of the
polyproline helix ensures that no intrachain hydrogen bonds are formed
and when three of these chains are wound together they are stabilised by
interchain H-bonding involving the small glycine residues positioned on
the inside of the triple helix. The presence of glycine at every third residue is
an absolute requirement for a stable triple helical coiled-coil structure. The
side-chains of the other amino acid residues are on the outside of the helix
and become involved in intermolecular interactions. The collagen molecule
or monomer that exists within a fibre is 300 nm × 1·5 nm and can therefore
be considered as a long, thin, rigid rod. Based mainly on electron
microscopic studies it has been established that these long molecules are
arranged in the fibre in a quarter-stagger fashion with an end overlap of
25 nm (Fig. 1). This arrangement accounts for the axial periodicity of
collagen fibres observed in the electron microscope, the dense stain

                     *Allen J. Bailey*

penetrating the fibre at the hole region. The alignment of the molecule in this manner is governed by the asymmetric distribution of the charged residues along the molecule, and the hydrophobic residues play a part in maintaining the organisation. However, despite the precise alignment of the molecules, the fibre has no tensile strength, a property which requires the formation of covalent bonds between the molecules to prevent slippage during mechanical stress.[21] The ultimate strength of the particular tissue depends, not only on the cross-linking of the molecules within a fibre but on the organisation of these fibres within the tissue. A wide variety of morphologies exists between various collagenous tissues from the uni-directional fibres in tendon, the precise 90° laminations in cornea, the concentric rings in bone and the random fibre arrangement in skin.

More recently another factor has had to be included in any discussions attempting to understand the basis of the mechanical properties of collagen. Detailed chemical evidence has now established that there are a number of genetically distinct collagens and that they are, to a certain extent, tissue specific. At the present time, six or seven different types have been described.

The classical collagen of skin, tendon and bone is now referred to as Type I collagen and consists of two identical $\alpha$-1-chains and a chemically different $\alpha$-2-chain (Table 2). Type II collagen is the predominant collagen in cartilage, and is also present in intervertebral discs; it is composed of three

TABLE 2

ISOMORPHIC FORMS OF COLLAGEN[20]

| Type | Molecular composition | Nature | Distribution |
|---|---|---|---|
| I | $[\alpha\text{-1(I)}]_2\alpha\text{-2}$ | Fibrous | Tendon, skin, bone, dentine |
| II | $[\alpha\text{-II}]_3$ | Fine fibres | Cartilage, vitreous humour, intervertebral disc |
| III | $[\alpha\text{-III}]_3$ | Fine fibres | Foetal dermis, cardiovascular system, viscera, muscle |
| IV | Unknown (possibly two chains plus 7S component) | Amorphous basement membrane | Lens capsule, kidney glomeruli, placenta, lung |
| V | $[\alpha\text{-B}]_2\alpha\text{-A}$ | | Placental membrane, cardiovascular system, lung |

identical α-chains. Type III collagen is a major component of foetal skin and the vascular system, and consists of three identical α-chains. Collagen Types I, II and III are known to be fibrillar in nature in these tissues. In contrast, Type IV collagen is the major collagenous component of the non-fibrillar, or amorphous, basement membranes. This type of collagen has a complex structure which has not yet been fully characterised.[22,23] It is highly likely that the composition of the basement membrane and its collagenous components may vary considerably between membranes, e.g. lens capsule, glomeruli and vascular. Type V collagen appears to be closely associated with basement membranes but it is not clear whether or not it possesses a fibrillar structure. More recently new collagens have been identified in cartilage which appear to be pericellular in nature and may therefore play a role in controlling the activities of the chondrocyte.[24]

To fully evaluate the role of collagen in muscle it is clearly necessary to appreciate the complexity of collagen itself. Primarily, the genetic type of collagen, the morphology of the fibrillar network and the nature of the covalent cross-links must be understood.

**Intramuscular Collagen**
Classical histological studies have identified collagen at various levels in the structure of a muscle. The entire muscle is surrounded by thick collagen fibres which extend to the tendon; this is referred to as the epimysium. These fibres are, on the whole, unidirectional. The muscle fibre bundles are also surrounded by a sheath of collagen fibres which is known as the perimysium. These are generally fine fibres arranged in a laminated structure around the bundles and are often referred to as reticulin. The individual myofibres also possess a connective tissue sheath, known as the endomysium. This sheath is made up of an amorphous basement membrane and an associated layer of fine fibres.[25]

By using both chemical and immunological techniques, muscle has been shown to contain Type I, Type III, Type IV and Type V collagen. The distribution was found to be reasonably specific, Type I collagen being the major component of the epimysium, Type III collagen predominating in the perimysium, and Types IV and V confined to the endomysium.[26] The tendon, as anticipated, contains primarily Type I collagen but immunofluorescent staining with anti-Type III collagen revealed that the bundles within the tendon are surrounded by a sheath of Type III collagen fibres and an inner sheath of Type IV basement membrane collagen.[26,27]

The relative mechanical strength of these different fibres, and hence their contribution to the transmission of muscular contraction, is not yet known.

However, considerable progress has been made towards understanding the nature of the covalent cross-links stabilising each of these individual collagens.[28,29]

It is generally accepted that following secretion of procollagen, excision of the procollagen peptides and subsequent fibre formation, the only enzymic process involved in stabilising the fibre is the oxidative de-amination of specific lysyl or hydroxylysyl residues. These residues are present in the short non-triple helical regions at the N and C termini of the molecule and their ε-amino groups are oxidised to aldehydes by the enzyme lysyl oxidase. The aldehydes then spontaneously react with an adjacent collagen molecule, and because of the precise quarter-stagger alignment, form an aldimine or keto-imine bond with the ε-amino group of a specific hydroxylysine 25 nm from the end of the molecule, i.e. at the end of the overlap region (Fig. 4). Despite various reports, no convincing evidence has

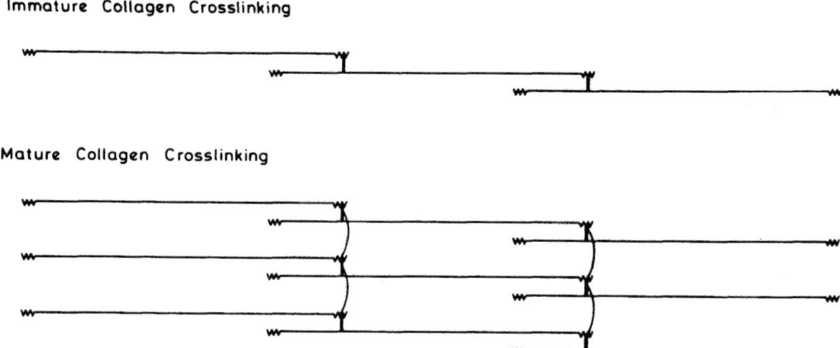



FIG. 4.   Diagrammatic representation of the location of the cross-links. (a) In immature collagen the cross-links are restricted to the end-overlap region involving the telopeptide aldehyde; (b) in mature collagen the above intermediate cross-links form interfibrillar cross-links with molecules in register.

been provided for cross-links at other sites. Similar reactions take place at these locations in the other fibrous collagens, Types II and III. This head-to-tail cross-linking builds up a long microfibril within the fibre and would confer considerable tensile strength on the fibre, but to ensure that slippage between microfibrils does not occur, interfibril cross-linking is necessary. Evidence to support the presence of such transverse as well as lateral bonds, has been forthcoming recently from a detailed analysis of collagen fibre from mature animals.[30] Earlier studies revealed that the reducible aldimine and keto-imine cross-links present in developing collagenous tissues rapidly disappeared with increasing age such that at maturity they were barely detectable.[31] In contrast, at this stage the fibres were extremely stable as assessed by tensile strength, isometric tension, swelling and solubility. A detailed analysis of isometric tension curves confirmed the increase in stable non-reducible cross-links. In this test, the fibre, which normally shrinks to one-quarter of its length on heating to about 65 °C, is restrained in clamps and the tension generated on thermal denaturation determined on a strain gauge. If the fibre is stabilised only by aldimine cross-links, the tension generated decreases on further heating due to the cleavage of the labile aldimine cross-link. With increasing age the tension generated increases dramatically and the extent of relaxation decreases, indicating a greater proportion of thermally stable bonds.[32] Unlike the aldimine and keto-imine cross-links, these stable bonds are not reducible and it has proved difficult to characterise them.

Some evidence as to the location of these bonds has been achieved by cyanogen bromide digestion, which cleaves proteins specifically at methionine residues, of collagen fibres from mature tissues. Radioactive labelling of the tyrosine in the peptide chain near the reducible cross-links followed by CNBr digestion, revealed that the label, originally only present in CB6-CB5 cross-linked peptide in young tissue, was present in a very high molecular weight component in mature tissue. Analysis of this component revealed it to be comprised primarily of $\alpha$-1CB6 and $\alpha$-1CB5 indicating a polymeric peptide CB6-CB5-CB6-CB5- etc.[30]

These results demonstrate that in mature tissues the cross-linking occurs through molecules in register. On the assumption that the molecules are aligned two-dimensionally in a quarter-stagger fashion to permit head-to-tail cross-linking (i.e. CB6-CB5) and three-dimensionally in a pentafibril, then the pentafibrils must be arranged in register to permit interaction between the end-terminals (i.e. CB6-CB5-CB6- etc.). In this way a network of lateral and transverse cross-links to stabilise the fibre can be generated involving no other reaction than those of the N- and C-terminal lysine-aldehydes. Presumably the transverse cross-link is a tri- or tetravalent cross-link derived from the reducible aldimine and keto-imine cross-link. Fujimoto *et al.*[33] have identified a potential trivalent cross-linking compound, 3-hydroxypyridinoline, and Eyre and Oguchi[34] have suggested that it is derived from the reaction of two keto-imine cross-links. Analysis of the polymeric CB6 from mature tendon or bone collagen failed to reveal this component, hence its role as a stabilising cross-link must await its identification in a cross-linked collagenous peptide.

The mechanical role of collagen in muscle therefore depends on the genetic type of collagen, the nature and extent of the cross-linking, the age of the tissue and the alignment of the fibres.


## MUSCLE PROTEINS AND MEAT QUALITY

On conversion of muscle to meat the myofibrillar proteins are fixed in rigor but the conditions to which the muscle is subjected pre- and post-rigor determine to a large extent the eating quality of meat. For example, pre-slaughter stress affects the water-binding capacity, pre-rigor chilling can result in super contraction of the myofibrils (cold shortening), the amount and maturity of the collagen affects toughness and during conditioning enzymes digest these structural proteins and increase tenderness.[35,36] In addition to the changes during processing, the meat scientist must also consider the changes in properties of the myofibrillar proteins and the collagen in the thermally denatured form following cooking. On heating, the myofibrillar proteins interact to form a firm gel and at higher temperatures the collagen shrinks exerting a considerable squeezing effect on the myofibrillar proteins. Both these changes result in an increase in the toughness of the meat but the reasons for the considerable variation in the extent of these changes poses further problems for the meat scientist. Some of these problems will be discussed in the following lectures (see Dutson and Tarrant, Ch. 17 and Ch. 14).

It is only from a detailed study of the properties of the structural proteins of muscle that the food scientist can achieve any control over the practical problems affecting the eating quality of meat.


## REFERENCES

1. LAWRIE, R. A., *Meat Science*, 3rd Edn. Pergamon Press, Oxford, 1979.
2. BENDALL, J. R., *Muscles, Molecules, and Movement*, Heinemann Educational Books Ltd, London, 1979.
3. OFFER, G. W., in *Companion to Biochemistry*, A. T. Bull, J. R. Lagnado, J. O. Thomas and K. F. Tipton (eds.), Longman, London, 1974.
4. HUXLEY, H. E., in *Structure and Function of Muscle*, 2nd Edn., Vol. 1, G. H. Bourne (ed.), Academic Press, New York, 1973.
5. HUXLEY, H. E., *Science*, 1969, **164**, 1356.
6. MAW, M. C. and ROWE, A. J., *Nature*, 1980, **286**, 412.
7. SQUIRE, J. M., in *Fibrous Proteins: Scientific, Industrial and Medical Aspects*, Vol. 1, D. A. D. Parry and L. K. Creamer (eds.), Academic Press, London, 1979.

8. TRINICK, J., *J. Mol. Biol.*, 1981, **151**, 309.
9. LOCKER, R. H., BAINS, G. J., CARSE, W. A. and LEET, N.G. *Meat Sci.*, 1977, **1**, 87.
10. TOYODA, N. and MARUYAMA, K., *J. Biochem. (Tokyo)*, 1978, **84**, 239.
11. WANG, K., MCCLURE, J. and TU, A. *Proc. Natl. Acad. Sci. (US)*, 1979, **76**, 3698.
12. TRINICK, J. and LOWEY, S., *J. Mol. Biol.*, 1977, **113**, 343.
13. OFFER, G., MOOS, C. and STARR, R., *J. Mol. Biol.*, 1973, **74**, 653.
14. COLLINS, J. H., *Nature*, 1976, **259**, 699.
15. KLEE, C. B., CROUCH, T. J. and RICHMAN, P. G., *Ann. Rev. Biochem.*, 1980, **49**, 489.
16. SEYMOUR, J. and O'BRIEN, E. J., *Nature*, 1980, **283**, 680.
17. WEBER, A. and MURRAY, J. M., *Physiol. Rev.*, 1977, **53**, 612.
18. PERRY, S. V., *Biochem. Soc. Trans.*, 1979, **7**, 593.
19. VIIDIK, A. and VUUST, J., *Biology of Collagen*, Academic Press, London, 1980.
20. BAILEY, A. J. and ETHERINGTON, D. J., in *Comprehensive Biochemistry*, Vol. 19B, M. Florkin, A. Neuberger and L. van Deenen (eds.), Elsevier Publ. Co., Amsterdam, 1979.
21. BAILEY, A. J., ROBINS, S. P. and BALIAN, G., *Nature*, 1974, **251**, 105.
22. BAILEY, A. J., SIMS, T. J., DUANCE, V. C. and LIGHT, N. D., *FEBS Letts.*, 1979, **99**, 361.
23. KUHN, K., WIEDEMANN, H., TIMPL, R., RISTELI, J., DIERINGER, H., VON, T. and GLANVILLE, R. W., *FEBS Letts.*, 1981, **125**, 123.
24. SHIMOKOMAKI, M., DUANCE, V. C. and BAILEY, A. J., *FEBS Letts.*, 1980, **121**, 51.
25. BLOOM, W. and FAWCETT, D. W., *A Textbook of Histology*, 10th Edn., W. B. Saunders, Philadelphia, USA, 1975.
26. DUANCE, V. C., RESTALL, D. J., BEARD, H., BOURNE, F. J. and BAILEY, A. J., *FEBS Letts.*, 1977, **79**, 248.
27. BAILEY, A. J., RESTALL, D. J., SIMS, T. J. and DUANCE, V. C., *J. Sci. Fd Agric.*, 1979, **30**, 203.
28. NICHOLLS, A. C. and BAILEY, A. J., *Biochem. J.*, 1980, **185**, 195.
29. HEATHCOTE, J. G., BAILEY, A. J. and GRANT, M. E., *Biochem. J.*, 1980, **190**, 229.
30. LIGHT, N. D. and BAILEY, A. J., *Biochem. J.*, 1980, **185**, 373.
31. ROBINS, S. P., SHIMOKOMAKI, M. and BAILEY, A. J., *Biochem. J.*, 1973, **131**, 771.
32. ALLAIN, J. C., LE LOUS, M., BAZIN, S., BAILEY, A. J. and DELAUNAY, A., *Biochim. Biophys. Acta*, 1978, **533**, 147.
33. FUJIMOTO, D., MORIGUCHI, T., ISHIDA, T. and HAYASHI, H., *Biochem. Biophys. Res. Commun.*, 1978, **84**, 52.
34. EYRE, D. and OGUCHI, H., *Biochem. Biophys. Res. Commun.*, 1980, **96**, 403.
35. BAILEY, A. J., *J. Sci. Fd Agric.*, 1972, **23**, 995.
36. DAVEY, C. L. and WINGER, R. J. in *Fibrous Proteins: Scientific, Industrial and Medical Aspects*, Vol. 1, D. A. D. Parry and L. K. Creamer (eds.), Academic Press, London, 1979.

# 14

# Muscle Proteins in Meat Technology

P. V. TARRANT

*Meat Research Department, Dunsinea Research Centre,
Castleknock, Co. Dublin, Ireland*

## INTRODUCTION

Skeletal muscle is, both qualitatively and quantitatively, the most important tissue of meat. Muscle proteins are of primary importance for the nutritive value and eating quality of meat. They determine the colour, tenderness and water-binding capacity of meat and contribute also to other quality characteristics. During processing the muscle proteins undergo changes which affect their functional properties and hence influence meat quality. A basic aim of the meat processor is to control the changes in the functional properties of meat proteins and so to preserve or improve the quality of the meat. This paper examines the influence of meat processing on the proteins of muscle and the consequences for the quality of meat products. The first part of the paper will review the relationship between muscle protein and some primary quality characteristics of meat and the second part will review the effect of some important meat processes on muscle proteins.

## MUSCLE PROTEINS AND MEAT QUALITY

With the exception of water, proteins are the major constituent of lean meat (Table 1). On the basis of solubility they may be broadly classified into water-soluble, salt-soluble and insoluble fractions (Table 2). The salt-soluble and insoluble fractions contain the fibrous proteins, actomyosin and collagen respectively, which comprise the important structural elements of muscle. The structural proteins determine toughness in raw meat

## TABLE 1

APPROXIMATE COMPOSITION OF MAMMALIAN SKELETAL MUSCLE (PERCENTAGE FRESH
WEIGHT BASIS)[1]

| Component | Percentage (range) |
|---|---|
| Water | 75·0 (65 to 80) |
| Protein | 18·5 (16 to 22) |
| Lipid | 3·0 (1·5 to 13·0) |
| Non-protein nitrogenous substances | 1·5 |
| Carbohydrates and non-nitrogenous substances | 1·0 (0·5 to 1·5) |
| Inorganic constituents | 1·0 |

## TABLE 2

APPROXIMATE COMPOSITION OF THE PROTEIN FRACTION OF
MAMMALIAN SKELETAL MUSCLE (PERCENTAGE FRESH WEIGHT
BASIS)[1]

| Component | Percentage |
|---|---|
| Myofibrillar proteins | 9·5 |
|   Myosin | 5·0 |
|   Actin | 2·0 |
|   Tropomyosin | 0·8 |
|   Troponin | 0·8 |
|   M-protein | 0·4 |
|   C-protein | 0·2 |
|   $\alpha$-Actinin | 0·2 |
|   $\beta$-Actinin | 0·1 |
| Sarcoplasmic proteins | 6·0 |
|   Soluble sarcoplasmic and mitochondrial enzymes | 5·5 |
|   Myoglobin | 0·3 |
|   Haemoglobin | 0·1 |
|   Cytochromes and flavoproteins | 0·1 |
| Stroma | 3·0 |
|   Collagen and reticulin | 1·5 |
|   Elastin | 0·1 |
|   Other insoluble proteins | 1·4 |

and together with the precipitated sarcoplasmic proteins they also determine cooked meat tenderness.

## The Collagen Component of Toughness

In mature beef animals, connective tissue comprises 2–8 % of muscle dry weight.[2] Collagen usually makes up 95 % or more of the fibrous elements of connective tissue: most of the remaining 5 % is elastin. Lean meat is permeated by connective tissue sheaths which surround entire muscles (epimysium), bundles of muscle fibres (perimysium) and individual fibres (endomysium). Their physiological function is to support the tissue and to transmit the force generated by the contractile proteins of the myofibre to the skeleton via the tendon. For this purpose connective tissue is tough and difficult to pull apart.

Collagen fibres are straight, inextensible and non-branching. Their molecular structure has been reviewed recently.[3,4] The basic recurring subunit of collagen is the triple-stranded tropocollagen molecule. The polypeptide chains of tropocollagen are arranged helically and held together mainly by hydrogen bonds. Collagen fibres are laid down in a well-defined criss-cross lattice which is orientated at an angle to the long axis of the muscle.[5]

Collagens are distinctive in containing a considerable amount of the amino acids proline and hydroxyproline, rarely found in other proteins. Together they constitute about 29 % of bovine intramuscular collagen. The polypeptide chains of adjacent tropocollagen subunits are bound together by covalent cross-links—notably by dehydrolysinonorleucine residues formed by an enzymatic reaction between two lysine residues of adjacent tropocollagen subunits. The nature of the covalent cross-links of collagen has been reviewed recently by Light and Bailey.[6]

In the young animal many of the covalent bonds linking tropocollagen molecules are relatively labile, being easily ruptured by pH changes, heat or denaturing agents. They are stabilised, however, as the animal ages, being gradually replaced by bonds which are much more resistant to thermal rupture.[7] Thus the percentage of intramuscular collagen solubilised by various treatments (collagenase, or water at 70° to 100 °C) is several times greater in veal than in mature beef[8-10] and a similar age effect has been demonstrated using the muscles of sheep and pigs.[11] The decrease in the proportion of labile to stable bonds, rather than the actual quantity of collagen present in meat, is the factor which determines the collagen contribution to meat toughness.

Examination of changes in collagen structure during conditioning of meat has not revealed transformation comparable in extent to those which affect the myofibrillar structure. Nevertheless, discreet changes in collagen structure which increase its thermal solubility may occur. A muscle lysosomal fraction was recently found to reduce the strength of isolated beef collagen fibres.[12]

**The Actomyosin Component of Meat Toughness**

The contractile protein complex is the other major contributor to meat toughness. By virtue of its biological function it undergoes rapid structural alteration during contraction in the live animal or shortly after death, with the result that its contribution to meat toughness is more variable and unpredictable than that of collagen. The contractile protein complex consists of the contractile proteins, actin and myosin, and the associated regulatory proteins including the tropomyosin–troponin complex, $\alpha$- and $\beta$-actinin, M-protein and C-protein (Table 2). They are all involved in muscle contraction or in the regulation of contraction. A detailed description of the contractile proteins is provided by Ebashi and Nonomura[13] and Briskey and Fukazawa[14] and the architecture of the myofibril was reviewed by Bendall.[15]

The thick myofilaments are composed of myosin, which constitutes about 55% by weight of the contractile protein complex (Table 2). There are 200 to 400 molecules of myosin in each thick filament, with each molecule being 1500 nm long and 13 nm wide.[16] Each myosin molecule consists of a heavy head, called heavy meromyosin (HMM), and a tail, called light meromyosin (LMM). HMM possesses both functionally active sites on the myosin molecule. One site contains the ATPase activity to produce chemical energy by splitting ATP, while the other site forms the cross-bridges that interact with the actin filaments during muscle contraction.

Actin is the second major protein of the contractile protein complex, contributing about 20% by weight. The thin myofilaments are composed of F-actin (fibrous actin) which is formed by the longitudinal polymerisation of globular actin monomers (MW 47 000). Two strands of F-actin are spirally coiled around one another to form the right-handed superhelix of the thin filament. Actomyosin is formed by the reversible linking of actin and myosin during muscle contraction and by the permanent cross-linking which characterises rigor mortis.

The critical feature, from the point of view of tenderness of meat, is the degree of overlap of the actin and myosin filaments.[17] Muscles are substantially tougher if allowed to go into rigor in the contracted state, compared with muscles at resting length or stretched muscles. In contracted muscle there is a large amount of overlap of actin and myosin filaments and a proportionally large number of cross-bridges are formed.

These cross-bridges are permanent in the absence of $Mg^{2+}$—ATP and contribute substantially to meat toughness. Attempts to improve meat tenderness by reducing the degree of overlap between the actin and myosin filaments have been made, most notably by the suspension of beef carcasses by the pelvic girdle (tenderstretch) rather than the Achilles tendon.[18]

**Meat Colour**

Myoglobin is mainly responsible for meat colour, accounting for about 90 % of the total pigment in beef muscle.[19] Haemoglobin of the red blood cells is the other pigment of significance to meat colour. The concentration of myoglobin varies considerably between different types of meat, being generally higher in beef (2–5 mg/g wet weight) and lamb (4–7 mg/g) than in poultry (1–2 mg/g) and pork (2–7 mg/g).[20] Important differences in myoglobin concentration also exist between muscles[19] and within individual muscles.[21] These are determined by the predominant muscle fibre type in any particular cut. Myoglobin content increases with exercise and age, resulting in darker meat from older animals. It can be reduced by dietary iron deficiency, an effect exploited in veal calf production to get whiter meat.

Myoglobin consists of a protein monomer (globin) and a single, iron-containing haem group. The protein is colourless, but variations in its primary structure between species may affect the rate of colour change of myoglobin from different meat species held under similar conditions.[20] The oxidation state and type of ligand (i.e. oxygen, water, nitric oxide) bound to the iron centre of the haem group determines the colour of myoglobin under most conditions.

Deoxymyoglobin is the native meat pigment in which iron is in the ferrous ($Fe^{2+}$) state. It exists under *anoxic conditions* such as occur in the interior of meat cuts. This form of myoglobin is purplish-red and is responsible for the corresponding colour in the interior of freshly cut beef. Metmyoglobin is the oxidised form of myoglobin in which iron is in the ferric ($Fe^{3+}$) state. It is a brown derivative and is generally undesirable in meat products. Oxidation of oxymyoglobin to metmyoglobin and the reverse process (reduction) occurs continuously in living muscle and for a significant time *post mortem*. For the preservation of fresh meat colour, the conditions favouring the pigmentation balance towards the reduced form must be maintained.[20] The *in vivo* site of metmyoglobin reduction is unknown, but probably involves NADH and a recently characterised metmyoglobin reductase.[22] Oxymyoglobin, the oxygenated form of myoglobin, is bright red in colour and is responsible for the familiar 'bloom' of fresh meats. Oxymyoglobin is a stable derivative and is relatively resistant to oxidation. If oxygen tension becomes low enough for partial deoxygenation to deoxymyoglobin, the rate of metmyoglobin formation is greatly enhanced and the latter will permanently replace red oxymyoglobin unless reducing conditions are present.

The bright red or pink colour of raw cured meats is due to the presence of nitric oxide myoglobin (nitrosomyoglobin) which is formed by a sequence of reactions between nitrite, myoglobin and ferrocytochrome.[23] Nitrosometmyoglobin is an intermediate product which is reduced to nitric oxide myoglobin by an enzyme system in the muscle mitochondria. The stability of the cured meat colour is enhanced if nitric oxide myoglobin is converted to nitric oxide myohaemochromogen. This is achieved by denaturation of the globin part of the molecule during cooking.[23]

The colour of fresh meat is determined by the relative proportions of the three states of myoglobin present at the cut surface of the meat. Upon exposure to the atmosphere myoglobin rapidly becomes oxygenated to a depth of a few millimetres below the cut surface. Factors affecting the subsequent rate of discolouration have been extensively studied, particularly in relation to meat packaging technology. Shopper discrimination against discoloured meat increases with increase in metmyoglobin content.[24] Autoxidation of myoglobin to metmyoglobin occurs much more rapidly in some beef muscles than in others.[25] This reaction is temperature-dependent[26,27] so that colour stability of fresh beef is greatest at temperatures just above freezing.[25] Denaturation of myoglobin by heat and pH enhances autoxidation and is responsible for the colour fading observed in PSE (pale, soft, exudative) pork.[20]

The susceptibility of myoglobin to oxidation at low partial pressures of oxygen is probably the most important consideration in designing a package.[28] At high partial pressures the concentration of oxygen in the surface layer will favour oxymyoglobin formation, and the metmyoglobin layer will form at a comparatively greater depth in the meat and be less visible. The reverse occurs at very low partial pressures of oxygen, at which the rate of myoglobin oxidation to metmyoglobin is maximal. Consequently, in designing gas-packaging systems for fresh beef, use has been made of anoxic atmospheres containing pure nitrogen or carbon dioxide.[29] Alternatively, fresh beef is packaged in air filled containers containing a proportion of carbon dioxide gas to suppress bacterial growth.[27]

Comparison of the optical properties of DFD (dark, firm, dry) meat with normal meat indicates that the major difference between them is in their opacity.[30] Muscle that is cut soon after slaughter is translucent and sticky to touch. Its dark, jelly-like appearance alters during rigor mortis, becoming progressively more opaque until the ultimate pH is reached, at which stage its appearance on cutting and oxygenation is moist bright red. The transition from translucent to semi-opaque is detectable at pH 5·9. In DFD meat this transition does not occur.[30] Light penetrates deeper into the meat, which has the characteristic reddish-purple colour of reduced myoglobin, because of reduced light scattering at the surface. DFD meat has a higher oxygen consumption rate[31,32] and a lower rate of inward diffusion of oxygen[32] which considerably increases the time for formation of a sufficiently thick layer of oxymyoglobin on the surface to appear red.

**Water-holding Capacity**
Lean meat contains about 75 % water, a small amount of which is tightly bound to the muscle proteins (water of hydration), the remainder being loosely bound and immobilised to varying extents. The loosely bound water is greatly influenced by the physical and chemical environment of the meat proteins and under the appropriate conditions part of it may exude from the meat as drip.

About 4 to 5 % of muscle water is directly bound to the surface of the protein molecules as hydration water which has a highly ordered structure and properties different from those of free water e.g. lower freezing point. The water of hydration remains tightly bound even during the application to the meat of severe mechanical or other physical force. It is not greatly influenced by changes in the structure or charge of the meat proteins[34,35] and consequently is not responsible for changes in meat water-holding capacity (WHC) during processing. Nevertheless, water of hydration is important in relation to meat processing because its removal during freezing or dehydration of meat (at very low temperatures or vapour pressures) greatly accelerates denaturation of the proteins.

The common changes in WHC of meat are related to the remaining 95 % of muscle water.[34,35] NMR data concerning the relative immobilisation of this water is controversial. However, the bulk of muscle water is considered to be partly immobilised by a weak attraction towards the charged groups of proteins and partly by the physical configuration of the myofibrillar proteins and the barriers created by the sarcolemma and connective tissue sheaths. There seems to be a continuous transition, from water which is strongly immobilised within the tissue and which can be expressed only with difficulty, to 'loose' water which exudes readily. It is not possible to get absolute values for the quantity of 'loose' water in meat because the amount is determined by the method (degree of force) used. Consequently, WHC is defined and expressed in terms of a particular method of measurement. Using the same method one can measure relative differences in WHC quite exactly.[35]

The importance of the spatial arrangement of the myofilaments was demonstrated by Hamm.[36] By increasing the electrostatic repulsion between adjacent myofilaments (by adjusting the meat pH away from the mean isoelectric point of the proteins) the network of myofilaments is enlarged and more water is immobilised between them. Conversely, by adjusting the pH of the meat towards the mean isoelectric point, tightening of the myofilaments is facilitated and water is more easily expressed from the meat and flows out under low pressure. The WHC of meat is at a minimum at pH 5·0 which corresponds to the isoelectric point of ac-tomyosin.[36] The myofibrillar proteins are close to their minimum WHC at normal meat ultimate pH and additives such as polyphosphates are used in the production of cured, cooked meat products to increase myofibrillar water-binding capacity. Retention of water between the myofilaments has been attributed to mechanical and chemical forces.[36]

The irreversible formation of actomyosin in rigor mortis also reduces the water-binding capacity of the myofibrils. Marsh[37,38] observed that the *post mortem* decrease in hydration in rabbit and whale muscle occurs even if the pH remains almost unchanged (as for example in the glycogen-depleted muscles of exhausted animals). He concluded that the hydrolysis of ATP *post mortem* causes a decrease of WHC in muscle. Hamm[39] estimated that two-thirds of the hydration drop in beef *post mortem* is due to the breakdown of ATP and one-third to the fall of pH. Addition of ATP at relatively high concentrations to homogenates of pre-rigor muscle was observed to restore WHC towards the *in vivo* level. This was attributed to the plasticising effect of ATP which prevents formation of actomyosin cross-links in the absence of calcium ions. The *post mortem* movement of water from the myofilament space into the sarcoplasmic space and possibly further into the extracellular space is supported by NMR studies[40] and by measurements of fibre diameter and extracellular space in pre-rigor and post-rigor muscles.[41] Hegarty[42] observed a decrease in muscle fibre diameter during rigor, which suggests that the cell membrane had become leaky and allowed fluid to escape. These changes are magnified in parts of the carcass which enter rigor before cooling much below body temperature, resulting in higher drip losses in the affected meat cuts.[43,44]

Meat WHC improves during ageing, partly due to a small increase in meat pH but even more so due to changes in the electrostatic charge on the myofilaments.[36] The latter is thought to increase through the gradual binding of $K^+$ and release of $Ca^{2+}$ by the myofibrillar proteins. For every divalent cation released, one negative charge on the proteins is freed to bind water molecules and force the chains apart.[1]

## CHILLING

For the production of good quality fresh meat it is essential for muscle to achieve a normal ultimate pH at a rate which avoids low muscle pH values at or near body temperature (37 °C). PSE pork is associated with a very rapid fall in muscle pH after slaughter so that the carcass may enter rigor within an hour of death, before chilling has commenced. Excessive denaturation of the sarcoplasmic proteins results from the high temperature and low pH conditions in the affected muscles. The normal cross-striations of skeletal muscle are obscured by what appear to be irregular bands of denatured protein.[45] Scopes[46] found that the particular sarcoplasmic protein denatured under these conditions was creatine kinase, and Penny[47] demonstrated that the actin and myosin filaments were also considerably denatured. The phenomenon of PSE may be entirely attributed to protein denaturation at high temperature and low pH, combined with a partial disruption of the sarcolemma; the first factor reduces the water-holding capacity of the muscle proteins and obscures the normal pink colour of the muscle pigment myoglobin,[48] and the second allows the solution of sarcoplasmic proteins to escape from the fibre and permeate the intrafibre space.[16]

Although the PSE condition is rarely observed in beef carcasses[19] substantial variations in the rate of pH fall occur between and within muscles, resulting in differences in colour and drip loss of meat cuts. In cooling carcasses of beef heifers, the mean rate of pH decline varies from 0·5 pH units/h in the fillet (*M. psoas major*) to less than 0·1 unit/h in the superficial part of the muscles of the round.[44] By comparison, PSE pork has a rate above 1 pH unit/h. Variations in the pH fall in beef muscle are partly due to inherent differences in muscle metabolism and function[49] and partly due to different cooling rates within the carcass. The faster rate of pH fall in the deep, slow-cooling parts of conventionally-chilled beef carcasses results in increased denaturation of myofibrillar and sarcoplasmic proteins and higher drip loss (Table 3).

The effect of the pH/temperature profile during chilling on meat proteins

## TABLE 3

DENATURATION OF SARCOPLASMIC AND MYOFIBRILLAR PROTEINS AND AMOUNT OF DRIP
LOSS, AS A FUNCTION OF THE DEPTH IN THE CHILLED BEEF ROUND. MEASUREMENTS WERE
MADE IN *M. adductor* TWO DAYS AFTER SLAUGHTER (ADAPTED FROM REF. 44)

| Depth in carcass (cm) | Myofibrillar ATPase activity ($\mu mol\ Pi/mg\ protein/min \times 10^2$) | $CPK^a$ activity (units/g $\times 10^{-2}$) | Percent drip (300 g, 30 min) |
|---|---|---|---|
| 1·5 | 7·1 | 53 | 10·4 |
| 5 | 6·3 | 37 | 18·4 |
| 8 | 4·4 | 10 | 21·7 |

[a] Creatine phosphokinase.

can be readily observed in a cross-section of the beef round. The deep meat lying close to the femur is lighter in colour, softer in texture and wetter than the outer parts.

### Hot Boning

Deboning of the hot carcass, followed by carefully controlled chilling of the vacuum packaged cuts, is a practical means of avoiding the large temperature gradients which develop in the cooling carcass.[50,51] The process of hot boning beef is now in commercial practice[52] and offers many advantages to the processor, including the elimination of chill loss by evaporation, reductions in energy costs and space requirements during chilling, better meat yields and the opportunity for selective ageing of particular cuts.

As the boned-out muscles are free to shorten, it is essential to avoid cold-induced muscle contraction and associated toughening of the meat. This can be achieved by holding the meat above 10°C until rigor has fully developed.[50-52] The meat can then be safely chilled to 0°C without risk of cold-shortening. Alternatively, carcasses may be electrically stimulated before hot boning to accelerate rigor development, thereby eliminating the need for a delay in chilling the hot packaged cuts. Taylor and colleagues[53] compared hot deboned beef using these two procedures and observed some significant differences. Hot deboning followed by delayed chilling reduced drip loss in the vacuum packaged cuts; the reduction in drip loss was smaller when electrical stimulation was used. The colour of large muscles which cool unevenly on the side was more uniform after hot deboning, but again the improvement was smaller after electrical stimulation.

Because pre-rigor meat has a higher respiration rate, residual oxygen in hot-boned, vacuum packed beef should be dispersed more quickly. Consequently, less metmyoglobin is formed on the surface of the meat, and after opening the pack the desirable oxymyoglobin pigment should be more prominent.[54,55]

## Cold Toughening

The commercial significance of cold toughening of meat during chilling was demonstrated in 1966 when the US importation of New Zealand frozen lamb was greatly reduced because of excessive toughness. This defect was caused by the introduction of a new, rapid, pre-rigor chilling and freezing process.[56] The problem was finally overcome by the application of electrical stimulation to carcasses to accelerate rigor development.[57,58]

Cold-toughening of meat (primarily lamb, beef and other red muscles) is attributed to muscle contraction and also to a slowing down of the normal autolytic degradation of the myofibrils during ageing. Cold-induced shortening was first reported by Locker and Hagyard in 1963 and since then, shortening phenomena have been conclusively demonstrated to result in increased toughness in meat.[17] Cold-shortening occurs when the muscle temperature drops below about 10 °C in the presence of ATP (i.e. pre-rigor). It is caused by the cold-induced release of bound $Ca^{2+}$, which activates the contractile actin–myosin ATPase of muscle. Although a comparatively slow phenomenon, cold-shortening proceeds by the normal physiological mechanism of contraction involving the sliding filaments. When rigor occurs the myofilaments are irreversibly fixed in the high overlap (contracted) position.

Several major muscles of beef and lamb carcasses are slack, and thus free to shorten, in conventionally-hung carcasses. Even in muscles which are firmly stretched on the skeleton, some sarcomeres shorten more quickly than others, due to unequal rates of cooling, and thus shortened nodes, alternating with lengthened portions, tend to form throughout the musculature. Although some tenderisation of cold-toughened meats occurs during conditioning, the degree of tenderness achieved, even after 14 days of ageing, does not equal that in properly chilled beef.[18]

## Electrical Stimulation

*Post mortem* electrical stimulation (ES) as a means of accelerating the development of rigor mortis is now widely practised in the beef and lamb chilling industry. ES is used to avoid cold-toughening and thaw-rigor toughening of meat, to facilitate grading of beef on the day after slaughter (in the US) and to increase meat tenderness. The latter effect has not yet been clearly established and appears to be independent of the elimination of cold-toughening. The effect of ES on the muscle proteins is mainly through increased protein denaturation resulting from the faster pH fall. This denaturation is similar but less severe than that observed in the PSE condition. PSE beef is occasionally observed as a result of ES (L. Buchter, personal communication). Additional effects of ES on the muscle proteins may include myofibrillar disruption consequent to violent muscular contractions, more effective release of muscle cathepsins, and alteration of the proportions of oxymyoglobin and myoglobin on the meat surface in the early *post mortem* period due to faster onset of rigor in stimulated beef.

Increased tenderness in ES carcasses is considered to be due to an acceleration of the ageing process, which begins soon after slaughter in well-stimulated muscles, at a temperature of about 20 °C.[59] In non-stimulated controls, rigor mortis sets in about 24 h post-slaughter and tenderisation begins at a temperature of about 7 °C. The different pH/temperature history of ES and control carcasses has been proposed as an explanation for the reduced ageing time required for ES beef.[59] At present there is no conclusive evidence that ES increases the activity of muscle proteinases, such as the $Ca^{2+}$ activated sarcoplasmic factor, or accelerates lysosomal release. Increased tenderness of ES beef might also be caused by physical disruption of myofibrils, decreasing their resistance to shearing. Savell *et al.*[60] examined micrographs of ES and control beef and observed contracture bands in the ES beef showing less well defined I-bands and Z-lines and sarcomeres on either side of the contracture bands seemed to be stretched or broken. The hypothesis that structural damage to muscle fibres causes improvements in tenderness was not supported by subsequent data.[61]

It has now been well established, particularly by the research group at Texas A and M University, that ES beef has a brighter lean meat colour at 18 to 24 h after death.[62] This effect may be caused by the faster rate of pH decline.[63] Normal oxygenation of myoglobin and normal light scattering at the meat surface is not achieved until the meat reaches a normal ultimate pH. ES carcasses are more likely to be in rigor at 18 to 24 h post-slaughter than non-ES carcasses, and consequently have a brighter colour. The improvement in meat colour, judged on the day after slaughter, was found to be mainly a transient effect. ES carcasses judged at 48 h or longer after slaughter were closer in colour to non-ES carcasses, but apparently not identical.[62] The residual colour effect sometimes observed in post rigor carcasses is another manifestation of protein denaturation, resulting in increased muscle opacity and light scattering.[30] Tang and Henrickson[64] reported that total pigment and total myoglobin concentrations were not affected by ES.

The phenomenon of 'heat-ring' may be observed at the rib-eye of beef carcasses which are graded at 18 to 24 h after slaughter. It consists of a dark colour on the outer part of the muscle in contrast to a normal, bright, cherry-red colour towards the muscle centre. 'Heat-ring' may delay or preclude federal grading of carcasses in the USA.[65] ES reduces or eliminates the incidence of this defect[62] which is caused by the outer, dark part of the rib-eye being in the pre-rigor condition when grading is carried out on the day after slaughter.

Just as hot deboning of beef results in reduced drip loss in vacuum packed beef on the grounds of altered pH/temperature gradients in the muscles during rigor development[50] ES is expected to increase drip loss in vac-pack beef. In a recent review of this subject Honikel[66] noted the wide divergence of experimental results and concluded that ES had no negative effect on water retention in general. Increased drip loss under particular circumstances may, nevertheless, occur.

## Meat Conditioning

In rigor mortis virtually all the possible binding sites in the area of overlap of actin and myosin are utilised. In contrast, during *in vivo* contraction only about 20 % of the binding sites are utilised.[1] Meat which is cooked and eaten while in rigor mortis is very tough, due to the irreversible formation of the actomyosin complex throughout the tissue. Subsequent tenderisation during conditioning of meat depends upon weakening of the myofibrillar structure by proteolysis.[67]

The relative importance of the proteolytic enzymes of muscle, principally the calcium-activated sarcoplasmic factor (CAF) and lysosomal cathepsins, during meat tenderisation by ageing has not been established. There is no evidence for hydrolysis of cross-bridges during conditioning of meat. Some lengthening of the sarcomeres has been reported and this is thought to result from a weakening of the Z-disc structure, particularly at its junction with the actin filaments.[67,68] CAF destroys the Z-disc structure of the myofibrils and rapidly hydrolyses many of the minor structural and regulatory proteins of the myofibril, but appears to have little action on actin or myosin.

Other structural components of the myofibre may also contribute to meat toughness. Recently, Locker *et al.*[69] proposed that a third set of filaments in the sarcomere, the 'gap filaments', set the limit to the tensile strength of the raw or heat-denatured myofibril. The gap filaments are thought to be highly elastic protein filaments which connect the thick myosin filaments in adjacent sarcomeres, passing between the actin filaments and through the Z-line. It was proposed that the observed increase in meat tenderness during ageing resulted from an attack by CAF on the integrity of these filaments. This theory has not yet been confirmed.[68]

The decrease in beef toughness during conditioning was found to parallel the proteolytic breakdown of troponin T[70] which is one of three components of the troponin complex of the myofibrils. Its breakdown is caused by the action of CAF, or cathepsin B, or both.[70] Because it is a minor protein located on the thin filament, the loss of troponin T is unlikely to have a major direct effect on toughness, but was found to be a good indicator of the rate and extent of toughness changes during conditioning.


# FREEZING

Freezing is an excellent means of preservation when properly carried out; nevertheless it is inevitably associated with some deterioration of meat quality. Voyle[71] identified the main quality problems as the increase in drip loss on thawing, an associated decrease in juiciness, and changes in meat texture due to sarcomere shortening and protein denaturation. Other than loss of nutrients in the exudate before cooking, Cutting[72] concluded that there was no published evidence that freezing meat had any deleterious effect on nutritive value. Loss of meat quality during freezing is associated with freeze damage to muscle proteins and membranes.

### Ice Formation

Freezing commences in mammalian muscle between $-1°$ and $-2°C$. As the temperature of the muscle is lowered below the freezing point, the ice content increases rapidly until a temperature of $-5°C$ is reached, and thereafter more slowly (Table 4). Solute is rejected by the growing ice and its concentration increases in the shrinking liquid phase. The temperature finally reaches the eutectic point where the solution is saturated and solute and ice solidify together. During freezing of muscle the concentration of various low molecular weight solutes change relative to each other because they have different eutectic points, until the final eutectic temperature has been attained.[76] Of major importance to muscle protein stability is the fact that a highly concentrated liquid phase can persist indefinitely at any point above the eutectic temperature, which was estimated at $-52°C$ for beef.[77]

Dissociation and denaturation of muscle proteins is believed to result from the higher solute concentration and accompanying changes in ionic strength and pH.[78] At common freezing temperatures ($-10°$ to $-20°C$) about 90% of the moisture freezes out, resulting in approximately a tenfold increase in the concentration of solutes. Theoretically, protein denaturation, aggregation or dissociation could result from the effect of salts on the secondary (i.e. non-covalent) forces which stabilise the tertiary and quaternary configuration of protein macromolecules.[76] Substances capable of reacting covalently with a protein may also undergo such reactions when pushed by the mass action of their increased concentrations.[76]

TABLE 4

AMOUNT OF ICE IN BEEF AND HADDOCK MUSCLES AT VARIOUS SUBFREEZING TEMPERATURES[73–75]

| *Temperature* *(°C)* | *% Water content frozen* | |
|---|---|---|
| | *Beef* *(74·5% $H_2O$)* | *Haddock* *(83·6% $H_2O$)* |
| $-1$ | 2 | 10 |
| $-2$ | 48 | 56 |
| $-3$ | 64 | 70 |
| $-4$ | 71 | 76 |
| $-5$ | 74 | 80 |
| $-10$ | 83 | 87 |
| $-20$ | 88 | 91 |
| $-30$ | 89 | 92 |

Some water remains strongly bound to proteins even when the temperature is below $-70°$. This so-called bound water has been estimated by Reidel[74,75] to comprise about 10% of the total water in cod and beef muscles. This represents about two molecules of water strongly adsorbed to each amino acid residue.[72]

Love and Elerian[79] have shown that in fish muscle the attainment of a temperature of about $-183°C$ causes an irreversible removal of some of the structurally-bound water of actomyosin, and increased toughness. Similarly beef muscle frozen in liquid nitrogen and thawed at 2 to 3°C showed ultrastructural damage which was not observed in muscle frozen conventionally.[80]

## Damage to the Cell Structure

The extent of myofibre damage is determined by the location and size of ice crystals, both of these factors are influenced by the rate of freezing. Exclusively extracellular ice crystals form in post-rigor muscle during slow freezing[73] probably because the extracellular fluid has a lower osmotic pressure than the intracellular fluid.[81] The current view is that a reversal of osmotic flow occurs during slow freezing. Water is drawn across the cell membrane from a concentrated solution within the cell to frozen water outside the cell, causing dehydration and shrinkage of the cells.[82] In frozen pork, Voyle[71] reported that the size of extracellular ice formations is determined by the rate of cooling. In general, the ice crystals occupied spaces of varying width between bundles of muscle fibres. In pork muscle cooled from $+5\,°C$ to $-5\,°C$ in 13·6 min, ice spaces were 30–90 $\mu$m in width, whereas in muscle cooled from $+5\,°C$ to $-5\,°C$ in 109 min these spaces measured 195–270 $\mu$m. On thawing, variable amounts of extracellular fluid is not readsorbed by the proteins of the muscle and is lost from the meat as drip.

The effect of frozen storage on the ultrastructure of bovine muscle was investigated using scanning electron microscopy.[81] Samples frozen slowly in a stationary air freezer at $-18\,°C$ and stored at that temperature for up to 26 weeks showed essentially no change in muscle ultrastructure.

Very rapid freezing causes formation of numerous ice crystals within each fibre, thereby reducing the translocation of water to the extracellular space. Ultrarapid freezing of small pieces of post rigor muscle produces ice crystals that are so small that they are beyond the resolving power of the light microscope. Cryogenic freezing gives better preservation of quality[73] but this was not confirmed by recent work done using the scanning electron microscope. Carroll and co-workers[80] found evidence of ultrastructural damage in liquid nitrogen-frozen samples of beef muscle and essentially no change in meat ultrastructure in samples frozen at $-18\,°C$ and stored for 26 weeks.

Storage temperature should be sufficiently low to prevent or retard the growth of ice crystals in the extracellular space. Such growth occurs when the temperature of the meat rises above its eutectic point and becomes increasingly undesirable at higher sub-freezing temperatures.[71] Storage temperatures which are insufficiently low or which fluctuate are associated with higher drip loss on thawing, but have no significant effect on the shear force of meat frozen after ageing.[83]

## Pre-rigor Freezing

Ramsbottom and Koonz[84] observed that intracellular ice crystals were formed in beef frozen pre-rigor but that extracellular crystals formed in beef frozen one day after slaughter. Pre-rigor freezing offers important advantages in preserving the protein infrastructure of meat from the damage associated with rigor development and post-rigor freezing. However, the activity of the contractile proteins in pre-rigor meat poses serious technological problems in relation to freezing. With small pieces of meat, rapid freezing is possible before significant cold-shortening and toughening of the meat can occur; with larger pieces or carcasses, current practice is to allow rigor mortis to develop fully before chilling to low temperatures.

## Thaw-contracture

In the case of meat frozen pre-rigor, activation of the contractile proteins may also occur during thawing, resulting in vigorous contraction and massive drip loss (losses of up to one-third of lean meat weight have been recorded) as soon as the meat is free of ice. This phenomenon, called thaw-contracture, has much in common with cold-shortening. The contractile actin–myosin ATPase of the muscle fibre is activated, due to the inability of the calcium pump of the sarcoplasmic reticulum to maintain the intracellular concentration of free calcium ions below $10^{-8}$M.[85] Thaw-contracture can best be avoided by holding the frozen, pre-rigor meat at $-3°C$ to $-5°C$ for some days before completion of thawing.[85] The conditioning period allows ATP hydrolysis to proceed and rigor to occur while the meat is still fixed in shape by the presence of ice crystals and unable to contract.

## Thaw Exudate

The drip formed during thawing of frozen meat consists of proteins and the soluble constituents of the sarcoplasm.[71] Electrophoretic and ultracentrifugal analysis of the exudate from a thawed beef quarter indicated that the proteins were of sarcoplasmic origin[86] and similar results were obtained with exudate from thawed whitefish.[87] Awad *et al.*[87,88] estimated that the protein lost in the thaw exudate amounted to 17 % of total muscle protein in beef stored for 8 weeks at $-4°C$ and 20 % in whitefish stored for 16 weeks at $-10°C$.

The amount of thaw exudate is dependent on the meat pH and degree of ageing and also by the freezing and storage conditions.[73] The amount of thaw exudate from beef, pork and mutton is greatest between meat pH 5·0 and 5·2[89] and decreases progressively until a minimum is reached at pH 6·4.[73,84] The quantity of exudate increases with time of frozen storage, particularly at temperatures near the freezing point ($-11°$ to $-4°C$), probably due to growth of ice crystal size.

**Protein Alterations in Frozen Muscle**

Most research into protein denaturation during freezing has been concerned with fish muscle and several excellent reviews have appeared.[73,78,90-93] The properties of the myofibrillar proteins are generally comparable in fish and mammalian species although isolated fish actomyosin and myosin preparations aggregate more readily.[78] The rate of denaturation of actomyosin can be used to estimate the rate of quality change in fish during frozen storage.[78,94]

The extractability of actomyosin from cod muscle which has been frozen rapidly and held below $-30\,°C$ does not decrease significantly. When muscle is stored between $-1\cdot5\,°C$ and $-20\,°C$ actomyosin is readily insolubilised.[73,95] According to the mechanism postulated by King[96] some actomyosin aggregates form by reaction with fatty acids. More actomyosin depolymerises and separates into G-actin and myosin in the high ionic strength liquid phase. Subsequently both proteins form insoluble aggregates. The aggregation process is thought to result from the formation of new ionic and covalent (disulphide) bonds as well as hydrogen bonds and hydrophobic associations.[78] Buttkus[97] postulated that sulphydryl–disulphide interchange is involved in the mechanism of myosin aggregation during frozen storage.

Myosin is considered to be the most sensitive fish myofibrillar protein with respect to freeze denaturation.[93] Increased ionic strength causes

TABLE 5

SUMMARY OF FACTORS CAUSING PROTEIN DENATURATION DURING FROZEN STORAGE OF FISH MUSCLE, AFTER SHENOUDA[93]

1. Changes in moisture
   (a) Damage due to formation of ice crystals.
   (b) Damage due to dehydration, i.e. migration of hydration water molecules to form ice crystals.
   (c) Damage due to increase in salt concentration.

2. Changes in lipids
   (a) Intact lipids, interact with myofibrillar proteins, role still unclear.
   (b) Oxidised lipids, adversely affect nutritional and functional properties of the proteins.
   (c) Hydrolysed lipids, FFA increases toughness and decreases protein extractability.

3. Production of formaldehyde by the action of trimethylamine oxidase, resulting in a decrease in the extractability of myofibrillar proteins.

reversible dissociation of myosin into heavy cores and light polypeptide chains, but with extended storage at high ionic strength an irreversible aggregate between these subunits takes place. Actin is relatively stable compared with myosin. Tropomyosin is considered to be the most stable of the myofibrillar proteins. The sarcoplasmic proteins of fish muscle are also comparatively stable during frozen storage.[98]

A summary of factors causing protein denaturation, and texture deterioration, in frozen fish is shown in Table 5, based on the review of Shenouda.[93] With the exception of the production of formaldehyde, which is a major problem in some fish species, these factors may also be of significance during frozen storage of mammalian and poultry muscles.


## COOKING


A lot of progress has been made in elucidating the changes induced in muscle proteins during cooking, some of which toughen meat while others increase tenderness. Beef undergoes two distinct phases of toughening during heating.[98] A three- to four-fold toughening occurs between 40 °C and 55 °C followed by a further doubling in toughness between 65 °C and 75 °C. The first phase is associated with denaturation and hardening of the contractile system and the second phase is closely associated with shrinkage of collagen. Heating tenderises meat in three distinct phases.[99,100] Up to 65 °C proteolysis occurs at specific sites in the myofibrils. This is in fact an acceleration of the ageing process in muscle. From about 70 °C the collagen is denatured and melts. Above about 100 °C the myofibrils begin to break down again, presumably from non-specific high temperature hydrolysis.

The relative importance of these toughening and tenderising changes in the meat proteins during cooking is determined by the nature of the meat, for example the extent of age-related cross-linking of the collagen and the degree of muscle contraction is important. In addition, the time and temperature of heating are of the utmost importance. Slow cooking at a prescribed temperature is often advocated as a means of tenderising.[101] This can be done at either of two temperature ranges, at 50°–60 °C which favours the ageing effect or at 80°–100 °C where the interstitial collagen is destroyed.[100] These two tenderising temperature ranges are well separated by a sharp minimum temperature of tenderisation at 66 °C to 68 °C. To hold meat in this temperature region would be the worst possible treatment because it favours collagen shrinkage and an increase in toughness.

                    *P. V. Tarrant*

## Muscle Structure

The microscopic striated appearance of the muscle fibres persists after heating in well-cooked meat.[102] After heating for 20 min at 60 °C the thick myosin filaments are clearly visible; they can still be discerned after heating for 45 min at 70 °C. After longer heating periods the myosin filaments become very indistinct[102] and actin filaments progressively lose their filamentous structure and continuity. The Z-lines become ill-defined and disorganised at the longer heating times and higher temperatures.

Heat shrinkage occurs in two phases in isolated muscle fibres examined under the microscope.[103,104] A decrease in fibre width begins at low temperatures (below 50 °C) followed by a decrease in length between 50 °C and 70 °C. The decrease in width may be due to coagulation of myofibrillar and globular proteins, while the shortening may be due to collagen shrinkage.[98] In general, the volume of the fibres decreases in the temperature range 45 °C to 80 °C. Giles[102] observed a 15 % decrease in sarcomere length during 20 min of heating at 70 °C; both the A-band and the I-band showed the same degree of shrinkage as the sarcomere as a whole. Sarcomere shrinkage closely resembled that of whole strips of meat.

## Effect of Heating on Collagen

The first change to occur during heating of collagen fibrils is a shortening to about one-third of the original length. This commences at about 56 °C and is usually complete in half the muscle collagen fibrils at 61 °C to 62 °C. Called collagen shrinkage, this comparatively rapid change is accompanied by an increased collagen solubility. The percent of beef collagen solubilised by heat increases gradually from about 60 °C to 100 °C.[23] At the latter temperature conversion to gelatin is marked. Gelatin formation is swift on pressure cooking at a 115 °C to 125 °C.[23] Gelatin is characterised by a low resistance to shear and high water-binding capacity compared with collagen. The major load-bearing structures of raw muscle are the myofibrils and connective tissue; prolonged cooking destroys the collagen of the latter[101] and the myofibrils are then held only loosely together and the force needed to shear cooked meat across the grain is due largely to myofibrils.[105]

Meat with a high content of connective tissue should be cooked in the presence of excess water, to facilitate the complete conversion of collagen to gelatin. The second connective tissue protein, elastin, is not susceptible to normal cooking temperatures. Thus, a cut that does not tenderise in spite of extensive cooking, is probably high in elastin fibre content. The only feasible method of tenderising such products is by using plant proteolytic enzymes.[1]

The term 'labile collagen' is used to describe that fraction of collagen which is rendered soluble on heating. The proportion of labile collagen decreases with age, due to an increase in heat-resistant covalent cross-links between amino acids in adjacent collagen molecules.[106] Field *et al.*[107] observed that the yield of labile collagen was significantly higher in tender beefsteak than in tough steaks. Epimysial tissue from yearlings yielded 22 % of labile collagen, compared to 2 % from tissue of old cows.

## Effect of Heat on Myofibrils

Tenderisation of meat by ageing continues during the early stage of cooking and causes a loss of myofibrillar tensile strength. The tenderising effect of ageing was found to be additional to that achieved by the breakdown of collagen during long cooking of beef muscle.[105] The toughness of contracted muscles persists in cooked meat. Davey and colleagues[105] reported considerably higher shear-force values in meat at 40 % shortening than in either unshortened meat or in meat shortened by 60 %.

The solubility of the myofibrillar proteins decreases greatly between 40 °C and 60 °C, accompanied by an unfolding of the protein chains and loss of ATPase activity.[98] Coagulation probably results from random associations between unfolded peptide chains, the chemical nature of which has not been established.[98] Heat coagulation of myofibrillar proteins is not due to oxidation of sulphydryl to disulphide groups as this reaction was not observed below 70 °C.[108] At temperatures above 80 °C a loss of sulphydryl plus disulphide groups of meat occurs, either by oxidation to cysteic acid or by loss of hydrogen sulphide. During heat sterilisation of meat the release of hydrogen sulphide may cause corrosion and discoloration of tins and their contents, and an unfavourable smell on opening the can.[109]

As expected, heat coagulation causes a decrease in myofibrillar water-holding capacity (WHC). Consequently, a large part of the muscle water becomes freely movable and is released from the tissue.[110] Decrease in myofibrillar WHC occurs primarily between 40 °C and 50 °C. At 60 °C the coagulation and subsequent release of juice is not yet completed.[110] Other workers reported the major decrease in meat WHC at somewhat higher meat temperatures.[111] Shrinkage of collagen occurs at about 60 °C and squeezes more water out of the coagulated myofibrillar structure, with a concomitant increase in toughness.[112] This effect is particularly important in the highly cross-linked collagens found in the meat of older animals.

                    *P. V. Tarrant*

**Sarcoplasmic Proteins**

Most of the sarcoplasmic proteins coagulate between 40 °C and 60 °C[110,111,113] although in some preparations heat denaturation of sarcoplasmic proteins is not completed below about 90 °C.[99] Electrophoretic studies suggest that the cathodic proteins are more thermostable than the anodic proteins.[114] The number and type of ionic charges on proteins affects their stability during heating.

The denaturation of myoglobin is of particular importance because it determines the change in meat colour during cooking from red to grey-brown. The principal pigment of cooked fresh meat is brown globin haemichromogen. Kvåle and Martens[115] noted that internal cooking temperatures below 50 °C yield beef with a raw colour; 50°–70 °C yield whitish beef (precipitated sarcoplasmic proteins) but with red juice; temperatures above 70 °C yield brown meat and clear juice, probably due to denaturation of myoglobin. At meat temperatures below 65 °C myoglobin denaturation may result from enzymic action or coprecipitation rather than from the temperature affect on globin *per se*.[23] Myoglobin is one of the more heat-stable of the sarcoplasmic proteins. Roberts and Lawrie[113] suggested that the initial resistance of myoglobin to heat denaturation and its subsequent loss of electrophoretic mobility between 75 °C and 88 °C could be developed as an index of the temperature attained by meat products.

Apart from the important effect of heat denaturation of myoglobin on cooked meat colour, heat denaturation of the sarcoplasmic fraction is of comparatively minor importance for the quality of cooked meats. Those sarcoplasmic proteins which exhibit proteolytic activity may play a role in tenderising meat before they are coagulated by heat during cooking.[116–119]


## COMMINUTION AND EMULSIFICATION


The common goal in the manufacture of comminuted meat products was defined by Schmidt[121] as the formation of a stable heat-set protein gel that will effectively bind legal limits of fat and water in an attractive and palatable meat product. A meat emulsion is a two-phased system, consisting of a continuous aqueous phase and a dispersed phase containing fat particles. Coarsely comminuted products contain meat chunks surrounded by an aqueous medium which is similar to the aqueous phase in emulsion-type products.

**Nature of the Aqueous Phase**

The continuous phase consists of a matrix of muscle fibres and connective tissue fibres suspended in an aqueous medium containing soluble proteins and other soluble muscle constituents.[1] The complex nature of the aqueous phase makes it difficult to determine the mechanism of binding of fat and water in practical experiments using whole meat. Considerable progress has been made using muscle protein fractions. The composition of the aqueous phase in manufactured meat products is greatly influenced by characteristics of the various machines used, about which there is little public literature.

In sausage emulsions, soluble proteins dissolved in the aqueous phase act as emulsifying agents by coating all surfaces of the dispersed fat particles. The soluble proteins may be either sarcoplasmic or myofibrillar. However, myofibrillar proteins are much more efficient emulsifying agents. Most sausage emulsions contain 2·5 to 3·0 % NaCl (0·5 to 0·6 M NaCl) to aid in extracting and solubilising the myofibrillar proteins so that they become available for coating fat particles. These proteins accomplish the function of emulsion stabilisation through the presence of reactive groups which are oriented across the water interface.[120]

Chopping under vacuum is used to enhance myofibrillar protein extraction, by reducing the incorporation of air into the sausage mix. The extracted protein which would otherwise surround the air bubbles is available for coating fat particles.[121]

Once the fat is coated the emulsion is stable for a period of some hours. Permanent stability is achieved by heat-coagulation of the proteins of the aqueous phase. A three-dimensional protein network holds the fat in suspension for an unlimited period of time and also holds water by capillary attraction.[122]

The mechanism of binding between chunks of meat in chopped or coarsely-comminuted products requires the formation of a concentrated emulsion between meat pieces. Additionally a degree of orientation by the salt-soluble proteins has been observed on the meat surfaces before and during the heat-initiated binding phenomenon. This appears to involve the interaction of synthetic protein filaments (formed from dissolved myosin during mixing prior to heating) with myosin filaments located within muscle cells on or near the surface of the pieces of meat.[122]

Hamm[34] proposed that an expanded network of myofilaments suspended in the aqueous phase may be more important for the stabilisation of fat in a sausage mix than the emulsification properties of the proteins. The fat particles are prevented from coalescing during heating by mechanical fixation within the meshes of the myofilament network. The more expanded this network is, the better the fat-binding and water-binding properties of the mix are. Grinding of meat for the production of sausages of the frankfurter type destroys the sarcolemma. Consequently, swelling of the myofibrillar system is no longer limited by the cell membranes. This is demonstrated by the observed effect of pH on interfilament spacing which is greater in skinned muscle fibres than in normal fibres.[123]

### Massaging

Histological examination revealed muscle fibre disruption after several hours of massaging of pork muscle in the presence of salt and phosphate. Further massaging resulted in the separation of myofibrils from the surface of the fibres.[124] Actin filaments and Z-discs were broken down leading to degradation of the sarcomere structure.[125]

Samples of the tacky exudate formed on meat surfaces as a result of massaging in the presence of salt and phosphate showed both solubilised protein and fragments from disrupted fibres. Samples without added salt showed disrupted fibre fragments only.[126] Cellular breakdown during massaging facilitates the extraction of the major myofibrillar proteins from the muscle chunks, which is beneficial in binding. Addition of soy protein appeared to enhance myofibrillar protein extraction by binding water, thus increasing the effective concentration of salt and phosphate.[121]

### Functional Properties of Muscle Proteins

MacFarlane *et al.*[127] compared the relative power of myosin, actomyosin and sarcoplasmic proteins for binding meat pieces together. In the presence of salt, myosin had the greatest binding power, and sarcoplasmic proteins exerted a deleterious effect on the binding power of myosin. This effect was attributed to the adsorption of denatured sarcoplasmic proteins onto the myofibrillar proteins, resulting in decreased extractability, as in the case of PSE muscle.[122] In the absence of salt the binding power of myosin is enhanced by the addition of sarcoplasmic proteins. The ionic contributions of sarcoplasmic proteins added to myosin improved its binding power in a manner similar to that of salt.

Myosin appears to be the only muscle protein that forms a rigid heat-set gel, and specifically the heavy chain portion of the myosin molecule can form a stable, filamentous three-dimensional matrix that gives processed emulsion meats the desired bind and texture.[121] The mechanism of gelation of myosin is not understood but is associated with irreversible changes in its quaternary structure that are caused by heating. Hamm[110] postulated that the stabilising bonds formed after the heat denaturation of myosin involved hydrogen and ionic interactions. Samejima *et al.*[128] studied some properties of gels formed by light meromyosin and heavy meromyosin and concluded that an intact myosin molecule offers the best functional properties. Although the greatest contribution to gelation of the aqueous phase of sausage emulsions comes from myosin, actin is also very important for the development of viscoelastic properties.[122]

Regarding the fat emulsifying capacity of muscle proteins, Schut *et al.*[129] reported no distinct difference between salt-soluble and water-soluble proteins of post rigor lean beef. However, emulsions prepared from salt-soluble proteins were more heat stable. This finding suggests that sarcoplasmic proteins are useful emulsifying agents (form strong mono-layer films at the lipid–water interface) but function poorly in heat-induced gelation.

Because they are insoluble, the connective tissue proteins are of limited functional value. Collagen can contribute to water-binding during comminution and emulsification. During heat processing, collagen shrinks and is partly converted to gelatin, retaining good water-binding capacity but lacking fat emulsifying ability.[1] Meat raw materials are classified by their 'binding capacity'. This term refers to water and fat retention in emulsions and to surface cohesion of meat chunks in chopped meat products. Meats with a high amount of connective tissue, or a high fat content, are low in binding ability and commercially are referred to as 'filler meats'.


**Pre-rigor Meat**

Emulsion stability increases as the amount of soluble protein available to act as an emulsifying agent increases. The amount of protein extracted is affected by several factors. More protein is extracted as the pH of muscle increases. This may partly explain the more stable emulsions formed at higher meat pH values. The rigor state of meat ingredients also influences emulsification. Pre-rigor meat is superior to normal post-rigor meat because up to 50% more salt-soluble protein can be extracted.[1] The myofibrillar proteins are more soluble at high pH values and in the presence of ATP. Consequently, more fat can be emulsified with the protein extracted from pre-rigor meat than with that extracted from the same quantity of post-rigor meat.

The introduction of NaCl into the myofibrils at high pH and ATP concentration prevents normal onset of rigor mortis in the fragmented fibres. This is probably due to a strong electrostatic repulsion between adjacent protein molecules associated with binding of the chloride ion.[34] By freezing pre-rigor salted ground beef the high water and fat binding properties can be preserved for several months.

Most of the advantages of pre-rigor meat can be gained by the pre-blending of post-rigor meat prior to emulsification. In this process the meat is chopped with ice, salt and cure ingredients, and held at $0°$–$4°C$ for up to 12 h before emulsification, thereby allowing more efficient protein extraction.


## Modified Proteins

Chemical and/or enzyme modification of muscle proteins has been studied as a means of improving their usefulness in meat products.[130] This approach may be useful in the future for increasing the functional properties of low-cost proteins which are of low technological value in their native state; for example fish protein concentrates and beef heart myofibrils.

Acylation reactions appear to have the most potential for chemically modifying food proteins. Promising results have been obtained by the addition of acetyl or succinyl groups to the side chains of basic amino acids. Acylation results in the conversion of certain cationic amino groups of the native protein to neutral or anionic residues. This in turn promotes unfolding of polypeptide chains with an increase in solubility and changes in other physicochemical and functional properties. For example, succinylation renders myosin water-soluble. Acylating beef heart myofibrils renders them more soluble in $0.2M$ NaCl than unacylated protein was in $0.6M$ NaCl.[131] Acylation is a possible way to produce low-salt meat products without loss of functional characteristics of the proteins.

For utilisation in foodstuffs, acylated proteins must be non-toxic and digestible. Brekke and Eisele[130] considered that correct choice of acylation agents and control of the extent of modification can minimise reductions in nutritional properties. The protein efficiency ratio and net protein retention values for the modified heart myofibrillar proteins were comparable to values reported for lean beef and casein.

Enzymic modification of beef and fish muscle protein has also been studied using proteolytic enzymes. After limited proteolysis, beef muscle homogenates show an increased capacity to emulsify lipids[132] followed by a decrease on further proteolysis. Combined chemical and enzymic modifications have also been studied.[130]

# REFERENCES


1. FORREST, J. C., ABERLE, E. D., HEDRICK, H. B., JUDGE, M. D. and MERKEL, R. A., *Principles of Meat Science*, W. H. Freeman and Company, San Francisco, 1975.
2. DAVEY, C. L. and WINGER, R. J., in *Fibrous Proteins: Scientific, Industrial and Medical Aspects*, D. A. D. Parry and L. K. Creamer (eds.), Vol. 1, 97, Academic Press, London, 1979.
3. GLANVILLE, R. W. and KUHN, K., in *Fibrous Proteins: Scientific, Industrial and Medical Aspects*, D. A. D. Parry and L. K. Creamer (eds.), Vol. 1, 133, Academic Press, London, 1979.
4. TRAUB, W. and PIEZ, K. A., *Adv. Protein Chem.*, 1971, **25**, 243.
5. ROWE, R. W. D., *J. Fd Technol.*, 1974, **9**, 501.
6. LIGHT, N. D. and BAILEY, A. J., in *Fibrous Proteins: Scientific Industrial and Medical Aspects*, D. A. D. Parry and L. K. Creamer (eds.), Vol. 1, 151, Academic Press, London, 1979.
7. BAILEY, A. J., *J. Sci. Fd Agric.*, 1972, **23**, 995.
8. GOLL, D. E., BRAY, R. W. and HOEKSTRA, W. G., *J. Fd Sci.*, 1964 *a*, **29**, 622.
9. GOLL, D. E., HOEKSTRA, W. G. and BRAY, R. W., *J. Fd Sci.*, 1964 *b*, **29**, 608.
10. GOLL, D. E., HOEKSTRA, W. G. and BRAY, R. W., *J. Fd Sci.*, 1964 *c*, **29**, 615.
11. HILL, F., *J. Fd Sci.*, 1966, **31**, 161.
12. KOPP, J. and VALIN, C., *Meat Science*, 1980–81, **5**, 319.
13. EBASHI, S. and NONOMURA, Y., Proteins of the myofibril. In *The Structure and Function of Muscle*, G. H. Bourne (ed.), Vol. 3, 285, Academic Press, New York, 1973.
14. BRISKEY, E. J. and FUKAZAWA, T., *Adv. Fd Res.*, 1971, **19**, 279.
15. BENDALL, J. R., *Muscles, Molecules and Movement*, Heinemann Ltd, London, 1969.
16. BENDALL, J. R. in *The Structure and Function of Muscle*, G. H. Bourne (ed.), Vol. 2, 243, Academic Press, New York, 1973.
17. MARSH, B. B. and CARSE, W. A., *J. Fd Technol.*, 1974, **9**, 129.
18. JOSEPH, R. L. and CONNOLLY, J., *J. Fd Technol.*, 1977, **12**, 231.
19. HUNT, M. C. and HEDDRICK, H. B., *J. Fd Sci.*, 1977, **42**, 716.
20. LIVINGSTON, D. J. and BROWN, W. D., *Food Technol.*, May 1981, 244.
21. TARRANT, P. V., HEGARTY, P. V. J. and McLOUGHLIN, J. V., *Proc. Roy. Irish Acad.*, 1972, **72B**, 229.
22. HAGLER, L., COPPES, R. I. JR. and HERMAN, R. H., *J. Biol. Chem.*, 1979, **254**, 6505.
23. LAWRIE, R. A., *Meat Science*, 3rd edn., Pergamon Press, Oxford, 1979.
24. HOOD, D. E. and RIORDAN, E. B., *J. Fd Technol.*, 1973, **8**, 333.
25. HOOD, D. E., *Meat Science*, 1980, **4**, 247.
26. GIDDINGS, G. G., *J. Fd Sci.*, 1977, **42**, 288.
27. MacDOUGALL, D. B. and TAYLOR, A. A., *J. Fd Technol.*, 1975, **10**, 339.
28. GOVINDARAJAN, S., *C.R.C. Crit. Rev. Fd Technol.*, 1973, **4**, 117.
29. O'KEEFFE, M. and HOOD, D. E., *Meat Science*, 1980–81, **5**, 27.
30. MacDOUGALL, D. B. and JONES, S. J., in *The Problem of Dark-Cutting in Beef*, D. E. Hood and P. V. Tarrant (eds.), Current Topics in Veterinary Medicine and Animal Science, Vol. 10, Martinus Nijhoff, The Hague, 1981.

31. BENDALL, J. R. and TAYLOR, A. A., *J. Sci. Fd Agric.*, 1972, **23**, 707.
32. HALL, J. L., LATSCHAR, C. E. and MACKINTOSH, D. L., *Kansas Agric. Exp. Stat. Tech. Bull.*, 1944, **58**.
33. LAWRIE, R. A., *J. Sci. Fd Agric.*, 1958, **9**, 721.
34. HAMM, R. in *Meat*, D. J. A. Cole and R. A. Lawrie (eds.), Butterworths, London, 1974, 321.
35. HAMM, R., *Die Fleischwirtschaft*, 1975, **10**, 1415.
36. HAMM, R., *Adv. Fd Res.*, 1960, **10**, 356.
37. MARSH, B. B., *Biochim. Biophys. Acta*, 1952 *a*, **9**, 127.
38. MARSH, B. B., *Biochim. Biophys. Acta*, 1952 *b*, **9**, 247.
39. MARSH, B. B., *Biochim. Biophys. Acta*, 1952 *b*, **9**, 247.
39. HAMM, R., *Biochem. Z.*, 1956, **328**, 309.
40. PEARSON, R. T., DUFF, I. D., DERBYSHIRE, W. and BLANSHARD, J. M. V., *Biochim. Biophys. Acta*, 1974, **362**, 188.
41. HEFFRON, J. J. A. and HEGARTY, P. V. J., *Comp. Biochem. Physiol.*, 1974, **49A**, 43.
42. HEGARTY, P. V. J., *Proc. 15th Eur. Meet. Meat Res. Workers*, 1969, **A5**, 50.
43. PENNY, I. F., *J. Sci. Fd Agric.*, 1977, **28**, 329.
44. TARRANT, P. V. and MOTHERSILL, CARMEL, *J. Sci. Fd Agric.*, 1977, **28**, 739.
45. BENDALL, J. R. and WISMER-PEDERSEN, J., *J. Fd Sci.*, 1962, **27**, 144.
46. SCOPES, R. K., *Biochem. J.*, 1964, **91**, 201.
47. PENNY, I. F., *Biochem. J.*, 1967, **104**, 609.
48. McLOUGHLIN, J. V. and GOLDSPINK, G., *Nature*, 1963, **198**, 584.
49. BENDALL, J. R., *Meat Science*, 1978, **2**, 91.
50. FOLLETT, M. J., NORMAN, G. A. and RATCLIFF, P. W., *J. Fd Technol.*, 1974, **9**, 509.
51. TARRANT, P. V., *J. Sci. Fd Agric.*, 1977, **28**, 927.
52. BUCHTER, L., *Proc. International Symposium on Electrical Stimulation and Hot Boning*, C. Valin (ed.), INRA, Theix, Beaumont, France, 1980.
53. TAYLOR, A. A., SHAW, B. G. and MacDOUGALL, D. B., *Meat Science*, 1980–81, **5**, 109.
54. CUTHBERTSON, A., *Institute of Meat Bulletin (UK)*, 1977, No. 97, 3.
55. MULDER, S. J., Central Institute for Nutr. and Fd Res., Zeist, Netherlands, 1979, Report No. R5890.
56. MARSH, B. B., WOODHAMS, P. R. and LEET, N. G., *J. Fd Sci.*, 1968, **33**, 12.
57. CARSE, W. A., *J. Fd Technol.*, 1973, **8**, 163.
58. CHRYSTALL, B. B. and DEVINE, C. E., *Meat Science*, 1978, **2**, 49.
59. VALIN, C. and ROMITA, A., *Rept. EEC Working Group on Electrical Stimulation and Hot Processing*, C. Valin (ed.), INRA, Theix, Beaumont, France, 1981.
60. SAVELL, J. W., DUTSON, T. R., SMITH, G. C. and CARPENTER, Z. L., *J. Fd Sci.*, 1978, **43**, 1606.
61. McKEITH, F. K., SMITH, G. C., DUTSON, T. R., SAVELL, J. W., HOSTETLER, R. L. and CARPENTER, Z. L., *J. Fd Protection*, 1980, **43**, 795.
62. TARRANT, P. V. and CASTEELS, M., *Rept. EEC Working Group on Electrical Stimulation and Hot Processing*, C. Valin (ed.), INRA, Theix, Beaumont, France, 1981.

63. McKeith, F. K., Smith, G. C., Savell, J. W., Dutson, T. R., Carpenter, Z. L. and Hammons, D. R., *J. Fd Sci.*, 1981, **46**, 13.
64. Tang, B. H. and Henrickson, R. L., *J. Fd Sci.*, 1980, **45**, 1139.
65. Savell, J. W., Smith, G. C. and Carpenter, Z. L., *J. Anim. Sci.*, 1978, **46**, 1221.
66. Honikel, K. O., *Rept. EEC Working Group on Electrical Stimulation and Hot Processing*, C. Valin (ed.), INRA, Theix, Beaumont, France, 1981.
67. Dutson, T. R., *Proc. Kellogg Foundation International Symposium on Food Proteins*, University College, Cork, 1981 (see Ch. 17 of this book).
68. Morrissey, P. A. and Fox, P. F., *Irish J. Fd Sci. Technol.*, 1981, **5**, 33.
69. Locker, R. H., Daines, G. J., Carse, W. A. and Leet, N. G., *Meat Science*, 1977, **1**, 87.
70. Penny, I. F. and Dransfield, E., *Meat Science*, 1979, **3**, 135.
71. Voyle, C. A. in *Meat Freezing—Why and How?* C. L. Cutting (ed.), Proc. Agricultural Research Council MRI Symposium No. 3, Meat Research Institute, Langford, Bristol, UK, April 1974, 6.1–6.6.
72. Cutting, C. L. (ed.), in *Meat Freezing—Why and How?*, Meat Research Institute, Langford, Bristol, UK, 1974.
73. Powrie, W. D., in *Low Temperature Preservation of Foods and Living Matter*, O. R. Fennema, W. D. Powrie and E. H. Marth (eds.), Marcel Dekker Inc., New York, 1973, 282.
74. Reidel, L., *Kältetechnik*, 1956, **8**, 374.
75. Reidel, L., *Kältetechnik*, 1957, **9**, 38.
76. Taborsky, G., in *Proteins at Low Temperature*, O. Fennema (ed.), Advances in Chemistry Series, Am. Chem. Soc., Washington, DC, 1979, 1.
77. Reidel, L., *Kältetechnik*, 1961, **13**, 122.
78. Matsumoto, J. J. in *Proteins at Low Temperatures*, O. Fennema (ed.), Advances in Chemistry Series 180, Am. Chem. Soc., Washington, DC, 1978, 205.
79. Love, R. M. and Elerian, M. K., *Proc. 11th Intl. Congr. Refrig.*, Munich, 1963, 887.
80. Carroll, R. J., Cavanaugh, J. R. and Rorer, F. P., *J. Fd Sci.*, 1981, **46**, 1091.
81. Callow, E. H., *J. Sci. Fd Agric.*, 1952, **3**, 145.
82. Grieve, P. W. and Povey, M. J. W., *J. Sci. Fd Agric.*, 1981, **32**, 96.
83. Khan, A. W. and Lentz, C. P., *Meat Science*, 1977, **1**, 263.
84. Ramsbottom, J. M. and Koonz, C. H., *Food Res.*, 1940, **5**, 423.
85. Bendall, J. R. in *Meat Freezing—Why and How?*, C. L. Cutting (ed.), Meat Research Institute, Langford, Bristol, UK, 1974.
86. Howard, A., Lawrie, R. A. and Bouton, P. E., *Spec. Rept. Fd Invest. Bd Lond.*, 1960, **68**, VIII.
87. Awad, A., Powrie, W. D. and Fennema, O., *J. Fd Sci.*, 1969, **34**, 1.
88. Awad, A., Powrie, W. D. and Fennema, O., *J. Fd Sci.*, 1968, **33**, 227.
89. Sair, L. and Cook, W. H., *Can. J. Res.*, 1938, **16D**, 255.
90. Connell, J. J. in *Low Temperature Biology of Foods*, J. Hawthorne and E. J. Rolfe (eds.), Pergammon Press, Oxford, 1968, 333.
91. Dyer, W. J. and Dingle, J. R., in *Fish as Food*, G. Borgstron (ed.), Academic Press, New York, 1961, Vol. 1, 275.

92. FENNEMA, O. R., in *Low Temperature Preservation of Foods and Living Matter*, O. R. Fennema, W. D. Powrie and E. H. Marth (eds.), Marcel Dekker Inc., New York, 1973, 150.

93. SHENOUDA, S. Y. K., *Adv. Fd Res.*, 1980, **26**, 275.

94. VYNCKE, W., in *Food Quality and Nutrition—Research Priorities for Thermal Processing*, W. K. Downey (ed.), Applied Science Publishers Ltd, London, 1977, 325.

95. JARENBACK, L. and LILJEMARK, A., *J. Fd Technol.*, 1975, **10**, 309.

96. KING, J. F., *J. Fd Sci.*, 1966, **31**, 649.

97. BUTTKUS, H., *J. Fd Sci.*, 1970, **35**, 558.

98. HAMM, R., in *Physical, Chemical and Biological Changes in Food caused by Thermal Processing*, T. Hoyem and O. Kvale (eds.), Applied Science Publishers Ltd, London, 1977, 101.

99. DAVEY, C. L. and GILBERT, K. V., *J. Sci. Fd Agric.*, 1974, **25**, 931.

100. DAVEY, C. L. and NEIDERER, A. F., *Meat Science*, 1977, **1**, 271.

101. LAAKONEN, E., *Adv. Fd Res.*, 1973, **20**, 257.

102. GILES, B. G., *Proc. 15th Eur. Meet. Meat Res. Workers*, 1969, 289.

103. HOSTETLER, R. L. and LANDMANN, W. A., *J. Fd Sci.*, 1968, **33**, 468.

104. LANDMANN, W. A., *Rept. 13th Eur. Meet. Meat Res. Conf.*, Amer. Meat Sci. Ass. 1967, 26.

105. DAVEY, C. L., NIEDERER, A. F. and GRAAFHUIS, A. E., *J. Sci. Fd Agric.*, 1976, **27**, 251.

106. VERZÁR, F., *Int. Rev. Connect. Tissue Res.*, 1964, **2**, 243.

107. FIELD, R. A., PEARSON, A. M. and SCHWEIGERT, B. S., *J. Agric. Fd Chem.*, 1970, **18**, 280.

108. HAMM, R. and HOFMANN, K., *Nature*, 1965, **207**, 1269.

109. HOFMANN, K. and HAMM, R., *Adv. Fd Res.*, 1978, **24**, 1.

110. HAMM, R. in *The Physiology and Biochemistry of Muscle as a Food*, E. J. Briskey, R. G. Cassens and J. C. Trautman (eds.), 1st edn., Univ. Wisconsin Press, Madison, 1966, 363.

111. LAAKKONEN, E., WELLINGTON, G. H. and Sherbon, J. W., *J. Fd Sci.*, 1970, **35**, 175.

112. LEDWARD, D. A., *J. Sci. Fd Agric.*, 1981, **32**, 521.

113. ROBERTS, P. C. B. and LAWRIE, R. A., *J. Fd Technol.*, 1974, **9**, 345.

114. LEE, A. and GRAU, R., *Fleischwirtschaft*, 1966, **46**, 1239.

115. KVÅLE, O. and MARTENS, H., *Food Quality and Nutrition: Research Priorities for Thermal Processing*, W. K. Downey (ed.), Applied Science Publishers Ltd, London, 1977, 537.

116. LAAKKONEN, E., SHERBON, J. W. and WELLINGTON, G. H., *J. Fd Sci.*, 1970, **35**, 178.

117. LAAKKONEN, E., SHERBON, J. W. and WELLINGTON, G. H., *J. Fd Sci.*, 1970, **35**, 181.

118. PAUL, P. C., BUCHTER, L. and WIERENGA, A., *J. Agr. Fd Chem.*, 1966, **14**, 490.

119. RANDALL, C. J. and MACRAE, H. F., *J. Fd Sci.*, 1967, **32**, 182.

120. BECHER, P., *Emulsions, Theory and Practice*, 2nd edn., Reinhold Pub. Corp., New York, 1965, 2.

121. SCHMIDT, G. R., *New Methods in Meat Processing. Proc. Meat Ind. Res. Conf.*, 1979, 31.

122. SCHMIDT, G. R., MAWSON, R. F. and SIEGEL, D. G., *Food Technol.*, May 1981, 235.
123. APRIL, E. W., BRANT, P. W. and ELLIOTT, G. F., *J. Cell. Biol.*, 1972, **53**, 53.
124. THENO, D. M., SIEGEL, D. G. and SCHMIDT, G. R., *J. Fd Sci.*, 1978 *a*, **43**, 488.
125. RAHELÍC, S., PRIBIŠ, VJERA and VIČEVIĆ, Z., *Proc. 20th Eur. Meet. Meat. Res. Workers*, Dublin, 1974, 133.
126. THENO, D. M., SIEGEL, D. G. and SCHMIDT, G. R., *J. Fd Sci.*, 1978 *b*, **43**, 483.
127. MACFARLANE, J. J., SCHMIDT, G. R. and TURNER, R. H., *J. Fd Sci.*, 1977, **47**, 1603.
128. SAMEJIMA, K., HASIMOTO, Y., YASUI, T. and FUKAZAWA, T., *J. Fd Sci.*, 1969, **34**, 242.
129. SCHUT, J., VISSER, F. M. W. and BROUWER, F., *Proc. 24th Eur. Meet. Meat Res. Workers*, Kulmbach, 1978, W12.
130. BREKKE, C. J. and EISELE, T. A., *Food Technol.*, May 1981, 231.
131. EISELE, T. A. Chemical, functional and nutritional properties of acylated beef heart myofibrillar protein, Ph.D. thesis, Washington State University, Pullman, 1980.
132. DUBOIS, M. W., ANGLEMIER, A. F., MONTGOMERY, M. W. and DAVIDSON, W. D., *J. Fd Sci.*, 1972, **37**, 27.

# 15

# Biological Nitrogen Fixation—Significance to the Food Industry


FERGAL O'GARA

*Department of Dairy and Food Microbiology,
University College, Cork, Republic of Ireland*


## INTRODUCTION

Green plants are a major source of the protein and fibre required to support life on this planet. Proteins of plant origin may be assimilated directly or, alternatively, used as a feedstock to produce animal proteins for the human diet.

Nitrogen is an essential component of protein and although the earth's atmosphere contains an abundance of dinitrogen gas (approximately 80 %), it cannot be used directly in this form for protein synthesis. Gaseous nitrogen must first be converted or fixed into an inorganic form such as $NH_3$ or $NO_3^-$ before it can be assimilated into amino acids—the building blocks of all proteins. Any limitation in the quantity of fixed nitrogen will consequently affect the production of adequate amounts of protein. In natural ecosystems, proteins can be degraded by the selective action of soil microorganisms during the natural decay cycle of plant and animal tissue. Additionally, the resulting amino acids may be further degraded to $NH_3$ and $NO_3^-$. All the inorganic N generated during the nitrification process is normally not available for future protein synthesis since the nitrate produced can act as a substrate for denitrifying microorganisms. Consequently, to maintain sufficient fixed nitrogen for global protein production, the inorganic N lost through denitrification must be counterbalanced by adequate rates of fixed N production (Fig. 1). It is relatively difficult to determine the actual amount of nitrogen fixed annually on earth, but it is estimated to be in the region of 175–190 million tonnes.[1,2] There is general agreement that this amount of fixed nitrogen is adequate to produce sufficient protein to support the existing world population (i.e.

                    *Fergal O'Gara*

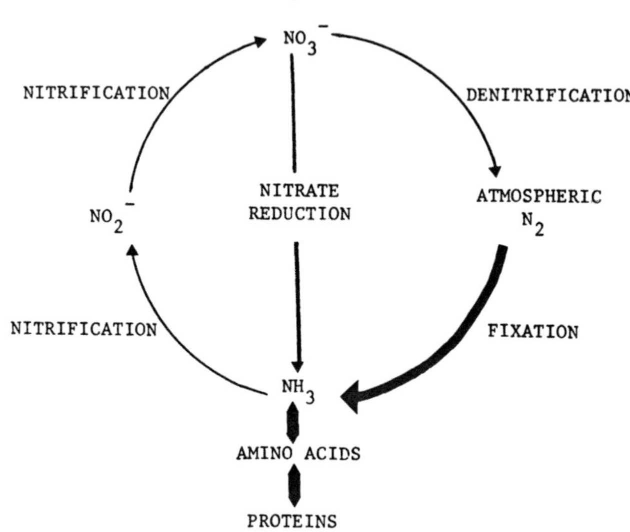



FIG. 1. Nitrogen interconversions in the biosphere. Atmospheric $N_2$ must first be converted to $NH_3$ by nitrogen fixation before it can be assimilated into proteins. $NH_3$ released following protein degradation may also be converted to $NO_3^-$ by the process of nitrification. This $NO_3^-$ may then be reconverted to atmospheric $N_2$ by the process of denitrification, or be reduced to $NH_3$.

assuming there were no problems in distribution, etc.). However, it is projected that the world population will increase by 40 % by the end of this century and will have doubled by the early decades of the next millennium, assuming there are no major global catastrophies (see Simantov, Ch. 2). In order to be in a position to produce enough protein to support this projected expansion in population, it is paramount that global fixed N production be increased. It has been estimated that an additional $75–150 \times 10^6$ tonnes of fixed N will be necessary to produce the additional food proteins required (reviewed in reference 3). If this increase is to be achieved, it is evident that fixed N output from existing technologies will have to be increased on a global basis in addition, perhaps, to developing new systems and technologies. In relation to present agricultural production, approximately two-thirds of the fixed nitrogen is produced by biological $N_2$ fixing systems while the remaining one-third $(50 \times 10^6$ tonnes) is synthesised commercially by the Haber process (Table 1).[4] Fixed nitrogen production by the Haber process is a very energy-intensive process. Hydrogen gas, which is generally obtained from fossil fuels, is a substrate molecule required for ammonia synthesis at high temperature

TABLE 1

ESTIMATES OF ANNUAL $N_2$ FIXATION ON EARTH[1]

| Crop | Cultivated area ($ha \times 10^6$) | Biological $N_2$ fixation ($tonnes/year \times 10^6$) |
|---|---|---|
| Legumes | 250 | 35 |
| Non-legumes | 1 150 | 9 |
| Permanent grassland | 3 000 | 45 |
| Forests | 4 100 | 40 |
| Fertiliser from chemical $N_2$ fixation | | 40 |

and pressure. Since fossil fuel products are required as a feedstock in the synthesis of ammonia fertiliser, it is not surprising that the production cost of N-fertiliser tends to be index-linked to that of fossil fuels and this has been the main reason for the escalation in N-fertiliser prices during recent years. A combination of other factors, particularly uncertainty surrounding the long-term availability of fossil fuels and environmental concern regarding the possible deleterious effects of increased production of oxides of nitrogen following greater use of synthetic N-fertiliser in the biosphere, have been responsible for focusing greater attention on ways of exploiting biological $N_2$-fixation which does not have many of the constraints associated with the Haber process for increasing fixed N-production.

## BIOLOGICAL NITROGEN-FIXING SYSTEMS

Nitrogenase, a unique iron sulphur protein, is the cardinal enzyme involved in biological $N_2$-fixation systems (Table 2). This enzyme is found only in certain procaryotic microorganisms and is inactivated by oxygen. The structural gene sequences coding for the enzyme from different nitrogen-fixing systems are highly conserved.[5] The functional enzyme is composed of two subunits (Fe-protein: Fe–Mo protein), both of which are required for activity. The Fe-protein is first reduced by specific electron transport proteins and then is able to transfer electrons to the Fe–Mo protein and reduce it. The numerous iron atoms of the Fe–Mo protein are capable of diverting electrons to nitrogen atoms bound to the Fe–Mo protein which, consequently, becomes reduced (reviewed in references 1 and 6). The necessary ATP and reducing power are produced from cellular energy-yielding pathways, principally from metabolism of carbohydrates which

### TABLE 2
SOME BIOLOGICAL SYSTEMS CAPABLE OF FIXING $N_2$

| *System* | *Organisms involved* |
| --- | --- |
| Symbiotic nitrogen fixation | *Rhizobium*/legumes |
| | Blue–green algae/*Azolla* |
| Associative nitrogen fixation | *Azotobacter, Spirillum* |
| Free-living fixers | *Azotobacter,* |
| | *Clostridium Bacillus,* |
| | *Rhodosprillum* |

may be derived indirectly from photosynthesis. Consequently, the biological reduction of $N_2$ is a renewable process which does not rely on non-renewable fossil fuel reserves.

Although nitrogenase is confined to procaryotic microorganisms, specific interactions of some of these microorganisms with eucaryotic partners extend the advantages of nitrogen fixation to some plant life important in agriculture. The symbiotic association between root nodule bacteria and legume plants represents the most important of these associations in agronomic terms (Fig. 2). Legumes, such as soybean, pea, bean, clover and alfalfa, are important sources of plant protein and have the major advantage of being able to grow without being supplemented with large amounts of synthetic N-fertiliser, in contrast to other food-protein crops, such as cereals. Specific interactions between root nodule bacteria and the root system of the host legume plant result in the formation of root nodules which can provide the plant with its nitrogen requirements. The basis of this symbiotic association is that the *Rhizobium* bacteria inside the nodule do not directly use all the ammonia formed by nitrogenase but, rather, excrete it into the plant tissue. Ammonia excretion was demonstrated in isolated bacteroids[7] and free-living *Rhizobium* spp., induced for nitrogenase activity *in vitro*,[8] following exposure to $^{15}N$ labelled gas. Even though root nodule bacteria are capable of assimilating ammonia when cultured under certain physiological conditions, the key $NH_3^+$-assimilation enzymes are undetectable or present at low levels in nitrogen-fixing bacteroids.[9,10] Consequently, a major portion of the $NH_3$ from nitrogenase is partitioned to the plant fraction where it is assimilated into amino acids by plant enzymes.[11] The bacteria in root nodules are energised by the provision of carbon substrate molecules derived during photosynthesis by the plant but the actual nature of the substrate molecules partitioned to the nitrogen-fixing bacteroids is not fully understood. In simple terms, the root nodule

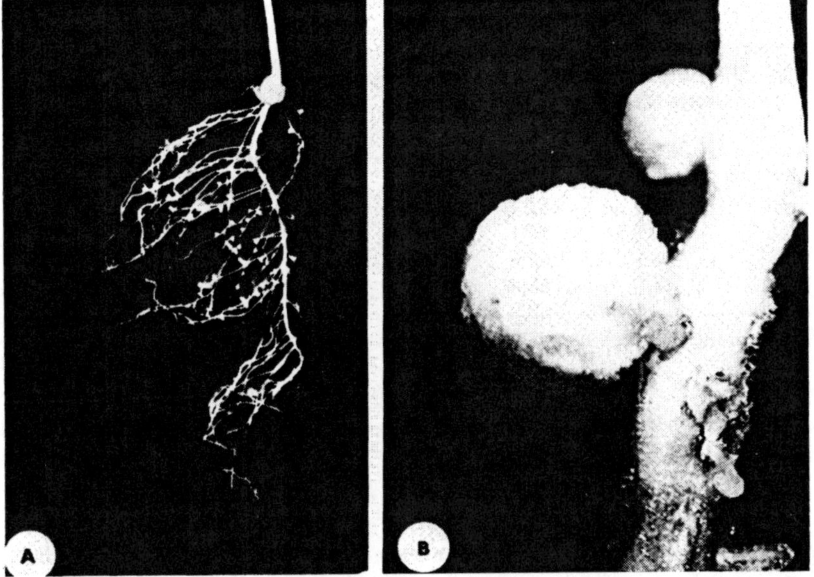

FIG. 2. Nitrogen-fixing root nodules on pea plant. (A) Entire root system showing numerous nodules. (B) Close up of an individual nodule.

bacteria may be visualised as specialised ammonia-excreting organelles in the root nodules that are supported by solar energy. Large amounts of food proteins are produced annually by these legume–*Rhizobium* symbiotic systems where the N-input comes from biological $N_2$-fixation. Soy beans now represent a significant source of protein and have the major advantage of being capable of growth without the use of large inputs of synthetic N-fertiliser since they can interact with nitrogen-fixing root nodule bacteria. World production of soy beans has more than doubled during the last decade and is currently in excess of $90 \times 10^6$ tonnes (Table 3). The US now produces approximately 65 % of world output but it is interesting to note that production in other countries, notably Brazil and some European countries, has increased significantly over the past decade.

Ammonia-excreting root nodule bacteria in symbiotic associations with legume plants make a major contribution to global protein production. However, this mechanism of $NH_3$ excretion by $N_2$-fixing procaryotes to support the N-requirements and growth of the plant partner in the symbiosis is not confined solely to the *Rhizobium*–legume symbiosis. Certain $N_2$-fixing blue–green algae (*Cyanobacterium*) are also capable of interacting with

*Fergal O'Gara*

### TABLE 3
#### WORLD SOY BEAN PRODUCTION

| Year | World production (tonnes × 10⁶) | US | % of Total China | Brazil | Others | Average world yield (kg/ha) |
|------|---------------------------------|------|-------|--------|--------|------------------------------|
| 1959 | 25·8 | 56·1 | 36·8 | 0·6 | 6·4 | 1 212 |
| 1964 | 28·1 | 67·7 | 24·6 | 1·0 | 6·5 | 1 177 |
| 1969 | 40·5 | 76·1 | 15·3 | 2·6 | 6·0 | 1 125 |
| 1974 | 54·3 | 60·8 | 17·4 | 14·5 | 7·2 | 1 387 |
| 1979 | 94·3 | 65·4 | 8·8 | 16·1 | 9·7 | 1 857 |

Source: United States Department of Agriculture Statistics.

plants and providing them with a source of biologically fixed N. In particular, blue–green algae in the Nostoc–Anabaena group exist in symbiotic associations in lichens and with liverworts and water-ferns.[12] The nitrogen-fixing blue–green algae, *Anabaena azollae*, can associate with the fern, *Azolla*, occurring as an endophyte in special cavities in the central surface of the dorsal lobe of each fern. The resultant crop is used as a forage and green manure of long recognised agronomic potential, especially in South East Asia.[13] For example, in Vietnam and Thailand, *Azolla* is used as a green manure for rice production.[14] Yields of 5 to 6 tonnes per hectare have been reported by exploiting *Azolla* as the major source of N-input[15] and current research in many countries is focused on the further exploitation of this system as a means of producing biologically fixed nitrogen for food production.[16,18]

## POTENTIAL FOR EXPLOITING BIOLOGICAL $N_2$-FIXATION

The importance of existing biological nitrogen-fixing associations, such as the legume and *Azolla* systems, is evident from the major contributions they are making to global food production. However, in attempting to exploit and further extend biological $N_2$-fixing systems, it has been recognised by investigators in many countries that biological $N_2$-fixing systems have inherent limitations which could perhaps be overcome by various modifications to the system through continued research efforts. For example, in the case of the legume system it has been recognised that nitrogen fixation can often be limited by the supply of energy from the host plant.[19] In addition, it has been demonstrated that the enzyme nitrogenase can divert a significant proportion of electrons required for $N_2$ reduction into a wasteful side-reaction to form $H_2$. Cell-free nitrogenase from all known sources produces $H_2$ concomitant with $N_2$ reduction and the energy expenditure in $H_2$ evolution by the nitrogenase system can be substantial.[20] Even though a number of investigators had observed relationships between $H_2$ metabolism and nitrogen fixation and demonstrated $H_2$ evolution and hydrogenase activity in various *Rhizobium*–legume associations,[21,22] it was not until recently that the significance of $H_2$ metabolism in relation to legume $N_2$-fixation was fully appreciated, mainly through the work of Evans and his co-workers.[23,24] In a survey of 77 samples of several different legumes inoculated with natural or commercial *Rhizobium* inocula, $H_2$ losses from nodules accounted for an average of 44 % of the electron flux through nitrogenase.[25] However, strains were identified that did not lose significant amounts of $H_2$ during nitrogen fixation and these strains were capable of recycling $H_2$ through an uptake hydrogenase system. $H_2$ metabolism is capable of supporting nitrogen fixation activity in free-living strains of *Rhizobium* induced for nitrogenase,[26] and also in isolated bacteroids.[27] The oxidation of $H_2$ is an important element in the maintenance of ATP supply in bacteroids and bacteroids formed without the unidirectional hydrogenase system lose $H_2$ to the atmosphere and cannot benefit from the potential energy advantages of the $H_2$-recycling process.[28] Available evidence suggests that *Rhizobium* strains possessing $H_2$ recycling capabilities increase the nitrogen content and yield of soybean compared with strains which are unable to recycle $H_2$.[29] Mutants defective in hydrogen recycling activity (Hup$^-$) have been isolated and compared with the wild type (Hup$^+$) strain in legume symbiosis. Maier *et al.*[30] described mutants of *R. japonicum* that lack the $H_2$ uptake capability in both free living and bacteroid forms. Soybean plants inoculated with these mutants fixed less $N_2$ than did plants inoculated with the Hup$^+$ parent, but the experiments were not conclusive, since it was not shown that the mutants were otherwise isogenic with the Hup$^+$ parent. Subsequently, Hup$^-$ mutants were isolated which spontaneously revert to the parent type at a frequency consistent with that of a single point mutation.[31] Soybean plants inoculated with these Hup$^-$ strains had lower dry weights and contained less total N than did plants inoculated with the parent Hup$^+$ strain in experiments conducted in growth cabinets. However, data are not yet available on the symbiotic properties of the revertants obtained from the Hup$^-$ mutant, information which would be important in specifically evaluating the rôle of $H_2$ metabolism during symbiosis. Nevertheless, these results, obtained with apparently isogenic lines of $H_2$ uptake-deficient *R. japonicum*, provide support for a beneficial rôle of the $H_2$ uptake phenotype in legume symbiosis.

It has been established recently that many *Rhizobium* strains in soils from major legume production areas in the US are Hup$^-$.[32] Consequently, the use of Hup$^+$ strains of *Rhizobium* in inocula to replace resident Hup$^-$ strains would appear to be a desirable development as a means of increasing nitrogen input into legumes for food production. The technology also exists to convert naturally occurring Hup$^-$ strains, which otherwise may have desirable symbiotic properties, to Hup$^+$ strains. This has been facilitated by the demonstration that determinants for hydrogenase activity in a strain of *R. leguminosarum* are genetically linked to determinants for nodulation ability, and are probably carried on a large molecular weight plasmid.[33] Although this particular plasmid was not self-transmissible, the determinants for nodulation ability and hydrogenase activity could be transferred to other strains of *R. leguminosarum* following recombination with another transmissible *R. leguminosarum* plasmid. The demonstration that certain key genes like *nif*[34] *nod*[35] and *hup*,[33] which are involved specifically with the nitrogen-fixing symbiosis, are located on plasmids, makes it more feasible to exploit biological nitrogen fixation in legumes by developing genetically improved strains of *Rhizobium*.

Research efforts aimed at providing improved *Rhizobium* strains for nitrogen fixation have received great impetus from the development and application of the novel methods of genetic engineering technology. However, producing improved strains and testing them under laboratory conditions represents only one of these phases in the overall process of enhancing nitrogen fixation in *Rhizobium*. An equally important consideration in this overall process is the ability of such 'engineered strains' to survive in soils and compete effectively with indigenous populations during the nodulation process of the host legume. It is well documented that *Rhizobium* strains used in inocula may fail to compete with the indigenous root nodule bacteria for nodule sites on the plant root.[36,38] For example, in field inoculation experiments with *R. phaseoli* a maximum of only 15% of the nodules on *Phaseolus vulgaris* were formed by the inoculum strain.[38] In similar experiments with soy bean, the *R. japonicum* inoculum strain could be recovered from 5% of the nodules when the inoculum had been applied at $3 \times 10^8$ cfu/cm row.[37] However, in these two latter studies, estimates of the relative numbers of indigenous *R. phaseoli* or *R. japonicum* in the particular soil under investigation are not available. Consequently, it is not possible to assess whether these low rates of recovery are due to a lack of competitiveness in these strains or to other factors.

Clearly, the inability to establish genetically engineered strains with desirable characteristics for enhanced nitrogen fixation in agricultural soil in production areas would impede overall efforts to increase nitrogen input through biological means. Many key factors, including the nature of events involved in competition and strain establishment, are poorly understood. However, unpublished studies by O'Gara and Higgins have shown that it is possible to select *Rhizobium* strains from natural populations which apparently have the ability to compete and become established in nodules following inoculation under field conditions. An inoculum of a genetically marked *R. leguminosarum* (CL100) strain was applied in a liquid form at the time of sowing to plots in which pea crops were being sown, under commercial conditions, at rates varying from $1-4 \times 10^8$ cfu/cm row, with a seeding rate of approximately one seed per 3 cm. Even though the *R. leguminosarum* (CL100) strain in the inoculum was entering soils which contained indigenous strains of *R. leguminosarum*, it nevertheless succeeded in forming the majority of the nodules on the pea plants (Table 4), clearly showing that strain CL100 was effective in competing with the indigenous strains of *R. leguminosarum* during nodule formation. The rates of nitrogen fixation, based on the acetylene reduction technique, by plants in inoculated plots were higher than those of plants in uninoculated control plots (Fig. 3). The overall pea yield from the inoculated plots

TABLE 4

ESTABLISHMENT OF *R. leguminosarum* CL100 IN PEA NODULES FOLLOWING SEED INOCULATION

| *Field experiment* | *Inoculation rate (cfu/cm row)* | *Indigenous population (cfu/g soil)* | *% Establishment* |
|---|---|---|---|
| 1 | $3.9 \times 10^8$ | $1.6 \times 10^3$ | 78 |
| 2 | $1.3 \times 10^8$ | $2.9 \times 10^2$ | 79 |
| 3 | $1.9 \times 10^8$ | ND | 92 |

ND. Not determined.

Establishment of *Rhizobium leguminosarum* CL100 in pea nodules following seed inoculation. Indigenous populations of *R. leguminosarum* were estimated by the most probable number method. The inoculum strain (CL100) was genetically marked (Str$^r$) and was monitored by transferring clones isolated from nodules to medium containing streptomycin and checking for growth. In addition, CL100 was also monitored by antibiotic fingerprinting analysis and by its plasmid profile.

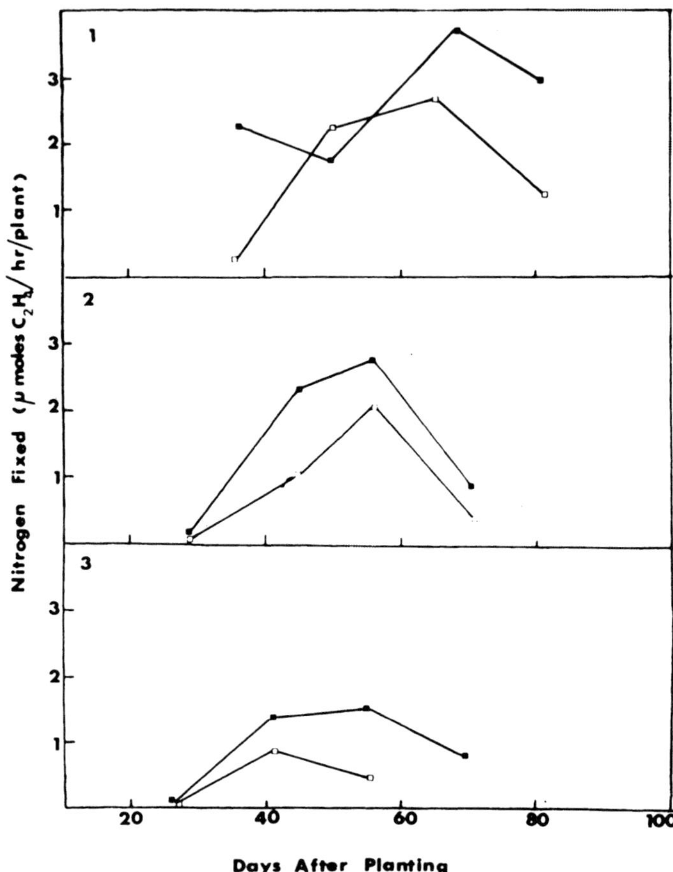



FIG. 3. Average rates of nitrogen fixation by inoculated pea plants compared with uninoculated controls during the growth season. Nitrogen fixation estimates were determined by the acetylene reduction technique on harvested plants and each value represents the mean of six determinations. Inoculated plants —■—■—■—; uninoculated control plants —□—□—□—. Experimental plots 1, 2 and 3 represent independent field experiments.

increased by an average of 14% compared with the uninoculated control plots—an increase which was statistically significant at the 0·01 level (O'Gara and Higgins, unpublished data). Based on the results of this inoculation programme, it seems likely that similar results could be achieved with other legumes, by ensuring that inefficient or low numbers of indigenous root nodule bacteria are replaced by adequate numbers of a competitive strain at the time of seed sowing. Such an approach would help ensure adequate N-input from biological N₂-fixation for the developing plant to produce consistent and high yields.

## GENETIC ENGINEERING OF NEW NITROGEN-FIXING PLANTS

The use of genetic engineering technology to improve existing biological nitrogen-fixing systems is only one of the possible applications of this powerful technology to enhance and further exploit biological $N_2$-fixation. Other approaches based on this technology are available; one which is continuing to receive a great deal of attention is the creation of new varieties of food crops which have been genetically engineered to fix their own nitrogen. A major objective in such an approach is to transfer nitrogen-fixing genes directly to plant cells so as to maintain them in a stable and functional manner. An alternative approach would be to 'engineer' a more promiscuous association between nitrogen-fixing bacteria and non legume plants, so that nitrogen fixation capabilities mediated through bacteria in the protective environment of nodules could be extended to some of our major food crops. The recent discovery of the apparently natural interaction between *Rhizobium* spp. and the tropical legume *Sesbania rostrata* resulting in the production of nitrogen-fixing nodules on the *stem* of the plant, widens the scope for achieving such an objective through genetic engineering technology.[39]

Research efforts to create new nitrogen-fixing plants by directly transferring nitrogen-fixing genes to plant cells have progressed significantly in recent years. The genes coding for the enzyme nitrogenase and some of the other ancillary proteins required to promote and regulate nitrogen fixation activity in *K. pneumoniae* have been isolated and studied in detail (reviewed in references 40, 41).

The close linkage of the *nif* genes to genes involved in histidine biosynthesis (*his*) on the *K. pneumoniae* chromosome facilitated the isolation of a transmissible plasmid carrying the *nif–his* region.[42,43] The plasmid pRDI has been used to study the expression of *nif* genes in different backgrounds and, when transferred to *E. coli*, this bacterium, normally incapable of fixing nitrogen, was converted to a nitrogen-fixing micro-organism. Plasmid pRDI was also the source of DNA for cloning the *nif* genes and a series of plasmids carrying fragments that cover the *nif* region have been constructed using recombinant DNA techniques.[44–46]

Similar recombinant plasmids carrying *nif* genes have been constructed in suitable vectors and then transferred into yeast cells in culture. Yeast cells, being eucaryotic organisms, are more closely related to higher plant cells than they are to bacteria and consequently serve as a useful intermediary model eucaryotic organism to study *nif* gene expression. The *nif* genes were replicated and maintained in the yeast cells for over 40 generations but attempts to demonstrate nitrogen fixation activity were negative.[47,48] However, the successful introduction of *nif* genes into a eucaryotic cell is a major breakthrough in crossing biological barriers with these agronomically important genes.

The introduction of *nif* genes into plant cells will be a much more challenging goal. A number of potential options are being explored but basically the strategy is to attach the *nif* genes to a suitable vector which can transform plant cells and be stably maintained. Some progress has been made in developing suitable vectors for introducing foreign genes into plant cells and perhaps the most promising method is the use of TI plasmid found in the bacterium *Agrobacterium tumefaciens*. This bacterium induces crown gall tumours on most dicotyledonous plants. The mechanism of infection is to insert a segment of its TI plasmid into the nucleus of cells resulting in the neoplastic transformation on plant cells.[49] The insertion of T-DNA is therefore a natural form of genetic engineering which confers several unusual properties on the transformed plant cell. Since the T-DNA region of the TI plasmid is known to intregate into plant nuclei, it is possible that a vector carrying this region could promote integration of linked genes. Plant cells transformed with a TI plasmid with an insertion of transposon Tn7 in its T-DNA region, have been found to contain this Tn7 DNA.[50] Therefore, *A. tumefaciens* cells, provided with the desired genetic information at the right position on the TI plasmid, may possibly be used in the future to achieve transfer of *nif* genes into plant cells. However, if stable transfer is achieved, attention must be focused on the expression of these genes and the problem of inactivation of nitrogenase proteins in an aerobic environment. It would appear that the transfer of *nif* genes to yeast may provide a useful model system to study these aspects of expression of nitrogen fixation activity in eucaryotic cells.

## REFERENCES


1. HARDY, R. W. F. and BURNS, R. C. *Nitrogen Fixation in Bacteria and Higher Plants*, Springs, New York, 1975.
2. DELWICH, C. D. Energy relations in the global nitrogen cycle. *Ambio*, 1977, **6** 104.

3. POSTGATE, J. R. in *Food Chains and Human Nutrition*, Sir K. Blaxter, ed., Applied Science Publishers Ltd., London, 1980.
4. BURRIS, R. H. in *Nitrogen Fixation*, W. E. Newton and W. H. Orme-Johnson, (eds.), University Park Press, Baltimore, USA, 1980.
5. RUVKUN, G. B. and AUSUBEL, F. M. *Proc. Natl. Acad. Sci., U.S.A.*, 1980, **77**, 191.
6. MORTENSON, L. E. and THORNEBY, R. N. F. *Ann. Rev. Biochem.*, 1979, **48**, 387.
7. BERGERSON, F. J. and TURNER, G. L. *Biochim. Biophys. Acta.*, 1967, **141**, 507.
8. O'GARA, F. and SHANMUGAM, K. T. *Biochim. Biophys. Acta.*, 1976, **437**, 313.
9. BROWN, C. M. and DILWORTH, M. J. *J. Gen. Microbiol.* 1975, **86**, 39.
10. ROBERTSON, J. G., WARBURTON, M. P. and FARNDEN, K. J. F. *FEBS Lett.* 1975, **55**, 33.
11. SCOTT, B. D., ROBERTSON, J. G. and FARNDEN, K. J. F. *Nature* 1976, **263**, 703.
12. FOGG, G. E., STEWART, W. D. P., FAY, P. and WALSBY, A. E. *The Blue–Green Algae*, Academic Press, London, 1973.
13. MILLBANK, J. W. in *Biology of Nitrogen Fixation*, A. Quispel, ed., American Elsevier Publishing Co. Inc., New York, 1974.
14. MOORE, A. W. *Bot. Rev.*, 1969, **35**, 17.
15. TRAN QUANY THUYET and DAO THE TUAN. Azolla. A green manure, *Agric. Prob.*, 1973, **4**.
16. PETERS, G. A., RAY, T. B., MAYNE, B. C. and TOIRA JR, R. E. in *Nitrogen Fixation, Vol. 11*, W. E. Newton and W. H. Orme-Johnson (eds.), University Park Press, Baltimore, 1980.
17. TALLEY, S. N. and RAINS, D. W. in *Nitrogen Fixation, Vol. 11*, W. E. Newton and W. H. Orme-Johnson, eds., University Park Press, Baltimore, 1980.
18. NEWTON, J. W. and CAVINS, J. F. *Plant Physiol*, 1976, **58**, 798.
19. HARDY, R. W. F. and HAVELKA, U. D. *Science*, 1975, **188**, 633.
20. BULEN, W. A. and LE COMTE, J. R. *Proc. Natl. Acad. Sci., U.S.A.*, 1966, **56**, 979.
21. DIXON, R. O. D. *Ann. Bot.*, 1967, **31**, 179.
22. HOCH, G. E., SCHNEIDER, K. C. and BURRIS, R. H. *Biochim. Biophys. Acta.*, 1960, **37**, 273.
23. EVANS, H. J., RUIZ-ARGUESO, T., JENNINGS, N. T. and HANUS, J., in *Genetic Engineering for Nitrogen Fixation*, A. Hollaender (ed.), Plenum Publishing Corp., New York, 1977.
24. EVANS, H. J., PUROHIT, K., CANTRELL, M. A., EISBRENNER, G., RUSSELL, S. A., HANUS, F. J. and LEPO, J. E. in *Current Perspectives in Nitrogen Fixation*, A. H. Givson and W. E. Newton, eds., Australian Academy of Science, Canberra, 1981.
25. SCHUBERT, K. R. and EVANS, H. J., *Proc. Natl. Acad. Sci., U.S.A.*, 1976, **73**, 1207.
26. O'GARA, F. and SHANMUGAM, K. T., *Proc. Natl. Acad. Sci., U.S.A.*, 1978, **75**, 2343.
27. RUIZ-ARGUESO, T., EMERICK, D. W. and EVANS, H. J. *Biochem. Biophys.* Res. Commun., 1979, **86**, 259.
28. EVANS, H. J., EMERICH, D. W., RUIZ ARGUESO, T., MAIER, R. J. and ALBRECHT, S. L. in *Nitrogen Fixation, Vol. 11*, W. E. Newton and W. H. Orme-Johnson (eds.), University Park Press, Baltimore, 1980.

29. ALBRECHT, S. L., MAIER, R. J., HANUS, F. J., RUSSELL, S. A., EMERICH, D. W. and EVANS, H. J. *Science*, 1979, **203**, 1255.
30. MAIER, R. J., POSTGATE, J. R. and EVANS, H. J. *Nature*, 1978, **276**, 494.
31. LEPO, J. E., HICKOK, R. E., CANTRELL, M. A., RUSSELL, S. A and EVANS, H. J. *J. Bacteriol*, 1981, **146**, 614.
32. LIM, S. T., ANDERSEN, K., TAIT, R. and VALENTINE, R. C. *T.I.B.S.* June 1980, p. 167.
33. BREWIN, N. J., DEJONG, T. M., PHILLIPPS, D. A. and JOHNSTON, A. W. B. *Nature*, 1980, **288**, 77.
34. NUTI, M. P., LEPIDI, A. A., PRAKASH, R. K., SCHILPEROORT, R. A. and CANNON, F. C. *Nature*, 1979, **282**, 533.
35. JOHNSTON, A. W. B., BEYNON, J. L., BUCHANON-WOLLASTON, A. V., SETCHELL, S. M., HIRSCH, P. R., and BERINGER, J. E. *Nature*, 1978, **276**, 635.
36. HAM, G. E. in *Nitrogen Fixation, Vol. 11*, W. E. Newton and W. H. Orme-Johnson, eds., University Park Press, Baltimore, 1980.
37. KUYKENDALL, D. L. and WEBER, D. E. *Appl. and Environ. Microbiol.*, 1978, **36**, 915.
38. BEYNOR, J. J. and JOSEY, D. P., *J. Gen. Microbiol.*, 1980, **118**, 437.
39. DREYFUS, B. L. and DOMMERGUES, Y. R., *FEMS Lett*, 1981, **10**, 313.
40. BRILL, W. J. *Microbiol. Rev.*, 1980, **44**, 449.
41. DIXON, R. A., KENNEDY, C. and MERRICK, M. in *Genetics as a Tool in Microbiology*, S. W. Glover and D. A. Hopwood (eds.), Cambridge University Press, Cambridge, 1981.
42. CANNON, F. C., DIXON, R. A. and POSTGATE, J. R., *J. Gen. Microbiol*, 1976, **93**, 111.
43. DIXON, R. A., CANNON, F. C. and KONDOROSI, A. *Nature*, 1976, **260**, 268.
44. CANNON, F. C., RIEDEL, G. E. and AUSUBEL, F. M. *Proc. Nat. Acad. Sci.*, *U.S.A.*, 1977, **74**, 2963.
45. CANNON, F. C., RIEDEL, G. E. and AUSUBEL, F. M. *Molec. Gen. Genet.*, 1979, **174**, 59.
46. PUHLER, A., BURKHARAT, H. J. and KLIPP, W. *Molec. Gen. Genet.*, 1979, **176**, 17.
47. ELMERICK, C., SIBOLD, L., GUERINEAU, M., DE MARSAL TANDEAU, N., CHOCAT, P., GERBAUD, C. and AUBERT, J. P. in *Current Perspectives in Nitrogen Fixation*, A. H. Gibson and W. E. Newton (eds.), Australian Academy of Science, Canberra, Australia, 1981.
48. ZIMIR, A., MAINA, C. V., FINK, G. R., MORRIS, D., NOTI, J., OSBORN, F. and SZALAY, A. A. in *Current Perspectives in Nitrogen Fixation*, A. H. Gibson and W. E. Newton (eds.), Australian Academy of Science, Canberra, Australia, 1981.
49. WILLMITZER, L., DE BEUCKELEER, M., LEMMERS, M., VAN MONTAGU, M. and SCHELL, J. *Nature*, 1980, **287**, 359.
50. HERNALSTEENS, J. P., VAN MONTAGU, M. and SCHELL, J. *EMBO Workshop on Plant Tumour Research 1978*, Noorwijkerhout, The Netherlands.

# 16

# Microbial Proteolysis of Milk Proteins

B. A. Law

*National Institute for Research in Dairying,
Reading, UK*

## 1. INTRODUCTION

Protein breakdown by microbial enzymes in milk and milk products is, on balance, a desirable process, upon which depends the growth and lactic acid production of the starter bacteria used in the manufacture of fermented dairy products, and also the development of typical flavour and texture in ripening cheese. In recent years microbial proteinases have also been exploited as alternative coagulants in cheese manufacture to rectify the shortfall in supplies of rennet (chymosin) from calf stomachs. There are, however, examples of detrimental effects of microbial proteolysis which affect the manufacturing properties of milk and the keeping quality of milk products. This paper attempts to review the current state of knowledge in these areas and to suggest ways of further exploiting the desirable consequences of protein degradation on the one hand, and of avoiding the problems of misplaced or mis-timed proteolysis on the other.

## 2. MICROBIAL PROTEOLYSIS IN CHEESE MANUFACTURE

### 2.1. Growth of Starter Lactic Acid Bacteria in Milk

All fermented diary products rely for their manufacture on growth to relatively high populations of streptococci and lactobacilli (added deliberately to milk as 'starter' cultures) whose immediate function is to convert the milk lactose to lactic acid. The types of starters used for different products are listed in Table 1. These bacteria are nutritionally fastidious and require supplies of some preformed amino acids for growth.

307

TABLE 1

LACTIC ACID BACTERIA USED IN STARTER CULTURES FOR THE
MANUFACTURE OF FERMENTED DAIRY PRODUCTS

| Culture | Product |
| --- | --- |
| Str. cremoris<br>Str. lactis<br>Str. diacetylactis | Most hard cheeses resembling<br>Cheddar and English territorials |
| Str. cremoris<br>Str. diacetylactis<br>Str. lactis<br>Leuconostoc spp. | Mould-ripened cheeses,<br>Semi-hard cheeses |
| Str. cremoris<br>Str. diacetylactis<br>Leuconostoc spp. | Cottage cheese, cream cheese,<br>cultured cream |
| Str. thermophilus<br>Lb. bulgaricus<br>Lb. helveticus<br>Lb. lactis | 'Cooked-curd' cheese<br>(e.g. Emmental), yoghurt |

Milk contains relatively low concentrations of free amino acids but the
starter bacteria possess a variety of transport systems which enable them to
utilise protein-bound amino acids.[1] Mills and Thomas[2] concluded that the
free amino acids normally present in fresh milk were sufficient to support
the growth of Str. cremoris to cell densities corresponding to only 8–16%
of those found in coagulated (fully grown) milk cultures. Peptides appeared
to provide a constant source of amino acids throughout growth, but the
milk proteins (including whey proteins) became increasingly important as
amino acid sources at high cell densities. Thus, proteolysis of milk proteins
by lactic acid bacteria is essential if they are to grow to sufficient levels for
the required acidification of fermented products. When one considers that
these products are the basis of the second most important worldwide
fermentation industry (the first is brewing), it is remarkable that relatively
little is known about the proteinases which are involved in the nitrogen (N)
nutrition of starters. Thomas et al.[3] established that the proteinases were
cell wall-bound, extracellular enzymes and Exterkate[4] subsequently described several proteinase activities in Str. cremoris HP which were
distinguished by their pH optima. Other strains of Str. cremoris differ
widely in the number and combinations of proteinases in their cell walls
(Table 2)[5] and one strain (AM1) produces a distinct extracellular, cell-free

proteinase[6] whose stability is not $Ca^{2+}$-dependent, unlike that of the cell-bound extracellular proteinase.[6,7] The necessity for $Ca^{2+}$ in the growth medium for the accumulation of catalytically active proteinases in the cell wall of *Str. cremoris*, and the repression of enzyme synthesis by high concentrations of amino acids,[6] may explain the slow initial growth in milk of starters which have been propagated in non-milk media (low in $Ca^{2+}$) supplemented with casein digests.

TABLE 2

SURFACE-BOUND PROTEINASES IN DIFFERENT STARTERS

| Strain | Proteinase | | |
|---|---|---|---|
| | *PI acid;*<br>*30°, 40°* | *PII neutral;*<br>*30°* | *PIII acid;*<br>*30°* |
| E8 | + | ± | − |
| HP | + | + | − |
| (HPprt⁻) | (±) | (−) | (−) |
| KH | + | + | − |
| $C_{13}$ | + | + | − |
| TR | + | + | + |
| AM2 | + | + | + |
| FD27 | + | + | + |
| AM1 | − | − | + |
| US3 | − | − | + |
| SK11 | − | − | + |

Source: reference 5. +, present; −, absent; ±, weak.

The importance of cell wall-bound proteinases for the growth of Group N streptococci in milk is well illustrated by the poor multiplication of proteinase-negative (prt⁻) variants of starter strains. These variants grow normally in media containing amino acids and peptides but lack one or more of the vital proteinases. For example, a prt⁻ variant of *Str. cremoris* HP was characterised by the loss of an acid and a neutral proteinase present in the parent strain (Table 2). Efstathiou and McKay[8] showed that the loss of proteolytic activity in *Str. lactis* was coincident with the loss of plasmid DNA and there is now increasing evidence that all cell wall-bound, extracellular proteinases are plasmid-coded in Group N streptococci.

The possession of cell-bound extracellular proteinases by the lactic streptococci has obvious advantages in that they produce peptides close to the cell and in proximity to the appropriate oligopeptide or dipeptide uptake systems whose size exclusion limits and specificities appear to be similar to those of the well-documented *E. coli* systems.[9-14] However, the complete lack of published data on the specificities of the proteinases themselves precludes any comment on their efficiency in providing peptides of the optimal size and composition for the uptake systems. Bond specificity data would also be valuable in order to clarify the relationship between the proteinases and the surface-bound peptidases recently reported in the starter streptococci.[15,16] These enzymes hydrolyse a wide range of synthetic peptides and peptide derivatives and display dipeptidase, tripeptidase, aminopeptidase and endopeptidase activities but, as with the proteinases, their physiological substrates are not known.

The final stage of protein utilisation for the supply of essential amino acids involves a large number of intracellular peptidases whose combined activities can cleave all known casein-derived peptide bonds.[17] These enzymes are also relevant to the rôle of starter bacteria in cheese ripening and will be discussed further when this aspect of proteolysis is considered (see Section 2.3).

Cell-bound extracellular proteinases and peptidases are not confined to the Group N streptococci. Argyle *et al.*[18] showed that an EDTA-sensitive proteinase was released from *Lb. bulgaricus* by lysozyme treatment. Shankar[19] showed that this organism had surface-bound proteinases which produced amino acids and peptides stimulatory for the growth of *Str. thermophilus* during yoghurt manufacture. The latter apparently used cell-free extracellular peptidases to hydrolyse peptides produced from milk proteins by *Lactobacillus*. However, this work requires confirmation since the evidence for the extracellular nature of the peptidases was not unequivocal. Intracellular peptidases in thermophilic starters have been described by Rabier and Desmazeaud,[20] Desmazeaud and Juge[21] and El Soda *et al.*[22] Their function in protein and peptide utilisation is presumably the same as that in the mesophilic starters. Peptide utilisation by lactobacilli is a well established phenomenon and has been widely observed[23,24] although the specificity of the uptake system has not been the subject of definitive studies equivalent to those concerning the mesophilic starter streptococci.

## 2.2. Microbial Rennets

The proteolytic cleavage of the $Phe_{105}$—$Met_{106}$ bond of $\kappa$-casein, which releases a charged macropeptide, destabilising the micellar casein suspension in milk, is the basis of curd formation during cheesemaking. Traditionally, this process is catalysed by calf rennet (chymosin) but shortages of this enzyme have stimulated the search for substitutes, many of which are of microbial origin. This subject has been so extensively reviewed[25-30] that a detailed account of the origins, properties and utilisation of microbial rennets would be superfluous in this chapter. However, an outline of the action of these proteinases is included for the sake of completeness.

The main difference between microbial rennets and chymosin is in the extent to which they catalyse general proteolysis in addition to the specific and desired breakdown of $\kappa$-casein. Because chymosin is very specific, it shows a very high ratio of milk-coagulating activity to proteolytic activity (Table 3). Bovine pepsin, a digestive proteinase, is less specific and shows a

TABLE 3

RATIOS OF MILK-COAGULATING ACTIVITIES TO PROTEOLYTIC ACTIVITIES FOR CHYMOSIN, BOVINE PEPSIN AND TWO MICROBIAL RENNETS

| *Enzyme* | *Coagulation: Proteinase*[a] |
|---|---|
| Chymosin | 1·40 |
| Bovine pepsin | 0·04 |
| *Mucor* rennet | 0·52 |
| *Endothia* rennet | 0·15 |

[a] Data adapted from reference 25.

much lower ratio. In general, bacterial milk-clotting enzymes (e.g. from *Bacillus subtilis, B. polymyxa* of *B. mesentericus*) are too proteolytic and cause bitterness and body defects in hard and semi-hard cheese. However, the fungal rennets, although showing higher clotting:proteolytic ratios than chymosin, have been used successfully for the manufacture of these types of cheese. The enzyme from *Endothia parasitica* is too proteolytic (Table 3) for ripened cheese and shows relatively broad bond specificity. However, it can be used if curd scalding temperatures are normally greater than 50 °C, so that the enzyme is inactivated. It may therefore find application in Emmental and Gruyere manufacture. Rennets from both *Mucor pusillus* var. *Lindt* and *Mucor miehei* are available commercially and have relatively high clotting:proteolytic ratios (Table 3). They are typical acid aspartate proteinases (like chymosin) and show strong bond specificity for aromatic and hydrophobic side chains. Preparations based on *Mucor miehei* proteinases are generally preferred because their $Ca^{2+}$ requirements are similar to that of chymosin; increased $Ca^{2+}$ concentrations are required for good clotting activity with *Mucor pusillus* rennet. The *Mucor* rennets have the disadvantage of being relatively heat stable so that they persist in cheese whey and preclude the use of this by-product in other foods unless it is subjected to suitable pasteurisation.[25]

## 2.3. Endogenous Microbial Proteinase in Cheese Ripening

Just as the formation of cheese curds is dependent on proteolysis, so their subsequent ripening to make mature cheese is dependent on the concerted action of a variety of proteinases and peptidases produced by the starter bacteria, and by the secondary microflora found in different cheese types. The precise rôle of proteolysis in cheese maturation is not known but it is considered essential for the conversion of the springy curd to the typically inelastic texture of mature cheese of all varieties. The amino acids and peptides produced in the later stages of proteolysis are also involved in the development of cheese flavours, the buffering of cheese and the nutrition and growth of secondary cheese microflora.

However, microbial proteolysis in cheese cannot be considered in isolation since a proportion of the coagulant remains in cheese curds to influence the overall pattern of protein breakdown. The action of microbial rennets has not been reported in great detail but they are generally thought to produce bitter peptides by hydrolysing bonds near hydrophobic side chains, other than the Phe—Met $\kappa$-casein bond. Although chymosin is more specific, it, too, acts on other bonds and is responsible for the formation of a large proportion of the pH 4·6-soluble N in cheese[31,32] as well as smaller peptides down to 1400 molecular weight.[33,34] Early rennet proteolysis is typified by the hydrolysis of the $Phe_{23}$—$Phe_{24}$ bond[35,36] or the $Phe_{24}$—$Val_{25}$ bond[37] of $\alpha_{s1}$-casein. $\beta$-Casein degradation occurs only slowly in cheese and its products appear late in cheese maturation; the most sensitive bonds are $Ala_{189}$—$Phe_{190}$ and $Leu_{191}$—$Tyr_{192}$. Endopeptidases from the starter bacteria also contribute to gross casein hydrolysis (i.e. hydrolysis leading to increased pH 4·6-soluble N). For example, an intercellular endopeptidase from *Str. diacetylactis* rapidly hydrolyses $\alpha_{s1}$-casein (but not whey proteins) and appears to be specific for peptide bonds involving the $\alpha$-amino group of hydrophobic residues (e.g. Leu or Phe[38]). Zevaco and Desmazeaud[39] demonstrated that this proteinase also hydrolysed $Pro_{186}$—$Ile_{187}$ and $Ala_{189}$—$Phe_{190}$ in $\beta$-casein, although activity was low. On the other hand, the enzyme efficiently degraded peptides derived from $\beta$-casein by chymosin action, by attacking the $Lys_{176}$—$Ala_{177}$, $Lys_{119}$—$Val_{120}$ and $Pro_{206}$—$Ile_{207}$ bonds. These authors concluded that the main starter proteinase activity in cheese was the degradation of those peptides released from casein by chymosin. This is consistent with reports by several research groups that cheese made with chymosin alone contains pH 4·6-soluble N but very little peptide and amino acid N, whereas cheese made with starter and chymosin contains relatively large amounts of free amino N. The work of Reiter *et al.*,[40] O'Keeffe *et al.*[33] and Visser,[32] has established that, in both Gouda and Cheddar cheeses, the proteinases of starter bacteria can slowly degrade whole casein to low molecular weight ($<$1400) peptides and amino acids. In normal cheese, chymosin-mediated gross proteolysis is so rapid (especially of $\alpha_{s1}$-casein) that it is doubtful whether this activity is significant. The most important rôle of starter proteinases and peptidases appears to be the degradation of large rennet-produced peptides to small peptides and amino acids. Recent evidence from studies with prt$^-$ variants suggests that the cell-bound extracellular proteinases of starters are significant in cheese proteolysis[41] but a significant proportion of the proteinase activity, and most of the peptidase activity, is intracellular and only released into the cheese matrix during the early ripening stage when the cells lyse.[42]

These intracellular peptidases have been widely studied and described in numerous papers. For example, Sorhaug and Solberg[43] used starch gel zymograms to demonstrate the presence of dipeptidases in *Str. lactis* capable of hydrolysing a wide range of substrates containing alanine, leucine, phenylalanine, tryptophan, histidine and glycine. Similar techniques with polyacrylamide gel electrophoresis (PAGE) revealed the presence of further amino- and dipeptidases in *Str. lactis*, *Str. diacetylactis* and *Str. cremoris* active against peptides and peptide derivatives containing arginine, methionine, tyrosine and proline.[44] Mou *et al.*[17] suggested that the peptidases of Group N streptococci were probably capable of the complete hydrolysis of casein to free amino acids, after initial proteolysis by chymosin and starter proteinases. Despite their predominately neutral to alkaline pH optima, these enzymes retain sufficient activity in cheese to function during maturation and can be recovered in active form from extracts of ageing cheese.[45]

*Rôle of Starter Proteinase in Bitter Defects in Cheese*

Bitterness in cheese is caused by an accumulation of peptides containing a high proportion of hydrophobic side chains.[46-48]

Opinions differ as to the importance of proteolysis by mesophilic starters in the production of bitter defects in cheese. Early hypotheses (e.g. reference 49) suggested that bitter peptides were produced by chymosin and

314                                    *B. A. Law*

that the so-called 'bitter' starters were those which had insufficient
peptidase activity to break down the bitter peptides to non-bitter peptides
and amino acids. However, the situation is more complex than this. While
it is true that chymosin produces bitter peptides from casein, the starter
proteinases can also do this and, indeed, can produce small bitter peptides
from non-bitter, casein-derived peptides (Fig. 1).[50,51] Lowrie and his co-
workers[50,51] suggested that this latter process is the single most important

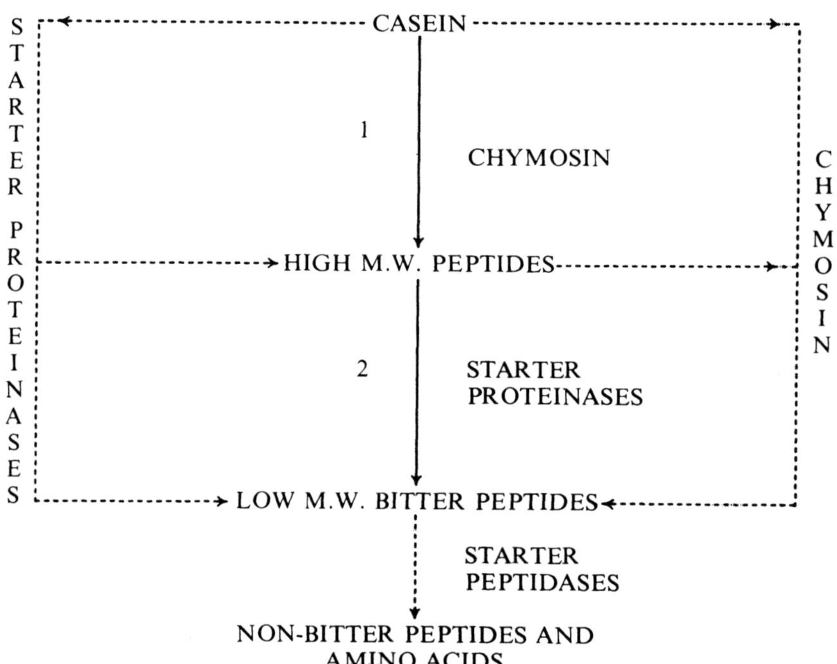


Fig. 1. Lowrie and Lawrence[50] model for bitterness development in Cheddar cheese. Bold arrows indicate the important stages (1 and 2). Broken lines indicate stages of lesser importance.

determinant in bitterness development and that starters which multiplied at
relatively high cooking temperatures during Cheddar manufacture (the
'fast' starters) were the most likely to give bitter cheese, simply because the
resultant high cell numbers contributed large quantities of bitter peptide-
producing proteinases. This hypothesis was supported with experimental
evidence showing that 'bitter' starters could be made to produce non-bitter
cheese if their number in curds was restricted by controlled bacteriophage

infections or by higher cooking temperatures. Conversely, the slow 'non-bitter' starters made bitter cheese if they were allowed to multiply to high cell numbers by altering the manufacturing process.[51] Direct evidence for the involvement of starter cell-wall proteinases in the development of bitterness was provided recently by the observation that proteinase-deficient variants of 'fast' starters produce less bitterness in cheese than their parent strains, even when total starter cell populations are high.[41]

The factors controlling bitter defects in Gouda cheese appear to be more complex since the starters generally reach high populations in curd at the relatively low cooking temperatures used for this variety. Stadhouders and Hup[52] showed that factors influencing the retention of chymosin in Gouda curd (e.g. cooking temperature, initial milk pH) also influence the tendency of the cheese to become bitter. They emphasised that some starter strains produce more bitter peptide-degrading peptidases than others. It is not known whether these are specific peptidases confined to non-bitter strains or general peptidases present at different levels. Chiba and Sato[53] identified both di-peptidase and amino-peptidase activity in fractions of cell-free extracts from starter streptococci capable of reducing bitterness but individual enzymes were not isolated. It appears, then, that proteolysis by mesophilic starters is important in producing the bitter defect in cheese but its contribution depends on the cheese variety in question.

*Proteolysis in Cheese by Non-starter Microorganisms*

Although the starter lactic acid bacteria die out during the first weeks of cheese storage, a variety of secondary microorganisms grow in the cheese (depending on type) and their proteinases augment those of the starter to influence the course of ripening. The main non-starter microflora of most hard and semi-hard cheeses are the lactobacilli, but although these are known to be weakly proteolytic,[54] little is known about their action in cheese. Desmazeaud and Juge[21] and El Soda *et al.*[22] have identified peptidases and proteinases in *Lb. casei*. The specificities of the peptidases are known: they hydrolyse a variety of di- and tripeptides containing lysine, leucine, tryptophan, phenylalanine, histidine, methionine and proline. Both the carboxypeptidase and endopeptidase described by El Soda *et al.*[22] have very restricted specificities and it is impossible at present to assess their contribution to cheese proteolysis. Early studies suggested that lactobacilli present in starter cultures did not influence cheese ripening, implying that strains which hydrolysed casein under laboratory test conditions did not do so in cheese. However, El Soda *et al.*[55] added cells and cell-free extracts of *Lb. casei* to Cheddar cheese curds and observed increases in soluble N due to their proteinase activity. The lactobacilli probably make very little contribution to normal proteolysis in hard and semi-hard cheese, but their enzymes may have applications in accelerated cheese ripening.[55]

Research on proteolysis in Emmental and Gruyere cheese has concentrated, in recent years, on the rôle of the propionibacteria. These organisms are essential for the characteristic eye formation in these cheeses but they also appear to play an important part in flavour development. *Propionibacterium shermanii* produces both proteinase and peptidase which are released into the cheese matrix by cell autolysis. Langsrud *et al.*[56,57] showed that proline-releasing peptidases predominated. This amino acid, together possibly with dimethyl sulphoxide (DMS), is thought to impart the typical sweet taste to Emmental cheese.[1]

*Brevibacterium linens* is a very proteolytic bacterium which, together with other coryneforms and yeasts, makes up the surface smear microflora of Gruyere, Limburg and Port Salut cheese. All microflora components appear to act in concert in breaking down casein near the cheese surface[58] but little is known about the yeast proteinases. In contrast, several studies have been published which describe intracellular and extracellular peptidases of *B. linens*. The organism is caseinolytic but not active against whey proteins.[59] Its ability to release 17 different amino acids in cheese[58] is attributable to the wide range of types and/or specificities of peptidases which it produces.[60-62] *B. linens* is particularly significant as a flavour-producing organism in cheese ripening because it can further metabolise the methionine which it produces to methanethiol.[63] This compound is important in the development of typical surface aroma and flavour in Trappist-type and Gruyere cheeses.[64-66]

Many cheese varieties owe their typical appearance, flavour and texture to the growth and metabolism of moulds, either on the the cheese surface or within the body of the cheese. These organisms are best known for their lipolytic activity and their ability to produce the methyl ketones necessary for Blue cheese flavour. However, their proteolytic action is also important; Kinsella and Hwang[67] considered that the high levels of mould-induced proteolysis (amounting to 50% of total N as TCA-soluble N in ripened cheese) were essential for the production of the very high free amino acid concentrations (10% of total N) which buffer Blue cheese to pH 6·5, aiding other enzyme activities and providing background flavour. The *Penicillium* moulds (*P. roqueforti* and *P. caseicolum*) produce a complete range of proteolytic enzymes which account for this high degree of casein breakdown in mould-ripened cheese (Table 4). Both organisms secrete extracellular acid (aspartyl) proteinase and neutral metalloproteinases[31]

## TABLE 4
PROTEOLYTIC SYSTEM OF *Penicillium* MOULDS USED FOR CHEESE RIPENING

|  | *Aspartyl proteinase* | *Neutral proteinase* | *Serine proteinase* | *Amino- peptidase* | *Carboxy- peptidase* |
|---|---|---|---|---|---|
| *P. roqueforti* | + | + | + | + | + |
| *P. caseicolum* | + | + | − | + | + |

Data summarised from reference 31.

which produce both high and low molecular weight ($> 1000$) peptides in cheese, but no amino acids. The starter peptidases probably produce some amino acids but the moulds also contribute extracellular acid and alkaline carboxypeptidases and alkaline aminopeptidases of very broad specificity.[68] Surface proteolysis by neutral enzymes of *P. caseicolum* is aided by the activity of lactate-utilising yeasts which raise the pH above 8·0 (see, for example, references 69 and 70). The proteinases of the surface mould are important to texture development in the high-acid soft cheeses such as Brie and Camembert. They appear to diffuse into the cheese matrix and cause softening from the outside.[71] However, washed-curd cheeses with higher pH (e.g. Meshanger) seem to depend for texture development on $\alpha_{s1}$-casein breakdown by chymosin and on a relatively high pH and moisture content.[72] The production of amino acids and ammonia by the surface flora is regarded as an integral part of flavour development in all types of surface-ripened cheese, regardless of the mechanism of texture development.

### 2.4. Use of Exogenous Microbial Proteinases to Accelerate Cheese Ripening

The time taken for cheese to acquire certain degrees of flavour intensity was probably less important than the period over which it remained edible when it was the product of a cottage industry. However, cheese manufacture is now a capital-intensive industry which benefits from a high rate of turnover and the running costs and interest charges involved in cheese storage represent a significant proportion of the total cost of converting milk into cheese. Any shortening of the time cheese is kept in store therefore represents a worthwhile saving provided that flavour development can be accelerated without impairment of flavour balance. Raising the cheese-ripening temperature is the most obvious method but, while this may speed up flavour-forming reactions, it also speeds up off-flavour formation and may promote the growth of ınwanted microbial contaminants such as moulds. More selective methi ds are therefore required and many have been investigated on a laboratory scale; examples range from the use of exogenous enzymes to the modification of the lactic acid starter bacteria.[73] Proteolysis is considered by most investigators to be so important to cheese maturation that much of the effort towards achieving accelerated cheese ripening has centred around methods of increasing the rate of this process. Notable examples are the studies of Nakanishi and Itoh,[74,75] Kosikowski and Iwasaki[76] and Malkii *et al.*[77] The more recent results of Sood and Kosikowski[78] indicate that suitably screened food grade proteinases can halve normal American Cheddar ripening times without off-flavour formation. Trials at the author's Institute, in which food-grade proteinases were added to Cheddar cheese curds, showed that those enzymes whose pH optima are close to the pH of cheese (fungal acid proteinases) catalysed excessive proteolysis during maturation.[79] Natural bacterial proteinase (*B. subtilis*) produced less proteolysis in cheese and the casein breakdown pattern on PAGE resembled that of normal cheese (Fig. 2). All the proteinase-supplemented cheeses had weak body and crumbly texture, though this tendency was not excessive in cheese treated with only 0·001 % (w/w) neutral proteinase. Acid proteinases caused bitterness even at very low concentrations and did not significantly enhance the development of typical Cheddar cheese flavour. Neutral proteinase, on the other hand,

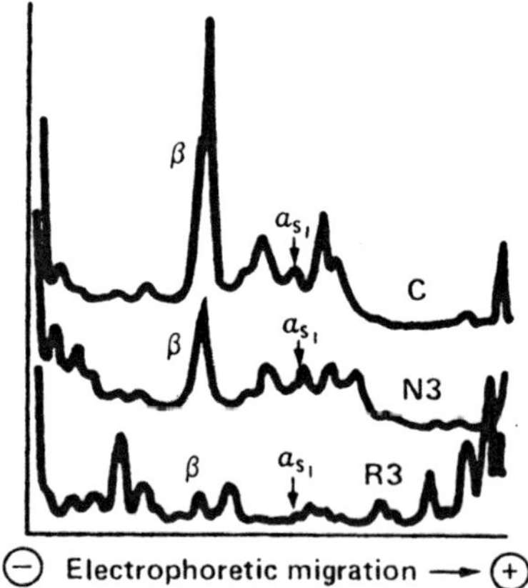


FIG. 2. Polyacrylamide gel electrophoretograms of 2 month-old cheeses treated with proteinases. The gels were scanned at $A_{590}$: C = untreated control; N3 = cheese treated with 2 units per gram bacterial neutral proteinase; R3 = cheese treated with 2 units per gram fungal acid proteinase.

could be added at levels such that flavour intensity was enhanced without the development of flavour defects (Table 5).

Many problems remain to be overcome if simple enzyme additions are to be applied commercially. For example, enzyme preparations are not yet available which can release flavour enhancing peptides and amino acids without excessive general proteolysis, leading to body defects. Greater

TABLE 5

MEAN FLAVOUR SCORES AND DEFECT SCORES FOR EXPERIMENTAL
CHEDDAR CHEESE TREATED WITH PROTEINASES

|     | Treatment | Cheddar intensity | Bitter defect | Rancid defect |
|-----|-----------|-------------------|---------------|---------------|
| (a) | N1        | 3·1               | 1·6           | 0·7           |
|     | N2        | 3·4               | 0·7           | 0·2           |
|     | N3        | 3·2               | 0·1           | 0·1           |
|     | R1        | 1·8               | 3·3           | 0·4           |
|     | R2        | 2·6               | 2·8           | 0·4           |
|     | R3        | 2·7               | 2·6           | 0·2           |
|     | C         | 2·2               | 0·1           | 0·2           |
| (b) | N2        | 4·6               | 1·0           | 0·5           |
|     | N3        | 4·8               | 0·2           | 0·3           |
|     | N2*       | 4·1               | 0·7           | 0·2           |
|     | N3*       | 4·2               | 0·1           | 0·2           |
|     | C         | 3·0               | 0·0           | 0·3           |

N1, N2, N3; 50, 10 or 2 units of neutral proteinase. R1, R2, R3; 50, 10 or 2 units of acid proteinase. C; untreated control. All neutral proteinase cheeses had mean Cheddar intensity scores which were significantly different ($P < 0.05$) from controls.
(a) Stored at 12 °C for 2 months.
(b) Stored at 12 °C for 4 months.
* Stored at 12 °C for 2 months, then 6 °C for a further 2 months.

exopeptidase activity in relation to endopeptidase may solve this problem. Also, it may be necessary to stop the accelerated ripening process if the demand for mature cheese falls; enzymes whose activity declines during maturation and/or at low temperatures will be useful for this purpose. Enzyme addition remains one of the biggest difficulties to be overcome. Proteinases added to cheese milk cause weakening of the coagulum and curd, large soluble N losses in whey and the whey itself is rendered unsuitable for many food applications unless the proteinases can be easily inactivated by heating. On the other hand, powdered enzymes added to cheese curds are difficult to distribute evenly and may present a health hazard to creamery personnel. All of these problems could be overcome if the proteinases could be encapsulated in such a way that they remained separated from milk proteins during cheese manufacture, became entrapped (like bacteria) in the forming curd and then were released in the cheese. Magee and Olson[80,81] described the encapsulation of carbohydrate-metabolising enzymes and their substrates in milk fat. However, their system has limited application since the enzymes are released when the fat melts ($>35\,^{\circ}$C) and most hard and semi-hard cheeses require a higher manufacturing temperature. Hardened fat may cause the opposite problem; the enzymes would be difficult to release into the cheese, a necessary step in 'delivering' proteinases in such a way that they can act on their substrates. Phospholipid liposomes or gelatin capsules may be more appropriate for proteinases since these compounds can be degraded in cheese to allow enzyme release.

## 3.  DETRIMENTAL EFFECTS OF MICROBIAL PROTEINASES IN MILK AND MILK PRODUCTS

It is now well established experimentally that degradative enzymes produced by Gram-negative psychrotrophic bacteria in raw milk have the potential to adversely affect the processing properties of the milk and to impair the quality of products. However, the extent to which this spoilage potential is realised in practice is not yet clear. Psychrotrophs and their enzymes should present no problems provided that milk producers and manufacturers are able to ensure that raw milk supplies are kept at a temperature at which the psychrotrophs multiply only very slowly ($4\,^{\circ}$C or lower). However, the cooled, bulk raw milk handling system can be abused in several ways that are likely to permit a build up of relatively high psychrotroph populations ($>10^6\,\text{ml}^{-1}$) in silo milk prior to heat treatment. These factors have been reviewed recently[82] and are outside the scope of the present discussion, but their existence serves to emphasise the gap between our knowledge of what can happen in the laboratory and what actually happens in commercial practice. Hopefully, the *Survey of Relationship of Raw Milk Hygienic Quality to Product Quality* (organised by the UK Dairy Trade Federation) will yield sufficient information to bridge this gap and point the way to further fundamental research. Unfortunately, there were no rapid sensitive assay methods available for psychrotroph enzyme for use in the survey; until such methods are developed, standards cannot be established for permissible enzyme levels in milk for manufacturing and correlations between milk and product quality will continue to be based on viable bacterial counts.

A wide range of psychrotrophic bacteria are proteolytic, although the *Pseudomonas* is the most commonly cited genus.[83,84] Because their enzymes are secreted outside the cell, they can act directly on milk proteins during raw milk storage. The extent of proteolysis is undetectable by simple, direct methods (e.g. soluble N), unless the number of psychrotrophs exceeds $10^8$ ml$^{-1}$. Counts in stored raw market milks rarely exceed $10^7$ ml$^{-1}$, but the effects of proteinases from this level of contamination can be seen on casein fractions separated by PAGE or starch gel electrophoresis (SGE). For example, Law *et al.*[85] showed that a strain of *Ps. fluorescens* growing to approximately $10^7$ ml$^{-1}$ in milk produced sufficient proteinase to degrade $\beta$- and $\kappa$-casein during 3 days' storage of raw milk at 7 °C. These changes were not detected as soluble N fractions but were clearly seen after SGE of milk samples in which the most notable feature was the appearance of *para-$\kappa$*-casein (Fig. 3), suggesting the presence of

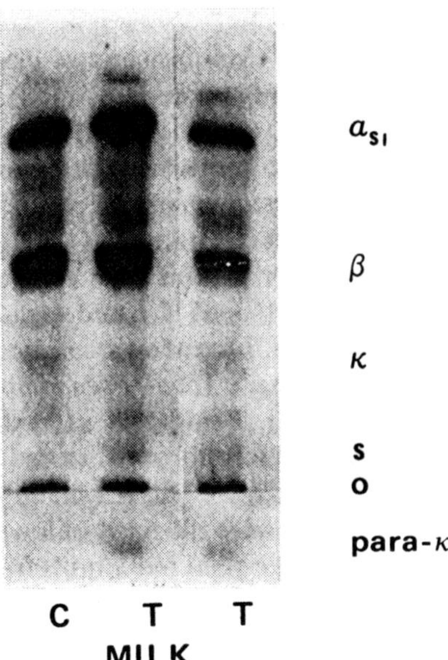


FIG. 3. Starch gel electrophoretogram of raw milk inoculated with a proteolytic strain of *Pseudomonas fluorescens* and stored for 3 days at 7 °C. C = Uninoculated control. T = Inoculated milk containing approximately $10^7$ bacteria per millilitre.

milk-clotting activity analogous to that of rennet. When milk, in which the *Pseudomonas* had grown, was subsequently UHT-sterilised and stored at 20 °C, those samples that had contained $>10^7$ and $8 \times 10^6 \, ml^{-1}$ *Pseudomonas* before sterilisation gelled after 10–14 days and 8–10 weeks, respectively (Table 6). The $\beta$- and $\kappa$-caseins were extensively degraded in gelled milks and some loss of $\alpha_{s1}$-casein was seen, but whey proteins seemed

TABLE 6

GELATION OF UHT-STERILISED MILK BY PROTEINASES FROM INCREASING NUMBERS OF PSYCHROTROPHS IN RAW MILK

| Numbers of Ps. fluorescens in stored raw milk (cfu/ml) | Time taken for gelation (Days) | UHT milk % casein breakdown at gelation time | | |
|---|---|---|---|---|
| | | $\alpha_{s1}$ | $\beta$ | $\kappa$ |
| $8 \times 10^5$ | $>56$ | 0 | 0 | Trace |
| $8 \times 10^6$ | 56 | 0 | 30 | 50 |
| $5 \times 10^7$ | 12 | 22 | 78 | 100 |

resistant to proteinase attack. Comparisons of electron micrographs of gelled and liquid UHT milks (Fig. 4) demonstrated aggregation of casein micelles in the gelled milks (cf. rennet action).

Adams *et al.*[86] reported that breakdown of casein fractions and whey proteins in whole milk by *Pseudomonas* proteinase rendered the milk unstable to heat treatment and it coagulated at the time of UHT treatment. Cousin and Marth[87] also observed this type of coagulation, although it occurred at lower temperatures (63–85 °C), and suggested that it was due to destabilisation of micelles by breakdown of both $\alpha_{s1}$- and $\beta$-casein under the influence of proteinase from *Pseudomonas, Flavobacterium, Micrococcus* and *Lactobacillus*. Feuillat *et al.*[88] demonstrated that the unidentified psychrotrophic microflora of stored raw milk produced peptides of 5000–20 000 molecular weight, the smallest of which were lost in the whey on cheesemaking and were the basis of lower yields of soft cheese made from stored milks. Average yield reductions from 14·88 % to 14·48 % were also required in cottage cheese manufactured with stored milks containing $>10^6 \, ml^{-1}$ *Ps. fluorescens* and curd structure was poor.[89] These changes were attributed to the proteolytic action of the *Pseudomonads* during storage of the milk. However, despite the apparent adverse effects of psychrotrophic proteinases in cheese manufacture in the

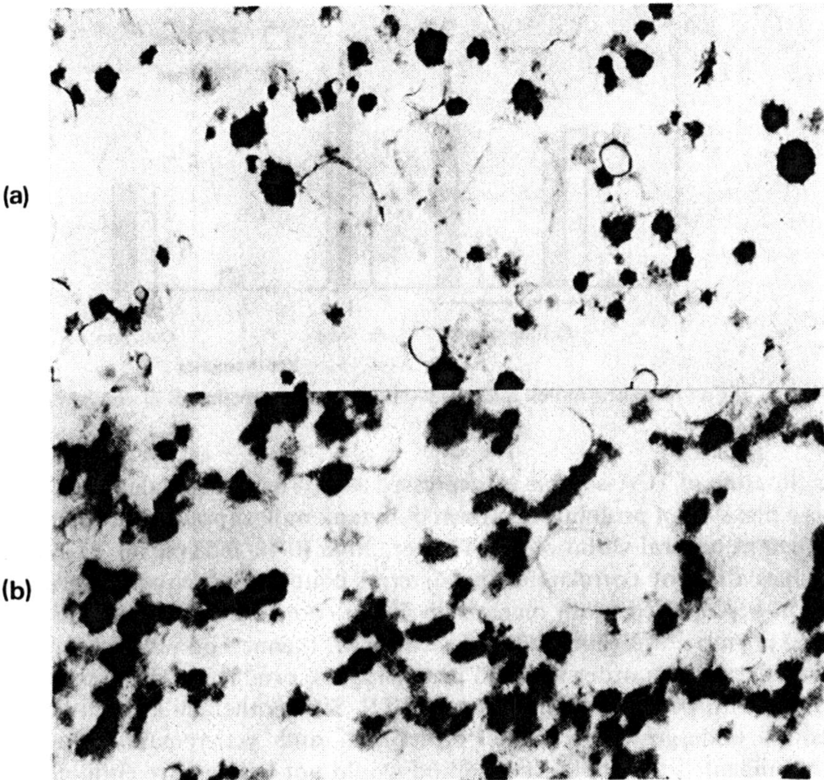



FIG. 4.   Electron micrographs of casein micelles in fresh UHT-sterilised milk (a) and in UHT-sterilised milk which has been gelled by the action of psychotroph proteinase (b).

vat, there have been no reports of flavour defects in cheese specifically associated with these enzymes. The proteinases are generally resistant to both pasteurisation and UHT treatments, although the degree of heat resistance between isolates differs (Fig. 5). Adams *et al.*[90] calculated that the proteinase of *Pseudomonas* sp. MC60 was about 4000-fold more heat resistant than the spores of *Bacillus stereothermophilus*. D-values (time required to reduce enzyme activity by 90 %) at 140 °C and 150 °C for the *Ps. fluorescens* AR11 proteinase (used in experiments on UHT-sterilised milks at the author's Institute) were approximately 60 s and 30 s, respectively.[91]

Until recently there has been no satisfactory published method available for the rapid and accurate determination of psychrotroph proteinases produced by populations of $<10^8$ ml in milk. Juffs[92] examined in detail the

                                    *B. A. Law*

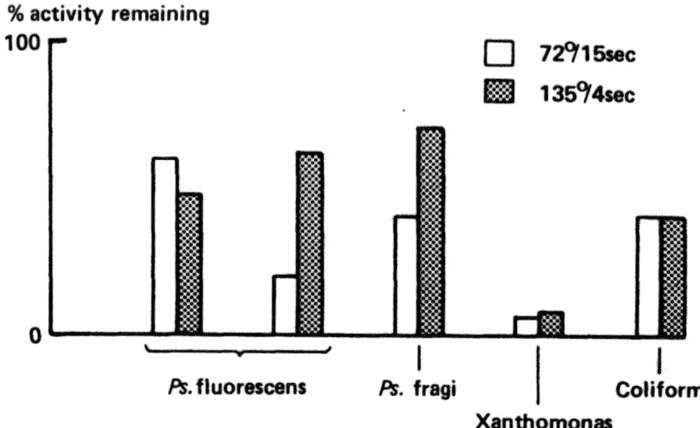



FIG. 5.    Effect of pasteurisation and UHT-sterilisation on proteinases of several raw milk psychrotrophs.

estimation of TCA-soluble N (expressed as a tyrosine equivalent value, TE) as a measure of proteolysis in farm bulk tank milk supplies, but found that inherent natural variations in TE were high ($0.40$–$0.58$ mg ml$^{-1}$) and the values did not correlate with bacterial counts. Subsequent storage of farm supplies led to an increase in TE in some cases but a threshold of $0.55$ mg ml$^{-1}$ was thought to be necessary if the method was to be applied commercially in order to avoid penalising the producers whose cows gave milks with particularly high natural TE. Since other milks with low TE could undergo considerable proteolysis and yet remain below the $0.55$ mg ml$^{-1}$ threshold, the method would not be sensitive enough to be useful. Preincubation of the milk samples for 20 h at 25 or 30 °C increased the sensitivity as a measure of TE above a universal threshold, but non-psychrotrophic bacteria made a major contribution to the increased proteolysis.[93] Juffs[94] later showed that TE, organoleptic quality and bacterial counts were related only in laboratory stored raw milk kept for 7–8 days at 5 °C and having viable bacterial counts of $10^7$–$10^8$ ml$^{-1}$. Post-pasteurisation proteolytic psychrotrophic contaminants were more clearly related to increased TE in stored pasteurised milks and the method may have an application in monitoring deterioration in this situation. The use of PAGE and SGE to detect proteolysis in milk has already been discussed in relation to reports of the effect of proteinases on milk quality; Juffs[95] examined SGE as a method of screening milks for proteolysis and although it has the advantage over the TE method that it specifically detects $\kappa$-casein breakdown and avoids problems of natural variations, only counts of $10^7$–$10^8$ bacteria per millilitre are sufficient to produce discernible proteolysis. Also, the method does not give quantitative results and may be too specialised for most commercial creamery laboratories.

The problem of inherently variable basal proteolysis levels can be overcome by the use of an exogenous artificial proteinase substrate added to the milk and incubated for a standard time. Cliffe and Law[96] have developed such an assay based on the ability of psychrotroph proteinases to release blue dye-labelled soluble peptides from an insoluble protein, Hide Powder Azure (HPA). The presence of proteinase activity from as few as $1.5 \times 10^6\,\text{ml}^{-1}$ psychrotrophic Gram negative rods can be detected by the formation of a blue colour in milk after incubation with this substrate. The colour intensity can be measured spectrophotometrically after the milk has been cleared by ether extraction and centrifugation. The method can be used to distinguish between raw milks which are likely to produce UHT milk with long or short storage lives.

# REFERENCES


1. LAW, B. A. in *Economic Microbiology*, Vol. 7. A. H. Rose (ed.), Academic Press, New York, 1981, (In press).
2. MILLS, O. E. and THOMAS, T. D. *New Zealand Journal of Dairy Science and Technology*, 1981, **16**, 43.
3. THOMAS, T. D., JARVIS, B. D. W. and SKIPPER, N. A. *J. Bacteriol.* 1974, **118**, 329.
4. EXTERKATE, F. A. *Netherlands Milk and Dairy Journal*, 1975, **29**, 303.
5. EXTERKATE, F. A. *Netherlands Milk and Dairy Journal*, 1976, **30**, 3.
6. EXTERKATE, F. A. *Archives of Microbiology*, 1979, **120**, 247.
7. MILLS, O. E. and THOMAS, T. D. *New Zealand Journal of Dairy Science and Technology*, 1978, **13**, 209.
8. EFSTATHIOU, J. D. and McKAY, L. L. *Appl. and Environ. Microbiol.* 1976, **32**, 38.
9. LAW, B. A., SEZGIN, E. and SHARPE, M. E. *J. Dairy Res.* 1976, **43**, 291.
10. LAW, B. A. *J. Dairy Res.* 1977, **44**, 309.
11. LAW, B. A. *J. Gen. Microbiol.* 1978, **105**, 113.
12. RICE, G. H., STEWART, F. H. C., HILLIER, A. J. and JAGO, G. R. *J. Dairy Res.* 1978, **45**, 93.
13. PAYNE, J. W. in *Peptide Transport in Protein Nutrition*, D. M. Matthews and J. W. Payne, eds., North Holland and American Elsevier, Amsterdam, Oxford, New York, 1975, p. 283.
14. PAYNE, J. W. in *Micro-organisms and Nitrogen Sources*, J. W. Payne, ed., John Wiley & Sons, Chichester, 1980, p. 211.
15. EXTERKATE, F. A. *Netherlands Milk and Dairy Journal*, 1981, **35**, 328.
16. LAW, B. A. *Journal of Applied Bacteriology*, 1979, **46**, 455.
17. MOU, L., SULLIVAN, J. J. and JAGO, G. R. *J. Dairy Res.* 1975, **42**, 147.

18. ARGYLE, P. J., MATHISON, G. E. and CHANDAN, R. C. *Journal of Applied Bacteriology*, 1976, **41**, 175.
19. SHANKAR, P. A. Thesis, University of Reading, 1977, pp. 58–62.
20. RABIER, D. and DESMAZEAUD, M. J. *Biochimie*, 1973, **55**, 389.
21. DESMAZEAUD, M. J. and JUGE, M. *Lait*, 1976, **56**, 241.
22. EL-SODA, M., DESMAZEAUD, M. J. and BERGERE, J.-L. *J. Dairy Res.* 1978, **45**, 445.
23. KIHARA, H. and SNELL, E. E. *Journal of Biological Chemistry*, 1960, **235**, 1409.
24. LEACH, F. R. and SNELL, E. E. *Journal of Biological Chemistry*, 1960, **235**, |3523.
25. AUNSTRUP, K. in *Economic Microbiology*, Vol. 5, A. H. Rose, ed., Academic Press, London, 1980, p. 50.
26. GREEN, M. L. *J. Dairy Res.* 1977, **44**, 159.
27. KONING, P. J. DE. *International Dairy Federation Bulletin*, 1980, **126**, 11.
28. MARTENS, R. and NAUDTS, M. *International Dairy Federation Bulletin*, 1976, **91**, 1.
29. MARTENS, R. and NAUDTS, M. *International Dairy Federation Bulletin*, 1978, **108**, 51.
30. PHELAN, J. *Dairy Industries International*, 1977, **42**, 50.
31. DESMAZEAUD, M. J. and GRIPON, J.-C. *Milchwissenschaft*, 1977, **32**, 731.
32. VISSER, F. M. W. *Netherlands Milk and Dairy Journal*, 1977, **31**, 210.
33. O'KEEFE, R. B., FOX, P. F. and DALY, C. *J. Dairy Res.*, 1976, **43**, 97.
34. VISSER, F. M. W. and DE GROOT-MOSTERT, H. A. E. *Netherlands Milk and Dairy Journal*, 1977, **31**, 247.
35. GRIPON, J.-C., DESMAZEAUD, M. J., LE BARS, D. and BERGERE, J.-L. *Lait*, 1975, **55**, 502.
36. HILL, R. D., LEHAV, E. and GIVOL, D. *J. Dairy Res.* 1974, **41**, 147.
37. CREAMER, L. K. and RICHARDSON, B. C. *New Zealand Journal of Dairy Science and Technology*, 1974, **9**, 9.
38. DESMAZEAUD, M. J. and ZEVACO, C. *Annales de Biologie animal Biochimie et Biophysique*, 1976, **16**, 851.
39. ZEVACO, C. and DESMAZEAUD, M. J. *J. Dairy Sci.* 1980, **63**, 15.
40. REITER, B., SOROKIN, Y., PICKERING, A. and HALL, A. J. *J. Dairy Res.* 1969, **36**, 65.
41. MILLS, O. E. and THOMAS, T. D. *New Zealand Journal of Dairy Science and Technology*, 1980, **15**, 131.
42. LAW, B. A., SHARPE, M. E. and REITER, B. *J. Dairy Res.*, 1974, **41**, 137.
43. SORHAUG, T. and SOLBERG, P. *Applied Microbiology*, 1973, **25**, 388.
44. DESMAZEAUD, M. J. and VASSAL, L. *Lait*, 1979, **59**, 327.
45. CLIFFE, A. J. and LAW, B. A. *Journal of Applied Bacteriology*, 1979, **47**, 65.
46. HARWALKER, V. R. *J. Dairy Sci.* 1972, **55**, 735.
47. MATOBA, T. and HATA, T. *Agr. Biol. Chem.*, 1972, **36**, 1423.
48. RICHARDSON, B. C. and CREAMER, L. K. *New Zealand Journal of Dairy Science and Technology*, 1973, **8**, 46.
49. CZULAK, J. *Australian Journal of Dairy Technology*, 1959, **14**, 177.
50. LOWRIE, R. J. and LAWRENCE, R. C. *New Zealand Journal of Dairy Science and Technology*, 1972, **7**, 51.
51. LOWRIE, R. J., LAWRENCE, R. C. and PEBERDY, H. F. *New Zealand Journal of Dairy Science and Technology*, 1974, **9**, 116.

52. STADHOUDERS, J. and HUP, G. *Netherlands Milk and Dairy Journal*, 1975, **29**, 335.
53. CHIBA, Y. and SATO, Y. *Japanese Journal of Dairy and Food Science*, 1980, **29**, 161.
54. TOURNEUR, C. *Lait*, 1972, **52**, 149.
55. EL-SODA, M., DESMAZEAUD, M. J., ABOUDONIA, S. and KAMAL, N. *Milchwissenschaft*, 1981, **36**, 140.
56. LANGSRUD, T., REINBOLD, G. W. and HAMMOND, E. G. *J. Dairy Sci.* 1977, **60**, 16.
57. LANGSRUD, T., REINBOLD, G. W. and HAMMOND, E. G. *J. Dairy Sci.* 1978, **61**, 303.
58. ADES, G. L. and CONE, J. F. *J. Dairy Sci.* 1969, **52**, 957.
59. FREIDMAN, M. E., NELSON, W. O. and WOOD, W. A. *J. Dairy Sci.* 1953, **36**, 1124.
60. FOISSY, H. *Milchwissenschaft*, 1978, **33**, 221.
61. TORGERSEN, H. and SORHAUG, T. *FEMS Letts*, 1978, **4**, 151.
62. SORHAUG, T. *Milchwissenschaft*, 1981, **36**, 137.
63. SHARPE, M. E., LAW, B. A., PHILLIPS, B. A. and PITCHER, D. G. *J. Gen. Microbiol.* 1977, **101**, 345.
64. CUER, A., DAUPHIN, G., KERGOMARD, A., DUMONT, J.-P. and ADDA, J. *Agr. Biol. Chem.* 1979, **43**, 1783.
65. DUMONT, J.-P. and ADDA, J. *J. Ag. Food Chem.* 1978, **26**, 364.
66. DUMONT, J.-P., ROGER, S. and ADDA, J. *Lait*, 1976, **56**, 595.
67. KINSELLA, J. E. and HWANG, D. *Biotechnology and Bioengineering*, 1976, **18**, 927.
68. GRIPON, J.-C. and DEBEST, B. *Lait*, 1976, **55**, 735.
69. SCHMIDT, J. L. and LENOIR, J. *Lait*, 1980, **60**, 272.
70. SCHMIDT, J. L. and LENOIR, J. *Lait*, 1980, **60**, 343.
71. LENOIR, J. *Revue Laitiere Francaise*, 1970, **275**, 231.
72. NOOMEN, A. *Netherlands Milk and Dairy Journal*, 1977, **31**, 75.
73. LAW, B. A. *Dairy Industries International*, 1980, **45**, 15.
74. NAKANISHI, T. and ITOH, M. *Japanese Journal of Dairy Science*, 1974, **23**, A121.
75. NAKANISHI, T. and ITOH, M. *Japanese Journal of Dairy Science*, 1973, **22**, A110.
76. KOSIKOWSKI, F. V. and IWASAKI, T. *J. Dairy Sci.* 1975, **58**, 963.
77. MALKKI, Y., ROUHIANEN, L., MATTSON, R. and MARKKONEN, P. United States Patent, 1977, 4,062,730.
78. SOOD, V. K. and KOSIKOWSKI, F. V. *J. Dairy Sci.* 1979, **62**, 1865.
79. LAW, B. A. and WIGMORE, A. S. *J. Dairy Res.* 1982, **49**, 137–146.
80. MAGEE, E. L. and OLSON, N. F. *J. Dairy Sci.* 1981, **64**, 600.
81. MAGEE' E. L. and OLSON, N. F. *J. Dairy Sci.* 1981, **64**, 616.
82. ROBERTS, A. W. *Journal of the Society of Dairy Technology*, 1979, **32**, 24.
83. LAW, B. A. *J. Dairy Res.* 1979, **46**, 573.
84. LAW, B. A., COUSINS, C. M., SHARPE, M. E. and DAVIES, F. L. in *Cold Tolerant Microbes in Spoilage and the Environment*, SAB Technical Series, Vol. 13, 1979, A. D. Russel and R. Fuller, eds., Academic Press, London, p. 137.
85. LAW, B. A., ANDREWS, A. T. and SHARPE, M. E. *J. Dairy Res.* 1977, **44**, 145.

86. ADAMS, D. M., BARACH, J. T. and SPECK, M. L. *J. Dairy Sci.* 1976, **59**, 823.
87. COUSIN, M. A. and MARTH, E. H. *J. Dairy Sci.* 1977, **60**, 1042.
88. FEUILLAT, M., LE GUENNEC, S. and OLSSON, A. *Lait,* 1976, **56**, 521.
89. MOHAMMED, F. O. and BASSETTE, R. *J. Dairy Sci.* 1979, **62**, 222.
90. ADAMS, D. F., BARACH, J. T. and SPECK, M. L. *J. Dairy Sci.* 1975, **58**, 828.
91. ALICHANIDIS, E. and ANDREWS, A. T. *Biochim. Biophys. Acta.,* 1977, **41**, 175.
92. JUFFS, H. S. *J. Dairy Res.* 1973, **40**, 371.
93. JUFFS, H. S. *J. Dairy Res.* 1973, **40**, 383.
94. JUFFS, H. S. *J. Dairy Res.* 1975, **42**, 31.
95. JUFFS, J. S. *J. Dairy Res.* 1975, **42**, 277.
96. CLIFFE, A. J. and LAW, B. A. *J. Dairy Res.* 1982, **49**, (In press.)

# 17

## Meat Proteolysis

THAYNE R. DUTSON

*Department of Animal Science,*
*Texas A & M University, Texas, USA*

### INTRODUCTION

Proteolysis in post-mortem muscle has long been considered to be one of
the possible mechanisms by which post-mortem tenderisation occurs.
However, direct evidence for specific proteolytic cleavage of muscle
proteins during post-mortem tenderisation has been elusive.[1] Indirect
evidence for the proteolytic breakdown of muscle *post mortem* has been
obtained by various authors in the form of increased free amino acids and
non-protein nitrogen.[2,3] Also, Sharp[4] demonstrated a continuous breakdown of protein and muscle tissue in aseptic rabbit and bovine muscle
during storage at 37 °C. However, the increases in free amino acids or non-
protein nitrogen are not related to the post-mortem increase in tenderness
and the amino acid increase is observed after most of the post-mortem
improvement in tenderness has occurred.[1,3] Thus, it was concluded[1,3] that
proteolysis, as measured by classical methods, is not related to changes in
tenderness.

Although classical indices of proteolysis may not be related directly to
post-mortem changes in tenderness, specific alterations in myofibrillar
proteins (before gross proteolysis has taken place) may occur at the time
post-mortem tenderisation is taking place in meat. Also, there is evidence to
indicate that, in an absence of differences in muscle shortening, higher
temperatures and lowered pH conditions produce increased tenderisation
of muscle.[5,6] Research by Goll *et al.*[7] indicated that the action of trypsin on
muscle tissue simulates the changes which occur in post-mortem muscle
during the period when tenderness increases. Research identifying two of

the enzyme systems which could most likely cause proteolysis in post-mortem muscle—the lysosomal enzyme system[8-12] and the calcium-activated factor system[13-15]—indicates that specific post-mortem proteolysis may be responsible for tenderness changes which occur in meat.


## PROTEOLYTIC ENZYME SYSTEMS IN MUSCLE

Two enzyme systems have been described in muscle tissue which have the ability to degrade specific muscle proteins, both *in vivo* and in post-mortem muscle. One of these systems operates at a neutral pH (neutral proteinases), the other at acidic pH values (acidic proteinases or cathepsins). These enzyme systems have recently been reviewed.[16-18]

### Neutral Proteinases

The major enzyme considered to be responsible for proteolytic breakdown in muscle at neutral pH values is the calcium-activated proteinase (CAF) which has been isolated and purified by Dayton *et al.*[19,20] This enzyme has been shown to cause partial degradation of myofibrils by removing $\alpha$-actinin from the Z-line and hydrolysing troponin-T, troponin-I, C-protein, tropomyosin, connectin and desmin.[21,22] Dayton and Schollmeyer[23] have demonstrated that CAF is localised only at the Z-line and at the sarcolemma.

The early work on CAF[20] involved an enzyme which had a requirement for calcium at concentrations ($> 1 \cdot 0$ mM) much greater than that found in muscle. However, more recently, a form of this enzyme, active at calcium concentrations in the range found in muscle cells ($\geqslant 5 \mu$M), has been discovered.[18,24] It appears that this proteinase is a low-calcium requiring form of CAF since this low-calcium requiring proteinase gives identical banding patterns on sodium dodecyl sulphate polyacrylamide slab gels as does the CAF enzyme, and an antibody of the 80 K dalton subunit of the CAF cross-reacts strongly with the purified low-calcium requiring proteinase.[24]

Indications are that this low-calcium form of CAF may be involved in the regulation of the activity of the proteinase. Activity of the proteinase may be reduced in living muscle by converting the proteinase to the high-calcium requiring form, thereby rendering it inactive in the cell where calcium concentrations are low. Activation of the enzyme system may occur by converting the high-calcium requiring form back to the low-calcium requiring form, enabling the enzyme to be active at cellular pH values and calcium concentrations.[24] Another possible factor for controlling the activity of the CAF enzyme system may be an endogenous inhibitor which has been isolated from skeletal muscle.[25,26] Preliminary localisation studies indicate that the inhibitor is also found in the region of the Z-line, which suggests that the inhibitor may function to control CAF activity *in vivo*.[24]

### Lysosomal Acid Proteinases

Lysosomes are small cytoplasmic organelles which contain many acid hydrolases and are surrounded by a membrane.[27] The activity of lysosomal hydrolytic enzymes has been shown to be increased by many treatments, including low pH and high temperature.[28,29] Low pH values and high temperatures destabilise the lysosomal membrane and allow the lysosomal enzymes to be accessible to components in the cytoplasm of cells.[28,29]

Bird[30] has shown that there are two groups of lysosomes in muscle tissue: one belongs to the muscle cells themselves while the other belongs to phagocytic cells within the muscle tissue. The lysosomes originating from muscle cells contain approximately 75% of the $\beta$-glucuronidase activity and 95% of the cathepsin and acid phosphatase activity found in muscle tissue. This refutes reports by Tappel[31] and Kohn[32] that the lysosomal enzymes found in muscle tissues are contributed by non-muscle cells of muscle tissues and that muscle cells themselves contain few or no lysozymes.

The major lysosomal cathepsins reported to be present in muscle cells are: cathepsins B and D;[33-35] cathepsin L;[17,36,37] and cathepsin H.[17] These enzymes have been shown to hydrolyse various muscle proteins: cathepsin B is active on myosin, actin, $\alpha$-actinin, troponin, tropomyosin, and collagen; cathepsin D is active on myosin, actin, myosin light-chain, troponin, tropomyosin, and collagen; cathepsin L is active on myosin, $\alpha$-actinin and actin; cathepsin H has very highly specific activity against myosin.

## POST-MORTEM MUSCLE PROTEOLYSIS AND MEAT TENDERNESS

Since the discovery by Locker[38] that meat tenderness depends greatly on the amount of muscle contraction and sarcomere shortening, many researchers have studied the relationship of sarcomere shortening to meat tenderness.[39-48] These researchers have confirmed a definite relationship between muscle shortening during onset of rigor mortis and the tenderness of meat.

Post-mortem temperature during the onset of rigor mortis has been shown to have a definite effect on meat tenderness, primarily related to the effect of cold on the shortening of muscle, i.e. cold-shortening.[3,45,49−59] However, there appears to be an additional factor related to post-mortem tenderisation of muscle at higher temperatures which is not associated with shortening of sarcomeres.[6] A study by Locker and Daines[5] has shown that, by altering the temperature of *Sternomandibularis* muscles in the final stages of rigor, markedly different degrees of tenderness can be produced, without a concomitant difference in muscle shortening. An experiment conducted in the author's laboratory has shown that, by holding carcasses at elevated temperatures during the first twelve hours *post mortem*, differences in tenderness can be achieved, even though no differences in muscle shortening occur.[6] In this experiment, both sides of each carcass were 'tenderstretched' to prevent any differences in muscle shortening due to the high temperature conditioning treatment. One side of each carcass was placed in a 1 °C cooler; the other side was conditioned at 22 °C for 4 h and at 16 °C for an additional 8 h. When samples were removed from these carcasses at 48 h *post mortem*, there was no difference in sarcomere length between the high temperature-conditioned side and the side chilled at 1 °C. There was, however, a significant difference, both in overall tenderness (as measured by a sensory panel), and shear force (as measured by a Warner–Bratzler shear device), with those sides held at the high temperatures being more tender.

Electrical stimulation is another treatment which improves the tenderness of meat and indicates that some mechanism other than muscle shortening influences tenderness. About 90 % of electrical stimulation treatments increase tenderness of meat without changing sarcomere length.[60] Thus, some mechanism other than the increase in sarcomere length is operative in tenderisation of muscle in a majority of electrical stimulation treatments.

Since high temperature conditioning of beef carcasses and electrical stimulation produce tenderness in the absence of differences in muscle shortening, it is possible that tenderness induced by these treatments could be due to autolytic proteolysis within the muscle, as both of these treatments decrease pH at a time when muscle temperature is elevated. Elevated temperatures and lowered pH values have been shown to effect the rupture of lysosomes and the release of lysosomal enzymes into the cytoplasm.[27,28] Experiments in the author's laboratory have shown that both high temperature conditioning[11,12] and electrical stimulation[61] increase the free activity of β-glucuronidase and cathepsin C (both lysosomal enzymes). Both increased fragmentation values which have been shown to be closely related to meat tenderness.[14,62] Thus, it appears that the tenderness produced by treatments involving low pH at higher muscle temperatures coincides with an increase in the free activity of lysosomal enzymes.

Parrish[13] and Olsen *et al.*[15] have indicated that the appearance of a 30 K dalton subunit in SDS gels of myofibrillar proteins of post-mortem, aged muscle accounts for up to 50 % of the increase in tenderness of *Longissimus dorsi* muscle. Other authors[63,64] have shown that a 30 K dalton component also appeared during post-mortem storage of chicken muscle. Parrish[13] and Olsen *et al.*[15] have postulated that, since CAF produced a 30 K dalton component by hydrolysis of troponin-T, it is the enzyme system responsible for increased tenderness during post-mortem ageing. However, Penny,[65] who also reported the appearance of a 30 K dalton component in bovine muscle during post-mortem ageing, was unable to show that CAF was the factor responsible for its production. Subsequently, Penny and Ferguson-Pryce[66] concluded that, although it is likely that CAF can degrade troponin-T, the major contribution to proteolysis of troponin-T during post-mortem ageing comes from the lysosomal enzyme system, since this system contains cathepsin B which is active in the acid pH range. Penny and Dransfield[67] also studied the relationship between toughness and the disappearance of troponin-T in muscle and concluded that the loss of troponin-T is a good indicator of the rate and extent of proteolysis or conditioning in beef. Using myofibrillar ATPase activities, Qualli and Valin[68] concluded that both CAF and a lysosomal proteinase could be involved in the ageing process.

High temperature conditioning has been shown to increase the tenderness of meat, increase fragmentation values, increase the rate of post-mortem pH decline, and to cause a more rapid release of lysosomal enzymes.[6,11,12,69] In view of these observations, experiments were conducted in the author's laboratory to determine if incubation at high temperatures and at different pH values would produce alterations in the myofibrillar proteins. This study was also designed to determine whether alterations in myofibrillar protein were greater when muscle was incubated at neutral pH (where CAF activity should be greater), or at acidic pH values (where lysosomal enzyme activity should be promoted). Results of this study indicate that high temperature-conditioned muscle samples exhibit a more rapid degradation of myosin heavy-chains with production of protein fragments at 50 K dalton, 95 K dalton, 135 K dalton and 150 K dalton and a more rapid degradation of troponin-T with production of a 30 K dalton fragment compared to the muscle samples maintained at normal coldroom temperatures.[70] Similar degradation patterns of myofibrillar proteins were observed in this study when muscle homogenates were incubated at pH 5·4. Little or no degradation of myosin heavy-chains and less degradation of troponin-T with production of a 30 K dalton fragment, was observed when muscle was incubated at pH 7·0 (optimum for CAF activity). Muscle fragmentation values were also greater for samples incubated at pH 5·4 than at pH 7·0.

Thus, when conditions in the muscle are favourable for promotion of lysosomal proteinase activity, greater alteration in myofibrillar proteins occurs than when conditions within the muscle are favourable for CAF activity. These data indicate that, although CAF causes some alterations in myofibrillar proteins, greater alterations are caused by lysosomal proteinases with concomitant alterations in muscle fragmentation and meat tenderness.

## REFERENCES


1. PARRISH, F. C., JR., *Proc. Recip. Meat Conf.*, 1971, **24**, 97.
2. DAVEY, C. L. and GILBERT, K. V., *J. Fd Sci.*, 1966, **31**, 135.
3. PARRISH, F. C., JR., GOLL, D. E., NEWCOMB, W. J., II, DE LUMEH, B. O., CHAUDHRY, H. M. and KLINE, A. E., *J. Fd Sci.*, 1969, **34**, 196.
4. SHARP, J. G., *J. Sci. Fd Agric.*, 1963, **14**, 468.
5. LOCKER, R. H. and DAINES, G. J., *J. Sci. Fd Agric.*, 1976, **27**, 193.
6. DUTSON, T. R., YATES, L. D., SMITH, G. C., CARPENTER, Z. L. and HOSTETLER, R. L., *Proc. Recip. Meat Conf.*, 1977, **30**, 79.
7. GOLL, D. E., STROMER, M. H., ROBSON, R. M., TEMPLE, J., EASON, B. A. and BUSCH, W. A., *J. Anim. Sci.*, 1971, **33**, 963.
8. ONO, K., *J. Fd Sci.*, 1970, **35**, 256.
9. ONO, K., *J. Fd Sci.*, 1971, **36**, 838.
10. DUTSON, T. R. and LAWRIE, R. A., *J. Fd Technol.*, 1974, **9**, 43.
11. MOELLER, P. W., FIELDS, P. A., DUTSON, T. R., LANDMANN, W. A. and CARPENTER, Z. L., *J. Fd Sci.*, 1976, **41**, 216.
12. MOELLER, P. W., FIELDS, P. A., DUTSON, T. T., LANDMANN, W. A. and CARPENTER, Z. L., *J. Fd Sci.*, 1977, **42**, 510.
13. PARRISH, F. C., JR., *Proc. Recip. Meat Conf.*, 1977, **30**, 87.
14. OLSEN, D. G. and PARRISH, F. C., JR., *J. Fd Sci.*, 1977, **42**, 506.
15. OLSEN, D. G., PARRISH, F. C., JR., DAYTON, W. R. and GOLL, D. E., *J. Fd Sci.*, 1977, **42**, 117.
16. DAYTON, W. R., LEPLEY, R. A. and SCHOLLMEYER, J. V., *Proc. Recip. Meat Conf.*, 1981, **34** (in press).
17. BIRD, J. W. C. and CARTER, J. H. in *Degradative Processes in Heart and Skeletal Muscle*, K. Wildenthal (ed.), Elsevier-North Holland Publishing Company, New York, 1981.

18. GOLL, D. E., *Proc. Recip. Meat Conf.*, 1981, **34** (in press).
19. DAYTON, W. R., GOLL, D. E., ZEECE, M. G., ROBSON, R. M. and REVILLE, W. J., *Biochemistry*, 1976, **15**, 2150.
20. DAYTON, W. R., REVILLE, W. J., GOLL, D. E. and STROMER, M. H., *Biochemistry*, 1976, **15**, 2159.
21. DAYTON, W. R., GOLL, D. E., STROMER, M. H., REVILLE, W. J., ZEECE, M. G. and ROBSON, R. M. in *Proteases and Biological Control*, E. Reich, D. B. Rifkin and E. Shaw (eds.), Cold Spring Harbor Lab., Cold Spring Harbor, New York, 1975.
22. ROBSON, R. M., *Proc. Recip. Meat Conf.*, 1981, **34** (in press).
23. DAYTON, W. R. and SCHOLLMEYER, J. V., *Exp. Cell Res.*, 1981 (in press).
24. DAYTON, W. R., SCHOLLMEYER, J. V., LEPLEY, R. A. and CORTES, L. R., *Biochim. Biophys. Acta*, 1981, **659**, 48.
25. OKITANI, A., GOLL, D. E., STROMER, M. H. and ROBSON, R. M., *Fed. Proc.*, 1976, **35**, 1946.
26. WAXMAN, L. and KREBS, E. G., *J. Biol. Chem.*, 1978, **352**, 5888.
27. HIRSCH, J. G. and COHN, Z. A., *Fed. Proc.*, 1964, **23**, 1023.
28. WEISMAN, G., *Fed. Proc.*, 1964, **23**, 1038.
29. STAGNIN, N. and DE BERNARD, B., *Biochim. Biophys. Acta*, 1968, **170**, 129.
30. BIRD, J. W. C., *Proc. Recip. Meat Conf.*, 1971, **24**, 67.
31. TAPPEL, A. L. in *The Physiology and Biochemistry of Muscle as a Food*, E. J. Briskey, R. G. Cassens and J. E. Trautman (eds.), University of Wisconsin Press, Madison, 1966.
32. KOHN, R. R., *Lab. Invest.*, 1969, **20**, 202.
33. SCHWARTZ, W. M. and BIRD, J. W. C., *Biochem. J.*, 1977, **167**, 811.
34. SPANIER, A. M., BIRD, J. W. C. and TRIEMER, R. E., *Fed. Proc.*, 1977, **36**, 555.
35. BIRD, J. W. C., SPANIER, A. M. and SCHWARTZ, W. M. in *Protein Turnover and Lysosome Function*, H. L. Segan and D. J. Doyle (eds.), Academic Press, New York, 1978.
36. KIRSCHKE, H., LANGNER, J., WIEDERANDERS, B., ANSORGE, S. and BOHLEY, P., *Eur. J. Biochem.*, 1977, **74**, 293.
37. OKITANI, A., MATSUKURA, U., KATO, H. and FUJIMAKI, M., *J. Biochem.*, 1980, **87**, 1133.
38. LOCKER, R. H., *Food Res.*, 1960, **25**, 304.
39. HERRING, H. K., CASSENS, R. G. and BRISKEY, E. J., *J. Sci. Fd Agric.*, 1965, **16**, 379.
40. HERRING, H. K., CASSENS, R. G. and BRISKEY, E. J., *J. Fd Sci.*, 1965, **30**, 1049.
41. HERRING, H. K., CASSENS, R. G., SUESS, G. G., BRUNGARDT, V. H. and BRISKEY, E. J., *J. Fd Sci.*, 1967, **32**, 317.
42. HOSTETLER, R. L., LANDMANN, W. A., LINK, B. A. and FITZHUGH, H. A., JR., *J. Anim. Sci.*, 1970, **31**, 47.
43. HOSTETLER, R. L., LANDMANN, W. A., LINK, B. A. and FITZHUGH, H. A., JR., *J. Anim. Sci.*, 1972, **37**, 132.
44. QUARRIER, E., CARPENTER, Z. L. and SMITH, G. C., *J. Fd Sci.*, 1972, **37**, 130.
45. BOUTON, P. E., FISHER, A. L., HARRIS, P. V. and BAXTER, R. I., *J. Fd Technol.*, 1973, **8**, 39.
46. DAVEY, C. L. and GILBERT, K. V., *J. Fd Technol.*, 1973, **81**, 445.
47. BUEGE, D. R. and STOUFFER, J. R., *J. Fd Sci.*, 1974, **39**, 396.

48. ABBAN, A. R., STOUFFER, F. R. and WESTERVELT, R. G., *J. Fd Sci.*, 1975, **40**, 1214.
49. LOCKER, R. H. and HAGYARD, C. J., *J. Sci. Fd Agric.*, 1963, **14**, 787.
50. MARSH, B. B. and LEET, N. G., *J. Fd Sci.*, 1966, **31**, 450.
51. MARSH, B. B. and LEET, N. G., *Nature*, 1966, **211**, 635.
52. BUSCH, W. A., PARRISH, F. C., JR. and GOLL, D. E., *J. Fd Sci.*, 1967, **32**, 390.
53. MARSH, B. B., WOODHAMS, P. R. and LEET, N. G., *J. Fd Sci.*, 1968, **33**, 12.
54. McCRAE, S. E., SECOMBE, C. G., MARSH, B. B. and CARSE, W. A., *J. Fd Sci.*, 1971, **36**, 566.
55. SMITH, G. C., ARANGO, T. C. and CARPENTER, Z. L., *J. Fd Sci.*, 1971, **36**, 445.
56. BOUTON, P. E., HARRIS, P. V., SHORTHOSE, W. R. and SMITH, M. G., *J. Fd Technol.*, 1974, **9**, 31.
57. HARRIS, P. V., *CSIRO Fd Res. Q.*, 1975, **35**, 49.
58. FIELDS, P. A., CARPENTER, Z. L. and SMITH, G. C., *J. Anim. Sci.*, 1976, **42**, 72.
59. SMITH, G. C., DUTSON, T. R., HOSTETLER, R. L. and CARPENTER, Z. L., *J. Fd Sci.*, 1976, **41**, 748.
60. DUTSON, T. R., SMITH, G. C., SAVELL, J. W. and CARPENTER, Z. L., *Proc. Eur. Meat Res. Workers*, 1980, **26**, II, 84.
61. DUTSON, T. R., SMITH, G. C. and CARPENTER, Z. L., *J. Fd Sci.*, 1980, **45**, 1097.
62. OLSEN, D. G., PARRISH, F. C., JR. and STROMER, M. H., *J. Fd Sci.*, 1976, **41**, 1036.
63. HAY, J. D., CURRY, R. V., WOLFE, F. H. and SANDERS, E. J., *J. Fd Sci.*, 1973, **38**, 981.
64. SAMEJIMA, K. and WOLFE, F. H., *J. Fd Sci.*, 1976, **41**, 250.
65. PENNY, I. F., *J. Sci. Fd Agric.*, 1974, **25**, 1273.
66. PENNY, I. F. and FERGUSON-PRYCE, R., *Meat Science*, 1979, **3**, 121.
67. PENNY, I. F. and DRANSFIELD, E., *Meat Science*, 1979, **3**, 135.
68. QUALLI, A. and VALIN, C., *Meat Science*, 1981, **5**, 233.
69. DUTSON, T. R. and YATES, L. D., *Proc. Eur. Meat Res. Workers*, 1978, **24**, II, E6.
70. YATES, L. D., DUTSON, T. R., CALDWELL, J. and CARPENTER, Z. L., *Meat Science*, 1981 (submitted).

# 18

# Protein Recovery from Food Factory Waste using Lignosulphonates


PER O. NETTLI

*Alwatech, Harbitzalleen 3, Olso, Norway*


It has been estimated that the UK meat industry discharges about 37 000 tonnes of BOD p.a. If 75% of this were recovered as a meat-meal containing 50% protein, the current value would be about £4·4 million p.a., while the world-wide recovery potential would be £200 million p.a. On the other hand, it can be estimated that if all the UK meat industry effluents were to be treated by conventional methods, the total quantity of surplus sludge requiring disposal could be $590 \times 10^6$ litres p.a. at 3% total solids.

For the UK poultry industry it has been estimated that about 8000 tonnes p.a. of protein-containing material with a value of £1·3 million p.a. is wasted in effluent, while in the UK fish industry a further 5000 tonnes of potential product is lost.

*The Alprecin Process* has been developed to recover this wasted material, and to provide effluent treatment systems which operate without the production of vast quantities of surplus sludge for off-site disposal.

It is essential that any recovery system meets certain basic criteria:

(i) any process to recover materials from effluents must be designed to operate on a constantly varying waste water, containing many substances in dilute suspension and/or solution;

(ii) the process should be capable of economical operation and not require highly specialised skills;

(iii) the chemicals used as precipitants must not render the recovered material unusable;

(iv) the finished product must be readily saleable without further processing.

The Alprecin Process meets all these criteria and the recovered product is currently being sold or reused by the producers in the UK, Sweden,

Belgium, Poland, Hungary and the USA, with plants soon to come on stream in Ireland. The chemical precipitant used—lignosulphonate (a by-product of wood pulping)—has been successfully tested in feeding trials and is an accepted additive for animal fodder by both the EEC and the US Regulatory Authorities.

The process is based upon the specific precipitation of proteins by lignosulphonic acid at low pH. A solution of lignosulphonate is added to the screened waste water which is then treated with sulphuric acid to reduce the pH to 3. Protein is precipitated and separated from the clarified effluent by dissolved air flotation in circular tanks, constructed from glassfibre-reinforced plastic. The effluent, which has only 25% of the original BOD and is low in suspended solids and fat, is neutralised and discharged to a municipal sewer for further treatment. The floated sludge is removed and can be mixed with surplus blood. If required, the sludge is then neutralised and heat-coagulated at 95°C with steam, followed by dewatering by a filterbelt or a decanter. The dewatered sludge can be dried, either alone or in admixture with other by-products, to produce an animal feed additive containing 45–70% protein. The recovered material is acceptable as an animal feed ingredient and is currently recovered in the UK and sold to an animal feed compounder. The current value is about £160/tonne.

Although, as with most commodities, the price of protein fluctuates widely, it would appear likely in view of the world protein requirements and supply, that the price will continue to rise in the long term.

If we compare the treatment of abattoir effluent by lignosulphonate precipitation to that in a conventional activated sludge plant, we find that every tonne of dry product recovered is equivalent to a saving of 5500 gallons of surplus sludge which would be produced by the conventional system. In financial terms this means an income of £160 instead of an expenditure of £125. Because of increasing transport costs and the reduced availability of tipping sites, the cost of sludge disposal can also be expected to rise.

In Tables 1 and 2 the economics of alternative treatment systems are considered in more detail for both abattoirs and poultry packing stations. We have, however, calculated in arbitrary units rather than in monetary values, although the figures are based on current UK costs. However, as the alternative systems to the Alprecin Process are hypothetical, some assumptions have been necessary.

First of all, we will consider an abattoir killing and jointing 700 cattle per day: efficient blood collection is practised and in the Alprecin Process the collected blood is mixed with the flotated sludge before heat-coagulation and dewatering. The dewatered sludge/blood mixture is then dried with

TABLE 1

ECONOMICS OF AN ABATTOIR EFFLUENT TREATMENT BY THE LIGNOSULPHONATE/ACTI-
VATED SLUDGE SYSTEM COMPARED WITH A SINGLE ACTIVATED SLUDGE TREATMENT

Basis for design:
Flow—910 m$^3$/day;
BOD$_5$—3 750 mg/litre (before screening).
Effluent standard required:
    BOD$_5$—20 mg/litre;
    Suspended solids—30 mg/litre.

The capital cost of the lignosulphonate plant is put at 100 arbitrary units. Costs and income are per annum and relate to this figure.

|  | *Lignosulphonate/ activated sludge* | *Activated sludge only* |
|---|---|---|
| Budget capital cost | 100 | 44·2 |
| Operating costs for chemicals | 5·1 | 0 |
| Electricity | 2·1 | 6·1 |
| Labour | 4·2 | 2·1 |
| Sludge coagulation and drying | 6·7 | 0 |
| Sludge disposal | 2·0 | 8·2 |
| Product recovery (tonnes)[a] | 1 456 | 0 |
| Value of product recovery | 40·0 | 0 |
| Net operating cost | 0 | 16·4 |
| Net income | 19·9 | 0 |

[a] The quantity of proteinaceous material recovered is based upon:
  (i)    1·2 kg dry solids produced per kg BOD$_5$ removed (75 % BOD$_5$ removal by lignosulphonate precipitation);
  (ii)   2·7 kg dry matter from blood recovered per animal killed;
  (iii)  1000 kg of dry screenings recovered per day with the removal of 20 % of the BOD$_5$;
  (iv)  the production of surplus activated sludge is based upon a sludge growth index of 0·6 × BOD$_5$ removed.

effluent screenings to about 8 % moisture before being mixed with other products by an animal feed compounder. The total annual recovery of protein, blood and screening would be about 1400 tonnes per year.

In Table 1 the cost relationship (in the UK) between the lignosulphonate precipitation process followed by an activated sludge plant is compared to a single stage activated sludge plant, either system being capable of reaching an effluent standard of 20 mg/litre BOD, and 30 mg/litre suspended solids.

The data show that the lignosulphonate process could be operated at a substantially lower cost than a conventional activated sludge plant; the difference in capital cost would be recovered in 2–3 years.

## TABLE 2

ECONOMICS OF A POULTRY PLANT EFFLUENT TREATMENT BY THE LIGNOSULPHONATE
SYSTEM COMPARED WITH A FERRIC CHLORIDE PROCESS.

Basis for design:
Flow—682 m$^3$/day;
BOD$_5$—2 500 mg/litre (after screening);
Fat—1 000 mg/litre;
BOD removal—75 %.

The capital cost of the lignosulphonate plant is put at 100 arbitrary units. Costs and income are per annum and relate to this figure.

|  | Lignosulphonate process | Ferric chloride process |
|---|---|---|
| Budget capital cost | 100 | 40 |
| Chemicals and electricity | 5·5 | 6·0 |
| Sludge coagulation and drying | 3·9 | 0 |
| Labour | 3·0 | 3·0 |
| Sludge disposal (3 % sludge) | 0 | 9·4 |
| Product recovery (tonnes)[a] | 348 | 0 |
| Value of product recovery | 30·9 | 0 |
| Net operating cost | 0 | 18·4 |
| Net income | 8·4 | 0 |

[a] The quantity of proteinaceous material recovered is based upon:
  (i)    1 kg dry solids produced per kg BOD$_5$ removed;
  (ii)   75 % removal of BOD$_5$.
The production of ferric chloride sludge is calculated upon the same basis.

In the second example (Table 2) the application of the lignosulphonate process to a poultry processing plant is compared with a system based on precipitation and flotation by ferric chloride. Both systems were designed to produce an effluent suitable for discharge to a municipal sewer. There was no provision for the recovery of blood after dewatering. The lignosulphonate process sludge was mixed with other by-products before cooking/drying. The cost of cookers was not included in the capital cost, as sufficient spare capacity existed. The example was in relationship to the UK and may be different for other countries.

The data show that the difference in annual costs, excluding capital repayment, was very much in favour of protein recovery and would enable the extra capital cost involved to be recovered in 3–4 years.

# 19

# Recovery of Protein from Food Factory Wastes by Ion Exchange


D. E. PALMER

*Bio-Isolates Limited,*
*Adelaide Street, Swansea, UK*


## INTRODUCTION

Ion exchange as a technology for the recovery of materials such as metals has been well known and employed industrially for many years. These exchangers are based on synthetic materials, usually of coal or oil origin, and their use is limited to the recovery of small ions, although attempts have been made to use them for the recovery of macromolecules. One of the main disadvantages of the early exchangers for the recovery of macro-molecules, such as proteins, is that such materials could not be recovered without denaturation. Additionally, adsorption capacity was generally small (Table 1).

The first attempts at producing ion exchangers suitable for the recovery of proteins and similar macromolecules from solution were restricted to hydrophilic exchangers based on natural cellulose. For a long time the technology was limited to laboratory use where it proved very valuable for protein isolation. A major obstacle to the scaling up of processes based on natural cellulose was that, because they were in the form of fibres, they had unsatisfactory hydraulic properties.

The second generation of hydrophilic ion exchangers were also based on cellulose, i.e. on regenerated cellulose. Work carried out in New Zealand and in the UK led to the development of a range of hydrophilic ion exchangers which had a granular structure and, consequently, improved hydraulic properties. The improved flow rates obtained with the new exchangers enabled protein fractionations to be performed up to ten times faster than previously, but without loss of resolution or efficiency. Some characteristics of ion exchangers are summarised in Table 1.

341

## TABLE 1
### APPLICATIONS OF ION EXCHANGERS

| | |
|---|---|
| | *'Small ion' Exchanger Employing Synthetic Resins* |
| Uses | De-ionisation, water treatment, metal recovery. |
| Limitation | Resins unable to successfully adsorb macromolecules, e.g. protein, with reasonable uptake and without damage to protein molecules. |
| | *Macromolecular (Hydrophilic) Ion Exchangers* |
| First Generation | Based on natural cellulose. Good capacity; protein undenatured but fibrous nature prohibited large-scale use. |
| Second Generation | Based on regenerated cellulose. Excellent capacity for protein; granular structure gives improved hydraulic properties and allows large-scale operation. |
| Third Generation | Based on modified, regenerated cellulose. High performance materials. |

# ION EXCHANGE SYSTEMS

Given a suitable ion exchange material, the other requirement for a successful ion exchange system is an effective means of separating solids from liquids. This can be achieved by using columns, fixed beds, fluidised beds, agitated columns or stirred tanks. The choice depends on the properties of the exchanger and/or the nature of the material to be treated. For instance, a stirred tank system is more successful than a fixed bed column for material with high suspended solids. When fat is present, an agitated column or stirred tank to assist in fat removal is more successful. More highly concentrated elutes may be obtained when columns are used.

# APPLICATIONS OF HYDROPHILIC ION EXCHANGE

## Abattoir Effluent Treatment

The new hydrophilic ion exchangers were first used, on a column system, in New Zealand for the treatment of abattoir effluent. In the UK, a continuous stirred tank reactor system, consisting of three sections with settling cones (Fig. 1), was developed. The effluent, together with the cellulose medium stream, is fed into a stirred tank where the protein is adsorbed and the suspension is allowed to flow into the first settling cone where the medium, with adsorbed protein, settles out and the treated liquor overflows. The medium with adsorbed protein is transferred to the second cone where the protein is deadsorbed by the addition of salt. The medium is allowed to settle and the protein solution overflows. The medium is transferred to the third cone where it is washed and regenerated, ready for recycling.

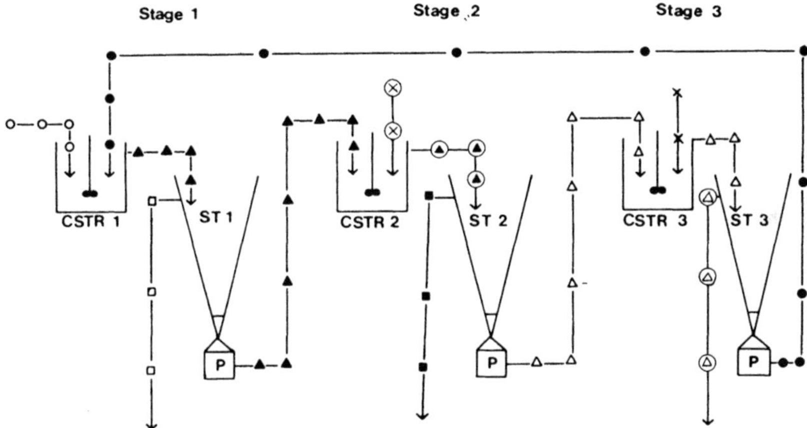


FIG. 1. Flow diagram of continuous ion exchange system. Stage 1: Protein adsorption. Stage 2: Protein recovery and media regeneration. Stage 3: Media washing. Effluent, containing protein (○); cellulose adsorbant medium (●); adsorbant plus protein stream (▲); regenerant, NaCl (⊕); desorbed protein plus adsorbant plus regenerant (⊛); concentrated protein stream (■); deproteinised effluent (□); regenerant plus cellulose adsorbant (△); wash water (×); regenerant plus wash water (⊗). CSTR: continuous stirred tank reactors. P: Pump. ST: Settling tanks.

Most of the protein is removed and recovered and since the BOD in the case of abattoir effluent is due mainly to protein, the recovery system also serves as an efficient effluent treatment system, capable of removing about 98 % of BOD.

However, the economics of the process for the treatment of abattoir effluent are doubtful unless effluent treatment charges increase. It seems likely that the application of the technique in abattoirs will, for some time, be restricted to the isolation of protein from hygienically collected blood, either for preparation of bulk quantities of albumin or globulins or for the isolation of high value proteins as reference proteins in biochemistry. This latter process is currently being perfected.

*D. E. Palmer*

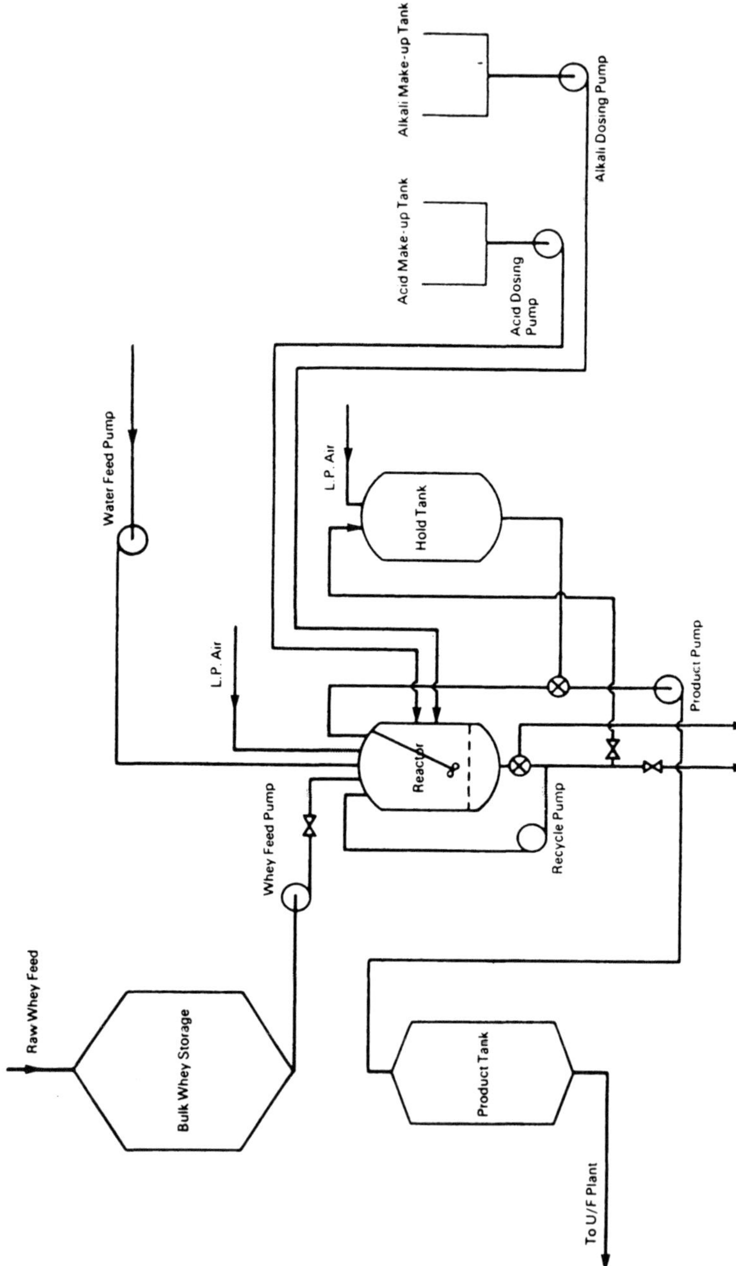


FIG. 2. Simplified flowsheet of whey protein recovery plant. (Courtesy: *Process Biochem.*, 1977, **12**(5), 24.)

## Whey Treatment

Cheese whey is an obvious source of high quality protein suitable for recovery by ion-exchange. Cheese or casein wheys contain the albumins and globulins of milk which together amount to about 20 % of the milk protein. Typically, whey contains: 0·6 % protein, 0·6 % minerals, 0·06 % fat and 4·6 % lactose. Total whey solids amount to 60 g/litre of which 6 g is high value protein. Approximately 70 % of the total proteins in human milk are classified as whey proteins. Consequently, the addition of whey protein derived from cows' milk is used to 'humanise' a number of proprietary baby foods.

A separation method used for whey treatment which is based on a filter bottom stirred tank reactor, has been developed. Although the process is essentially simple, the history of its development is a long one, following the classical stages of bench, pilot scale and finally large pilot or pilot production scale.

A simplified flow sheet is shown in Fig. 2. The reactor is a vessel fitted with a sieve bottom, the reaction taking place above the sieve. The process sequence is illustrated in Fig. 3. Whey is drawn into the reactor, the pH is adjusted and the protein is adsorbed onto the medium (stages 1 and 2). Deproteinised whey is removed through the sieve and, depending on the purity required, the protein-laden medium is washed free of impurities (stages 3 and 4). Water is added and the pH adjusted to release the protein into solution (stages 5 and 6).

After complete desorption, the protein solution is concentrated and spray dried to produce a white protein powder. As extremes of pH are not used, little protein denaturation occurs.

The pilot plant has been used for several months at an industrial site to produce substantial quantities of protein for evaluation and to yield design information for further scale-up. A larger scale plant is currently working continuously in a creamery in South Wales. In the latter operation the pure eluted protein solution is concentrated by ultrafiltration and pumped under high pressure to a specifically designed spray dryer. Drying conditions are such that the protein remains undenatured and of high quality and with good free-flowing properties.

The compositional analysis of a routine batch of powder (BIPRO DA) is shown in Table 2. It contains 96 % protein on a dry basis, is almost free of lactose and fat and has a low mineral content. The product has foaming properties similar to egg white and gels in ~4 min at 60 °C, which is comparable with egg white.

*D. E. Palmer*

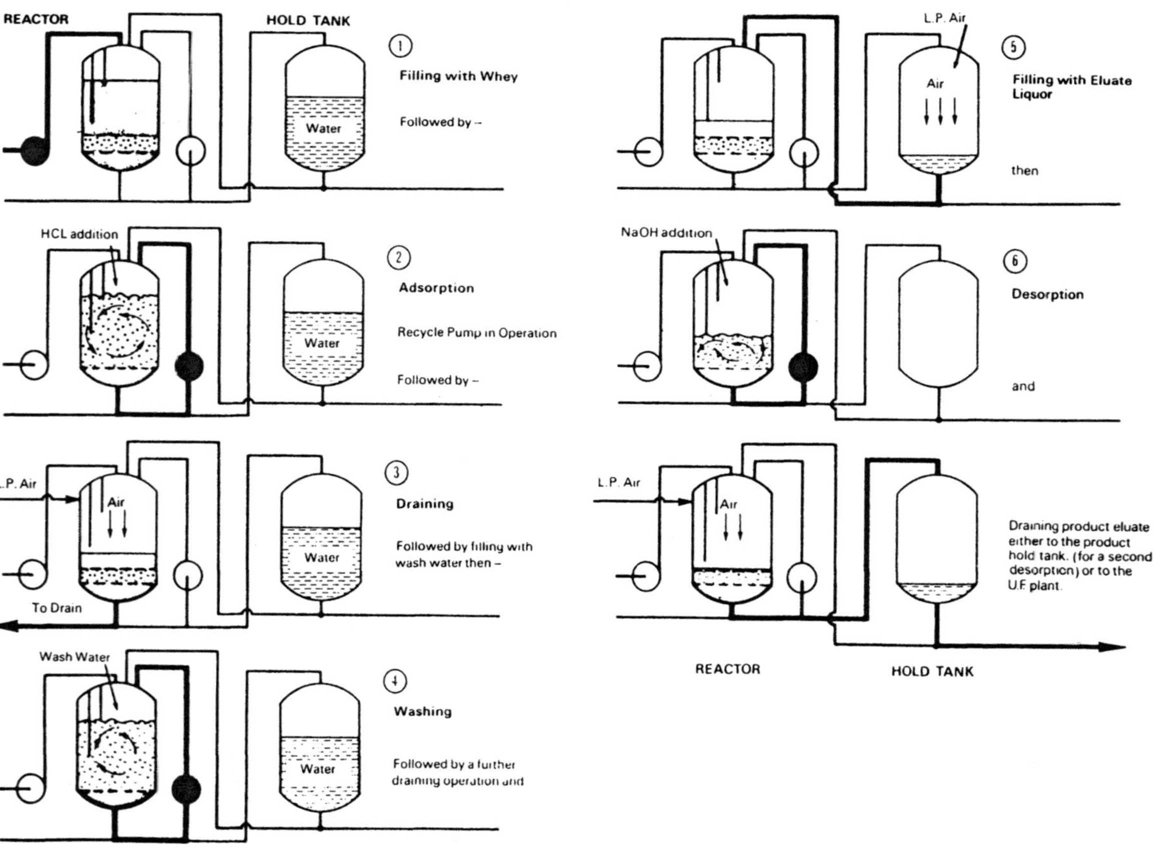



FIG. 3.  Milk whey pilot plant process sequence. (Courtesy: *Process Biochem.*, 1977, **12**(5), 24.)

TABLE 2
ROUTINE ANALYSIS OF BIPRO DA[a]

| Constituent | % |
| --- | --- |
| Protein | 92 |
| Moisture | 4 |
| Fat | 0·5 |
| Ash | 3 |
| *Expansion Ratio:* Typically 550–650 | |
| *Gelation:* Typically 4 min at 60 °C | |

[a] Product from large-scale operation.


## Economics of Production of Whey Protein Isolate

The process has been in operation for more than a year and enough work has been done to provide meaningful production costs, which are summarised in Table 3. The operating costs, i.e. direct isolation costs, total £500 to £600 per tonne. Labour costs are not included because this may

TABLE 3
SUMMARY OF PRODUCTION COSTS FOR WHEY PROTEIN ISOLATE

| | | | | |
| --- | --- | --- | --- | --- |
| Daily capacity (litres) | 40 000 | 70 000 | 140 000 | 280 000 |
| Batch volume (litres) | 4 000 | 7 000 | $2 \times 7\,000$ | $4 \times 7\,000$ |
| Annual cycles | 3 000 | 3 000 | 3 000 | 3 000 |
| Annual feed (litres) | $12 \times 10^6$ | $21 \times 10^6$ | $42 \times 10^6$ | $84 \times 10^6$ |
| Protein in feed (g/litre) | 6 | 6 | 6 | 6 |
| % Recovery | 75 | 75 | 75 | 75 |
| Annual production (tonnes) | 54 | 90 | 180 | 360 |
| *Production Costs, £ per tonne of Protein Powder* | | | | |
| Water | 49 | 52 | 52 | 52 |
| Power | 38 | 32 | 32 | 32 |
| Media | 80 | 83 | 74 | 74 |
| Process chemicals | 86 | 88 | 88 | 88 |
| Maintenance | 50 | 30 | 20 | 20 |
| Total isolate production cost | 303 | 285 | 266 | 266 |
| Ultrafiltration cost | 102 | 70 | 58 | 58 |
| Drying energy cost | 200 | 200 | 200 | 200 |
| Total operating costs | 605 | 555 | 524 | 524 |

## TABLE 4
ESTIMATE OF TOTAL COSTS AND PROFITABILITY OF PRODUCTION OF WHEY PROTEIN ISOLATE

| | (A)* | | | | (B)* | | | |
|---|---|---|---|---|---|---|---|---|
| Daily capacity (litres) | 40 000 | 70 000 | 140 000 | 280 000 | 40 000 | 70 000 | 140 000 | 280 000 |
| Capital cost (£k) | 250 | 310 | 500 | 810 | 200 | 260 | 300 | 510 |
| Sales volume (tonnes/year) | 54 | 90 | 180 | 360 | 54 | 90 | 180 | 360 |
| Operating costs (£/tonne) | 605 | 555 | 524 | 524 | 605 | 555 | 524 | 524 |
| Depreciation (£/tonne)[a] | 463 | 344 | 278 | 225 | 370 | 289 | 167 | 142 |
| Labour (£/tonne)[b] | 648 | 466 | 466 | 233 | 648 | 466 | 466 | 233 |
| Total costs (£/tonne) | 1 716 | 1 365 | 1 268 | 982 | 1 623 | 1 310 | 1 157 | 899 |
| Annual total costs (£k) | 93 | 123 | 228 | 354 | 88 | 118 | 208 | 324 |
| Annual value of sales (£k)[c] | 178 | 297 | 594 | 1 188 | 178 | 297 | 594 | 1 188 |
| Gross margin (£k/year) | 85 | 174 | 366 | 834 | 90 | 179 | 386 | 864 |
| Volume treated (litres/year) | $12 \times 10^6$ | $21 \times 10^6$ | $42 \times 10^6$ | $84 \times 10^6$ | $12 \times 10^6$ | $21 \times 10^6$ | $42 \times 10^6$ | $84 \times 10^6$ |
| Gross margin (p/litre) | 0·71 | 0·82 | 0·87 | 0·99 | 0·75 | 0·85 | 0·92 | 1·03 |
| Gross margin (p/gallon) | 3·2 | 3·7 | 3·9 | 4·5 | 3·4 | 3·8 | 4·1 | 4·6 |

[a] Depreciation over 10 years.
[b] Labour charges—5 men at £7 000 pa for 40 000 lpd plant; 6 men at £7 000 pa for 70 000 lpd plant; 12 men at £7 000 pa for others.
[c] Sales price of powder—£3 300 per tonne (range £3 000–£4 000 depending on application).
* (A) Based on higher capital cost. (B) Based on lower capital cost.

TABLE 5
MARKET VALUE OF SOME PROTEIN PRODUCTS (SEPTEMBER, 1981, UK)

| Product | % Protein | UK price/tonne | Average price/tonne protein |
|---|---|---|---|
| Soy concentrate | 60–70 | 1 012 | 1 560 |
| Soy isolate | 90 | 1 550 | 1 780 |
| Milk concentrates | 60 | 2 866 | 5 015 |
|  | 85 | 4 167 | 5 147 |
| Whey protein (by ultrafiltration) | 60–80 | 1 750–2 800 | 3 000–3 500 |
| Egg white, commercial grade | 85 | 3 300–4 900 | 4 000–6 000 |
| Whole egg, spray dried | 50 | 3 200 | 6 400 |

vary within different companies, depending on circumstances; neither are capital costs since capital policy will influence capital charges. In Table 4 an ex-factory cost and estimate of profitability have been derived based on certain assumptions of the operating and capital costs, sales value, etc. It can be shown that with a large-scale operation the ex-factory cost should be £900 to £1700 per tonne, depending on scale, and the sales value of the product should be £3000 to £4000 per tonne from our trials and market research, bearing in mind the current prices of proteins (Table 5).

TABLE 6
ESSENTIAL AMINO ACID COMPOSITION OF CERTAIN PROTEINS (g/100 g OF PROTEIN)

| | Whey protein | Egg | Casein | Soy |
|---|---|---|---|---|
| *Iso*-leucine | 6·55 | 6·45 | 5·80 | 5·15 |
| Leucine | 14·00 | 8·30 | 9·50 | 7·85 |
| Lysine | 10·90 | 7·05 | 7·60 | 6·20 |
| Methionine | 2·35 | 3·40 | 2·95 | 1·35 |
| Cystine | 3·15 | 2·25 | 0·40 | 1·35 |
| Phenylalanine | 4·05 | 5·80 | 5·40 | 5·10 |
| Tyrosine | 4·80 | 4·05 | 5·70 | 3·40 |
| Tryptophan | 3·20 | 1·50 | 1·30 | 1·25 |
| Valine | 6·85 | 7·15 | 6·80 | 5·30 |
| Threonine | 6·70 | 5·15 | 4·00 | 4·10 |
| Total | 62·55 | 51·10 | 49·45 | 41·50 |

*Nutritional Properties of Whey Protein Isolate*

The product (BIPRO DA) contains a high concentration of all the essential amino acids (Table 6) in biologically available form (digestibility, 99%; biological values, 94%; net protein utilisation, 93%; protein efficiency ratio, 3·2).

*Functional Properties of Whey Protein Isolate*

In the food industry, protein is frequently much more valued for its functional properties than for its nutritional value. The functional properties of proteins that are of value to the food manufacturer are blandness, solubility, foaming ability, heat gelation, water binding and fat emulsification. BIPRO DA has no flavour but enhances other food flavours; it is easy to incorporate into other foods, even low pH fruit squashes; it has good foaming, water binding and fat emulsification properties and sets at moderate temperatures. Whipping and gelation properties of whey protein isolated by the ion exchange method are similar to those of egg white (Tables 7 and 8). Because of its nutritional and

TABLE 7

WHIPPING PROPERTIES OF EGG AND DAIRY ALBUMIN (11% PROTEIN)

|  | *Specific foam value (ml/g)* | *Drainage at 30 min (ml)* |
|---|---|---|
| Egg white | 10·0 | 12 |
| Dairy albumin | 10·2 | 8 |
| Egg white[a] | 3·5 | 2 |
| Dairy albumin[a] | 3·1 | 4 |

[a] 100 g protein solution plus 100 g sucrose.
Adapted from Burgess and Kelly, *J. Food Technol.*, 1979, **14**, 325.

functional properties, the whey protein isolate has many uses in the food industry (Table 9). Products, which include foam whips, sponges and meringues, have been made successfully in our kitchens. It is likely that BIPRO DA will serve as an ingredient that can be used by the housewife in home cooking as well as being used in the baking industry.

## Other Protein Wastes

Other waste streams that have been treated successfully by ion exchange

TABLE 8

GELLING PROPERTIES OF EGG AND DAIRY ALBUMIN (11 % PROTEIN)

| Protein | pH | Gelation temperature (°C) | Water-holding capacity (%) | Hardness | Springiness |
|---|---|---|---|---|---|
| Egg albumin | 8·5 | 60 | 97·8 | 7·2 | 0·59 |
| Dairy albumin | 7·0 | 58 | 98·4 | 8·4 | 0·65 |
| | 8·0 | 56 | 99·0 | 5·2 | 0·60 |
| | 9·0 | 56 | 98·2 | 5·2 | 0·58 |

Adapted from Burgess and Kelly, *J. Food Technol.*, 1979, **14**, 325.

TABLE 9

APPLICATIONS FOR BIPRO DAIRY ALBUMEN

*Successful applications, currently confirmed*
Companies have carried out trials of the protein in several products and have reported as follows:

| | |
|---|---|
| Cakes | Cherry, angel, and madeira cakes have been made without eggs, without loss of volume, with enhanced flavour and marginally less expensive. |
| Pastry | The flaky characteristic is more stable using BIPRO. |
| Confectionery | Mallows are more stable and have a longer shelf life than when made with eggs. |
| Protein supplement | Included in diets for athletes in Europe |

*Applications being developed*

| | |
|---|---|
| Baking | Protein enriched breads with increased crustiness. Genoese cakes, macaroons and cheese cake. |
| Confectionery | Protein confectionery bars for the Third World. |
| Meat foods | Better bound beefburgers. |
| Soft drinks | Protein enriched fruit squashes and colas. |
| Protein packs | Containers of BPIRO to be sold in health food shops. |

Eggless cakes for people allergic to eggs.
Eggless custards, omelettes, meringues.
Protein enriched fruit jellies, etc.
Protein enriched soups and beef extracts and gravy thickeners.

*Other possibilities*
Protein enriched home-made jams.
Better home-made icecream.
Margarine, marmalade and melons can be enriched.
Oranges, peaches and pears can be enriched with protein syrups, as can also apple pie.
Porridge, cereals, muesli and potatoes can be enriched to make complete energy and protein foods.

### TABLE 10
RECOVERY OF PROTEINS FROM SOME FOOD PROCESSING WASTES BY ION EXCHANGE

| *Waste stream* | *Protein N (g/litre)* | | *% Removal of protein* |
|---|---|---|---|
| | *Before* | *After* | |
| Potato effluent | 0·78 | 0·035 | 95·5 |
| Distillery effluent | 0·48 | 0 | 100·0 |
| Animal by-product | 0·52 | 0·036 | 93·0 |

### TABLE 11
TYPICAL ANALYSIS OF PROTEINS ISOLATED BY ION EXCHANGE FROM SOME PROCESS EFFLUENTS

| *Process effluent* | *Percentage* | |
|---|---|---|
| | *Protein* | *Ash* |
| Soy | 92 | 4 |
| Potato | 94 | 1·5 |
| Rapeseed | 95 | 3 |

include potato effluents, distillery effluents and animal by-product treatment effluents. The efficiency of protein removal is very good (Table 10) and at some future date it is likely that high quality, high value proteins may be isolated from these streams. The compositions of protein products recovered from some of these wastes are summarised in Table 11. Other sources to be further examined include pea, sunflower, palm oil and leaf.

# Index

Blue–green algae, 297, 298
Bond energies, 63
Bovine sources, 241–2
Boyer process, 233, 234
Breadmaking, 85–93
Breast feeding, 141
*Brevibacterium linens*, 316
BV (biological value), 123, 126

C-protein, 250
Calcium-activated proteinase, 330–1,
       333–4
Calcium-activated sarcoplasmic factor
       (CAF), 273
Calorie
   intake, 44, 111
   requirements, 37, 109–13, 118
Cardiac arrhythmia, 116
Carotene, 202, 203
Casein, 65, 155–6, 157, 161, 162,
       163–5, 184, 217, 310, 311,
       313, 321
   acid, 180–3
   edible grades, 180
   fibres, 217
   lactic, 180–1
   manufacture in Ireland, 31
   micelles, 165–8
      stability, 168–72
   mineral acid, 181
   proteolysis, 70
   rennet, 183–4
   texturisation, 217
Caseinates, 185, 216
Cattle farming in Ireland, 14
Cereal(s)
   consumption in Ireland, 9
   production in Ireland, 7–11
   proteins, 85
   snack food, 216
Charge modification, 65
Cheese
   manufacture, 65
      Ireland, in, 31
      microbial proteolysis, 307–20
      proteolysis by non-starter
          microorganisms, 315–17

Cheese—*contd.*
   manufacture—*contd.*
      starter proteinase in bitter
          defects, 313–15
   ripening
      exogenous microbial proteinases
          to accelerate, 317–20
      microbial proteinase in 312–17
      mould ripening, 316
   whey, ion exchange, 345–52
Chemical properties, 51
Chilling effects on meat proteins,
       269–74
Chymosin, 310, 311, 313
Climatic effects, 114
Coagulation, 74
Collagen, 253, 254, 255, 257, 258, 263
   cross-linked, 281
   fibres, 253, 263
   heating effects, 280
   intramuscular, 255–8
   labile, 281
   structure, 263
Comminution, 282–6
Connective tissue, 285
   proteins, 252–5
Contractile protein complex, 264
Co-precipitates, 184
Creameries in Ireland, 26
Cross-linking, 256, 257
Curd
   formation, 180
   separation, 182
*Cyanobacterium*, 297

Dairy industry in Ireland, 26–32
DEAE-cellulose chromatography, 239
Denaturation, 84
Deoxymyoglobin, 265
Depolymerisation, 84
*Dermatophagoides pteronyssinus*, 138
Developing countries, 36, 42, 46, 47,
       108, 112, 129
Diet in Ireland, 3–6
Dispersability, 72
Dissociation–reassociation cycle, 168